**Waiwai**

# Waiwai

## Water and the Future of Hawai'i

Edited by Kamanamaikalani Beamer

University of Hawai'i Press
Honolulu

Printed in the United States of America

First printed, 2025

**Library of Congress Cataloging-in-Publication Data**

Names: Beamer, Kamanamaikalani, editor.
Title: Waiwai : water and the future of Hawai'i / edited by Kamanamaikalani Beamer.
Other titles: Water and the future of Hawai'i
Description: Honolulu : University of Hawai'i Press, 2025. | Includes bibliographical references and index.
Identifiers: LCCN 2024029665 (print) | LCCN 2024029666 (ebook) | ISBN 9798880700523 (hardback) | ISBN 9780824899318 (trade paperback) | ISBN 9780824899332 (epub) | ISBN 9780824899349 (kindle edition) | ISBN 9780824899325 (pdf)
Subjects: LCSH: Water resources development—Hawaii. | Water-supply—Government policy—Hawaii. | Water conservation—Hawaii. | Environmental responsibility—Hawaii.
Classification: LCC HD1694.H3 W35 2025 (print) | LCC HD1694.H3 (ebook) | DDC 333.91/1609969—dc23/eng/20241104
LC record available at https://lccn.loc.gov/2024029665
LC ebook record available at https://lccn.loc.gov/2024029666

Cover art caption and credit: Top photograph of two children, bottom photograph of a stream in Hawai'i. Photographs courtesy of Kamanamaikalani Beamer.

University of Hawai'i Press books are printed on acid-free paper and meet the guidelines for permanence and durability of the Council on Library Resources.

# Contents

## Acknowledgments

*Waiwai* would not be possible without our amazing co-authors, mahalo for all you do for wai!

*For those past who brought me to my wai:*

My tūtū, Papa and Dambie, Granny and Grandpa, Mom, Uncle Hank, Aunty Laurie, Tenny, and ka poʻe i aloha i ka ʻāina i hala. All of the forms of Kāne. Working daily to mahalo all you have given me from our times together and continuing to quench myself from the wai you provided.

*For those who experience and share our wai today:*

My ʻohana, co-authors, and UH Press team—Keliʻi, my keiki, Kapono and Dayna Beamer, broader Beamer ʻOhana, Adam Zaslow, Gini Kapali and Kapali ʻOhana and all the ʻohana out there fighting to mālama our wai. You are my wai, I deeply aloha and cherish each of you.

*My hoa aloha ʻāina*—Kapua Sproat, Kanekoa Schultz, Keliʻi Kotubetey, Kawika Winter, Donovan Preza, Kaliko Maʻiʻi, Kaeo Duarte, Kalā Hoe, Kalehua Krug, Camille Kalama, Malia Akutagawa, Mehana Vaughan, Kaleo Manuel, Wayne Tanaka, the entire ʻĀina Aloha Economic Futures hui, Kamakakūokalani, Kaʻohewai, and the Oʻahu Water Protectors—it's an honor to walk alongside of you. The Pōʻai ke Aloha ʻĀina lab, Kawena, Pua, Mahina, Emily, and Kelsy, for all your efforts in pushing this work forward! Go team! All the aloha ʻāina of today who have inspired our movement and stood for our wai across Hawaiʻi, in the loʻi and loko iʻa, in boardrooms, in classrooms, and the streets, you are my heroes.

*My kumu*—Ano, Uncle Calvin Hoe, Anakalā Eddie, Neil Hannahs, Dr. Jon Osorio, Dr. Carlos Andrade, Dr. Jon Goss, Chuck Lawrence, Jon Kaʻimikaua, Dr. Kanalu Young, Dr. Lilikalā Kameʻeleihiwa, Dr. Haunani Kay Trask, and Bill Tam. Mahalo for teaching me about kuleana and courage.

*For those who will experience the wai I will never touch:*

*My keiki and pulapula to come*—Halialoha, Kūlokuwiwoʻole, Kaulupōlani, Waikā a me Manono—and all the aloha ʻāina of tomorrow, it's a joy and a kaumaha to carry this forward; and you can—with our own flair, e kāmau.

To the unnamed uncles and aunties who climb toward the far limbs of branches to secure rope swings alongside mauka ponds across our islands—we live because of you—Ola i ka wai!

The ones who taught keiki to mālama ʻāina and love the kahawai—Ola i ka wai! The ones who wrote songs to honor our places and wai—Ola i ka wai! The ones that taught us to stand and fight for the future of wai and the future of Hawaiʻi—Ola i ka wai!

To the wai and places that provide life on our earth—habitat and ecosystem services—stories and memories for its peoples—may we cherish, fight for you, and never forget—Ola i ka wai!

# Waiwai

# Introduction

*Kamanamaikalani Beamer*

> The Ali'i's (ancestral class of rulers) role was to govern and listen to the people, the Koa's (warriors) role was going to battle—to fight if need be for what is Pono (proper, righteous). The Mahi'ai (farmers) were to farm and care for the 'āina (land, literally that which feeds) and the wai (water) so that resources were abundant for future generations. These roles were split up, and it worked well for our people. But today if you like be one Mahi'ai—one farmer, you also gotta be one Koa—you gotta go fight for get your water, and you gotta be one ali'i cause you gotta organize people and systems to achieve Pono. *Because whoever controls the water controls the future.*
>
> —Calvin Hoe

I was merely an apprehensive teenager when Uncle Calvin Hoe was a guest lecturer in my high school Art History course. Uncle Cal was a mahi'ai (Native Hawaiian taro farmer), a master crafter of hula implements, and a lead activist for water justice in the famed Waiāhole water struggle on O'ahu. While most of high school remains a blur, I never forgot the words Uncle Cal spoke to us that day. It is a moment I have looked back on many times in my life. His words still resonate with me and remain true in my approach toward achieving justice for our most precious resource—*wai—our water—our future.*

All the wonder and awe of our collective existence is only possible because of water. Reflect for a moment on those foundational memories of your life: the times spent with family in your youth gathered around home-cooked meals, the moment you crushed a grape between your teeth for the very first time and felt the sweet juice burst in your mouth. Maybe you can recall that first, delightful taste of a perfectly ripe mango just picked from your aunty's tree. Perhaps you have memories of being in a warm bath as a toddler as a loved one drenched your skin with a sopping towel, or collecting the cold, sweet mountain water of a natural spring. If you were raised in a city and danced under the falling droplets of a burst fire hydrant or swam in the community pool—you shared a communal space with others in water. Maybe you recall the first time at the beach you found the courage to dunk your head under the waves, anxiously holding your breath and likely the hand of your father, mother, or elder sibling. None of these tender memories exist without water. In fact, if you were to recall the most trying, adverse times of your life—the loss of a precious loved one, a failed attempt to advance your career, or a painful ending to love that once seemed everlasting, you might find your cheeks dampened by water flowing in the shape of tears.

Ola i ka wai. Life because of wai (water).

This ancestral Hawaiian idiom edifies our Indigenous understanding of water's vital importance to our world. It is a well-known yet often taken for granted fact that life on our planet would not exist without the continual blessing of wai. Ola i ka wai! Life because of water! Our bodies are comprised/composed mostly of water. Wai cleanses us when we are unclean, quenches our thirst, and when we feel down, the ritual of immersing our bodies in wai or kai (ocean) often brings immediate comfort and healing. Teardrops fall from the face of every human regardless of culture, religion, race, gender, or sexual orientation.

Although water is fundamental and completely integral to every aspect of our lives, ironically, much of contemporary society undervalues our single greatest resource. Perhaps the importance of water can be likened to a family member whose complete contribution to our lives is not entirely grasped until they have passed. To put it frankly, many people don't fully appreciate water until it is gone. But let us be clear, there is no culture, no family, no food, no wine, no community, no life—no us—without water. And when water is spent, lost, degraded, and polluted, most of the adverse effects are disproportionately felt and lived by the most vulnerable and least politically powerful among us. In this sense, every allocation, management decision, and personal choice humans make for water can flow toward or away from justice.

Our ancestors clearly understood the value of water, so Indigenous communities are often perplexed by the astonishing ways in which modern society undervalues, wastes, and even destroys sources of water for future generations. Perhaps contemporary societies are less mindful of our water resources because we live in a time where most of our water comes from a faucet, a bottle, or a ditch. The familiar image "Blue Marble," one of the first color photographs taken of our planet from space, is ingrained in our modern psyche and gives an impression that our wai resources are seemingly vast and limitless. A simple Google search for an image of our planet can enable the perception that our water resources are incredibly vast and at a scale that appears somewhat endless. However, it can be surprising that 97 percent of the water on our planet is too salty to drink and that merely 3 percent of our planet's water is fresh, of which likely half that amount is frozen in our world's glaciers. In addition, the impacts and effects of the capitalist-induced climate crisis are creating increased instability for our world's water resources. Glaciers are rapidly melting, and many places across our planet are experiencing extreme climatic systems causing drought or torrential flooding in areas where weather patterns used to be milder and easier to predict. At the same time, sea levels are rising and inundating coastal lands, nearshore freshwater systems, and coastal aquifers. The effects of these systemic changes across our planet are not as easily seen on Google Earth images. While some business interests and a few key politicians would just have us bury our heads in the sand, denying science and reason, they are ultimately wagering the survival of future generations for short-term profit and growth. Since 2012, global organizations such as the World Economic Forum have consistently listed the water crisis as one of the top five risks to our global economy. Consequently, there is increased attention by governments and private interest in our world's freshwater resources. Many economic and political theorists predict that the growing scarcity of our world's water resources will be the impetus for future wars, and extreme drought conditions already affect the ability of many communities across the world to access water. States like California, Nevada, and Colorado have all attempted to deal with the lingering effects of the twenty-three-year drought on the Colorado River. While here at home in Hawai'i, we have witnessed prolonged periods without rainfall on our island summits like the Koʻolau on Oʻahu or Waiʻaleʻale on Kauaʻi. In this period of drastic climate change, our islands and the world's water future are highly uncertain.

*Waiwai: Water and the Future of Hawaiʻi* seeks to make Hawaiʻi's complex water history transparent. This book illuminates the challenges our islands face while offering solutions to secure the future of water in Hawaiʻi. While each essay in this volume provides key insight into Hawaiʻi's water past and glimpses of our future, the themes and lessons from this book are also of value to a global audience. Most of Hawaiʻi's water challenges are similarly seen across the globe. Islands can be seen as model systems for understanding social and natural systems because of their scale and complexity.[1] As Peter Vitousek once said at a First Nations Futures program, our earth is in fact a delicate and habitable island in the vast and hostile sea of space. Water dynamics and its synergies with social and natural systems are more rapidly observable on islands than in larger continental systems. This makes it possible to understand the interconnectedness of water allocation and management with past failures and potential solutions to many of the complex ecological and economic challenges found in Hawaiʻi and across the world.

The word "wai" has multiple translations, each of which offers a glimpse into the deep significance of wai from a Hawaiian worldview. "Wai" is used in the phrases "ʻO wai kou inoa?" (What is your name?) and "ʻO wai ʻoe?" (Who are you?) These usages suggest that each of us has a wai, a water or essence we bring with us. "Wai" combined with "lua" (two) into "wailua" is translated as "spirit" or "ghost," suggesting a metaphysical aspect to wai that moves and continues even outside of our bodies. When "wai" is repeated in the Hawaiian language as "waiwai" (water, water), it is translated as "wealth." This ancient insight holds true today, as those who control water also control wealth. Access to clean, fresh drinking water is, for many, seen as a human right. While for others, privatization and more accurate accounting for the real costs of providing water for society are described as necessary for a more secure global water future.

## Who Speaks for Wai?

Humans do not create water. Over the course of history, human civilizations have flourished or failed because of their relationships with water. The chapters in this book discuss myriad implications for society when we position the environmental health of water as connected to the economic success and overall well-being of society. Perhaps one of the best measures of an enlightened society is how it values or devalues water. What are the legal regimes and frameworks that care for water in contemporary times? How are we doing in Hawaiʻi today? When it comes down to it, how do we manage and mitigate our actions with regard to water? What does water justice look like in our islands today? Who speaks for wai?

This book enables the reader to see the downflow effects of different regimes of water resource management. In ancestral Hawaiʻi and for many ʻŌiwi (Indigenous Hawaiians) today, wai is an akua (god) and not an object that can be owned or possessed. In earlier times, the mōʻī (Native sovereign) held the duty to care for and manage water from ma uka (uplands) to ma kai (seashore) for the benefit of present and future generations, but wai was free to all, and access to wai was an essential part of governance. This practice became the foundation of the Hawaiian Public Trust doctrine. Later Indigenous knowledge and ancestral insight guided Hawaiʻi's water management regimes through the formation of the Hawaiian Kingdom, which had to grapple with capitalism and foreign imperial interests. Hawaiian knowledge systems were nearly erased from Hawaiʻi's water governance following the rapid rise of sugar barons in the post-overthrow period. During these years (1893–1978), Hawaiʻi's water was nearly privatized by sprawling cash crop plantations. The pendulum began to swing back when, in 1973, William S. Richardson became the first Native Hawaiian since the overthrow to serve on the Hawaiʻi Supreme Court. Armed with his knowledge of Hawaiian customary practice and the legal precedents set during the Hawaiian Kingdom, Chief Justice Richardson's rulings on pivotal water cases reasserted earlier public trust principles from the Hawaiian Kingdom culminating with the formal reestablishment of the Public Trust doctrine into the Hawaiʻi State Constitution of 1978. This complex history of shifting power dynamics over water management regimes makes understanding how Hawaiʻi's water is managed today challenging to understand. This book seeks to demystify the complex and often highly specialized facets of water resource management to allow for greater understanding, relevance, advocacy, and community management of Hawaiʻi's wai. This book is urgently needed because water resource management has become so highly specialized that many do not understand the structures that govern Hawai'i's water resources. Most would not recognize who the decision-makers are for water, and perhaps know even less about how decisions governing our islands' most precious resource are made. In far too many instances, powerful business and development interests benefit because of this limited public understanding and engagement. Much of the content you will read in this book has previously been known only by a handful of experts who often make decisions in boardrooms or, at times, behind closed doors. This book seeks to change the state of Hawai'i's water resource management. Each of the chapters is authored by an esteemed colleague or community hui who has spent countless hours, sometimes decades, working to understand and advocating for wai. I must be honest and say that some of this work is very personal. I have spent eight years of my life serving on the Hawai'i State Commission for Water Resource Management (CWRM) as one of seven trustees over Hawai'i precious freshwater resources. I have volunteered countless hours carefully listening to community testimony, reading reports and case files, working with CWRM staff, and drafting and debating policy. I have spent days hiking alongside dry stream beds and flown in helicopters to the backs of remote valleys to assess the impacts of centuries-old ditch systems diversions on the public trust. I have sat in hearings and listened to the passionate testimonies of people from almost every Hawaiian island expressing their frustrations and fears. I have also been deeply inspired by their courage and relentless commitment to aloha 'āina and water justice for their communities, streams, and aquifer systems. This work has not been easy; I have been called names, I have taken political punches, I have been the subject of conspiracy theories, and I have even occasionally been thanked for my efforts. As an 'Ōiwi who understands the occupation of our islands and the significant odds against us, I found myself trying to shift a tainted structure. This work was like forcing a pendulum to swing toward justice. There were times it was joyous, as every time we made a decision to restore a stream, it returned flow to community and 'āina. I also understood that there were dozens of activists and community visionaries before me who fought across our islands and in the Hawai'i Constitutional Convention of 1978 for the establishment of the Hawai'i State Water Resource Management Commission who made this work possible.

Most significantly, even before my time as a Hawaiʻi state water resource commissioner, I was a keiki born and raised in our islands, who followed uncles up ma uka to be with streams. I learned to "jump cliff," to "jack-knife" and "cannonball." I learned to find the hīhīwai and invasive prawns. Like many keiki in the generations before (and hopefully after) me, I developed a deep love for the wai and kahawai (streams). They were places of joy for our family to gather and replenish our bodies, minds, and souls. I learned from these early experiences that something about being in and around wai brought out the best in people. I witnessed uncles who struggled to succeed in modern-day conventions like holding nine-to-five jobs—those who were unable to tolerate the daily struggles of surviving amidst the vast socioeconomic injustice in Hawaiʻi—be at home and at peace in the wai. People battling substance abuse and self-medicating became patient and welcoming, full of smiles and joy when their bodies were immersed in a cold stream. Something as basic as a rope swing tied to a mango tree became a portal into raucously shared laughter and subtle moments of zen. These places taught keiki courage, physical control of their bodies, and to kilo (observation skills) the water cycle. As I grew, I carried these lessons with me. I spent much of my twenties and early thirties learning from kūpuna (Hawaiian elders who care for ancestral knowledge) and friends around me about wai, loʻi kalo (taro systems), streams, and estuary biota. I worked alongside many to restore ancestral systems of agriculture and aquaculture in places across our islands. I have participated in rallies and stood in solidarity with courageous community groups like Kaʻohewai and the Oʻahu Water Protectors on critical issues like the navy's poisoning of the aquifer at Red Hill. I have also spent much of my academic career seeking to uplift aloha ʻāina as liberatory practice. This work has allowed me to learn about the deep interconnectedness of water with life and the economy. It brought me across continents and oceans, where the innovations of other places have deeply inspired me. This work enabled me to establish international partnerships on the development of the circular economy and to engage with strategies such as those authored by Kate Raworth on Doughnut Economics.[2]

All of this has led me to understand that we must adapt our water management tactics and strategies if there is to be a future for Hawaiʻi. Innovations for our water governance can also be economic innovations for our islands. And WE MUST INNOVATE. Hawaiʻi is facing multiple crises, and water will be of the utmost importance for life on these islands. *Waiwai: Water and the Future of Hawai'i* is a resource book for the community to understand the many water challenges we face today and the innovations that will guide us into the future. This book provides critical insight into the challenges decision-makers must navigate and the ramifications of those decisions. It traces the complex sets of historical problems (the occupation of the Hawaiian Kingdom) as well as the indefinite future of our island's water resources in the face of the climate crisis and increased economic pressures for life in our islands to survive.

Ancestral Hawaiian knowledge teaches us that water is the cornerstone of life; contemporary scientists, policymakers, legal experts, cultural activists, and perhaps even academics could agree with this ancient insight. Yet, water allocation and management in Hawai'i since the overthrow of the Hawaiian Kingdom have often been decided behind closed doors to benefit a few. In many cases, business interests and those with access to power networks and political kickbacks have controlled the management of Hawai'i's water for over 130 years at the expense of stream ecology, taro farmers, and a sustainable future. The exploitation of Hawai'i's water resources, however, is not merely economic. U.S. military installations located in Pu'uloa (Pearl Harbor) had constructed the world's largest underground fuel storage tanks. The U.S. Navy built them directly above one of O'ahu's most important aquifers. These tanks remain exempt from the basic environmental standards that your local gas station must follow and have leaked at least 180,000 gallons of jet fuel into O'ahu's most precious aquifer since their construction in the mid-1940s. Of course, the official position of the U.S. Navy before 2022 had been that these tanks are essential for "national security." However, at least one government official has questioned that logic by stating in 2014, "if we're poisoning our aquifers, that's not a form of defense."[3]

In truth, even when done in the most prudent and safe manner, every allocation of water is like withdrawing from the savings account of future generations with the hope that nature will somehow replenish the public trust. If anything can assist in better management of and care for our precious water resources, it is providing the public with access to knowledge about Hawai'i's potential water futures and the means to restore our relationships to wai.

## Ancestral Circularities

Water in the Hawaiian Islands has always been precious. Our place in the center of the Pacific does not allow us to simply divert water from neighboring states or nations. Essentially, our resources are incredibly finite; we have only what our islands and akua (the gods) have provided. ʻŌiwi have long understood the primacy of water for the abundance and diversity of life within both natural and human systems. In many ways, wai was the element that shaped much of the development of our social, economic, and governing systems since our Maoli (Oceanic, Pacific) ancestors navigated to our majestic high volcanic islands. Our ahupuaʻa (culturally appropriate, ecologically aligned, place-specific units with access to diverse resources[4]) most often maximized unique ecosystem niches based on the connectivity of water systems that ran ma uka to ma kai. Ancestral understandings of the circular patterns of the water cycle inspired even the nature of our social governance structures. Water was the essential component for establishing the Ancestral Circular Economy.[5] Wai was the element that inspired a Kālaiʻāina (the political process of redistributing lands in ancestral Hawaiian society)—it is literally and figuratively the element that carves and shapes our landscapes over time. The fluvial systems, flow, and connectivity of wai are the veins of the ʻāina, the lifeblood of that which feeds us. Water transports materials and nutrients from the uplands to the sea while replenishing springs and aquifer systems. The observations of cyclical patterns of the forms of Kāneikawaiola were what gave our kūpuna the inspiration to create social and political systems that mimicked its patterns. The political process of a Kālaiʻāina then can be likened to the cycling and reforming of water from gas to liquid or ocean evaporations to raindrops on mountain summits. A Kālaiʻāina was a unique innovation that enabled the rebalancing and redistribution of resources with every new mōʻī. It required the leadership class to be accountable to the community and competing leadership interests. It also made it difficult for one family or line to consolidate so much wealth and authority that it could rise to the totalitarian levels of power and influence we see in the 1 percent of our time. Perhaps this is an important political and economic insight that we can learn from today—that the redistribution of resources and wealth from the stranglehold of the elite was essential for a society to thrive for centuries in the most physically isolated islands on our planet. Perhaps for humanity to survive the climate crisis, we must establish systems that behave like water and seek ways to redistribute wealth while drastically altering societies' relationships with the natural world. It is clear that our ancestral understandings of water and its natural systems manifested themselves in our social, political, and economic structures. While similar structures of today's society function in a linear fashion, this was not the case in times of old. The water cycle and system itself are described as being the realm of Kāne, one of the major Hawaiian deities that were manifested in the water cycle. Ancient ancestral chants capture this powerful understanding. Here is one such example:

**"He Mele No Kane"**

He ui, he ninau:
E ui aku ana au ia oe,
Aia i hea ka wai a Kane?
Aia i ka hikina a ka La,
Puka i Haehae;
Aia Ali'i ka Wai a Kane.
E ui aku ana au ia oe,
Aia i hea ka Wai a Kane?
Aia i ka ulana-ka-la,
I ka pae opua i ke kai,
Ea mai ana ma Nihoa,
Ma ka mole mai o Lehua;
Aia i laila ka Wai a Kane.
E ui aku ana au ia oe,
Aia i hea ka Wai a Kane?
Aia i ke kuahiwi, i ke kualono,
I ke awawa, i ke kahawai;
Aia i laila ka Wai a Kane.
E ui aku ana au ia oe,
Aia i hea ka Wai a Kane?
Aia i kai, i ka moana,
I ke kua-lau, i ke anuenue,
I ka punohu, i ka uakoko,
I ka alewalewa:
Aia i laila ka Wai a Kane.
E ui aku ana au ia oe,

Aia i hea ka Wai a Kane?
Aia i luna ka Wai a Kane,
I ke ouli, i ke ao eleele,
I ke ao panopano,
I ke ao popolo hua mea a Kane la, e!
Aia i laila ka Wai a Kane.
E ui aku ana au ia oe.
Aia i hea ka Wai a Kane?
Aia i lalo, i ka honua, i ka wai hu,
I ka wai kau a Kane me Kanaloa
He waipuna, he wai e inu,
He wai e mana, he wai e ola.
E ola no, ea!

**"The Water of Kane"**

A query, a question,
I ask you:
Where is the water of Kane?
It is there powered by the arrival of the Sun
Where the light first appears at Haehae [place-name, easternmost spot of Hawai'i];
There is the water of Kane.
I ask you again:
Where is the water of Kane?
It is where the water is woven by light [ka ulana a ka la]
In the bank of clouds formed over the ocean,
Appearing at Nihoa island,
Down to the base of Lehua island,
There is the water of Kane.
I ask you again:
Where is the water of Kane?
It is on the mountains, on the upper ridges of the peaks,
It is in the valleys, in the streams;
There is the water of Kane.
I ask of you ask again:
Where is the water of Kane?
It is seaward in the ocean,

In the rains that fall over the open sea, in the rainbow,
It is in the low-lying rainbow, in the cloud that refracts light,
In the buoyant high clouds;
There is the water of Kane.
I continue to ask you:
Where, where is the water of Kane?
It is above us, the water of Kane,
It is in the approaching of a storm, in the dark clouds,
In the thick deep blue-black clouds,
In the purplish-blue and reddish-brown clouds of Kane;
There is the water of Kane.
I ask of you again:
Where is the water of Kane?
It is below us, deep in the ground, in the water that gushes out as springs,
In the aquifers of Kane and Kanaloa,
Appearing as springs, as water to drink,
As water that enables us to exercise agency—
As water that enables life!
Let there be life! Ea [breath, sovereignty]!

The chant "He Mele no Kāne" (a song for Kāne, akua of the water system) demonstrates the vast knowledge of the intricacies of the water cycle that was needed to compose this chant. It documents scientific processes of orographic lift while also illustrating the connection between the sun as the source for powering the water system and the cycle over the ocean. It tells us how the water droplets responsible for rainbows take on varying colors depending on the time of day and the location of the sun. It documents varying cloud formations and their relative hue in relation to their capacity for precipitation. Perhaps most astonishingly, it documents these unique aspects of the water cycle in a style that can be memorized by children and elders alike. What, then, is "He Mele no Kane"—song?, poetry?, prayer?, science? Perhaps it is a combination of all of these. A unique gem of what Indigenous knowledge can cradle and teach our world about human's relationship to water and the fierce reverence with which this relationship needs to be respected and maintained. What would it be like to drink and breathe in a society that deified the natural systems that were responsible for maintaining life? What kinds of reverence and space might one hold for water and its relationship to society?

Each of the contributors to *Waiwai: Water and the Future of Hawai'i* makes an urgent plea for people to rethink and restore their relationships with water. Perhaps one of the most readily achievable individual actions we can do to repair our fractured relationship with water is to speak the truths found in this book to power. We can use the information found in this collection to stand against short-term thinking and greed, which have diverted flow away from humans' innate relationship with water. In many ways, our material and cultural future are dependent on our relationship with water. While water's relationship with Hawai'i and our world is rapidly changing, *Waiwai: Water and the Future of Hawai'i* is a resource for us to understand that change.

The following chapters are authored by some of Hawai'i's and our world's premier legal, scientific, and cultural minds, as well as prominent community organizers. Each of our authors and the individuals interviewed for this book have demonstrated a significant commitment to wai advocacy and pushed Hawai'i's understanding of the nuances of water management and governance toward a better future. We have worked diligently to present their unique and specialized knowledge about water, justice, climate change, and our islands' future without technical and scientific jargon. Our book is divided into four parts, and each part begins with a brief introduction written by me. Part 1 is "Ola i ka Wai: Foundations of Water Law and Governance," where we explore the foundational economic and legal precedents established through wai. Part 2, "Threats to Hawai'i's Water Futures," features discussions on our current climate crisis, sea level rise, and the Red Hill catastrophe. Part 3 is titled "Fluid Relationships: Management and Stewardship" and features essays on stream restoration and community relationships with water. In Part 4, "Let Wai Flow," I discuss some of the conclusions and significant interventions that need to take place for the future of Hawai'i's water.

In our opening part, "Ola i ka Wai: Foundations of Water Law and Governance," we provide the reader with the historical and legal constructs to understand how Hawai'i has reached this pivotal moment through a collection of thoughtful essays on law and governance. In our first chapter, "Wai Governance and the Ancestral Circular

Economy," Kamanamaikalani Beamer, Pua Souza, and Kawena Elkington discuss water's role in shaping ancestral governance and the economy of early Hawai'i. Kawena Elkington, Pua Souza, Ikaika Lowe, and Kamanamaikalani Beamer, in chapter 2, "Law and Water Management in the Hawaiian Kingdom," provide an analysis of water management in the Hawaiian Kingdom and the effects of the illegal overthrow on Hawai'i's water resources. "Reframing Wai as Waiwai: The Public Trust as Paradigm in Hawai'i Nei" by law professor D. Kapua'ala Sproat and Mahina Tuteur offers us a powerful glimpse into the future implications for water and the Public Trust doctrine in Hawai'i by highlighting efforts in Wai'oli, Kaua'i, that elevate the "duty to aloha 'āina." In the final chapter of that part, William Tam offers his perspectives on the formation of the Hawai'i State Water Code and the politics of creating the Hawai'i State Commission on Water Resource Management.

In part 2, "Threats to Hawai'i's Water Futures," we focus on two of the converging crises affecting our islands' wai—the climate crisis and the Red Hill emergency at Kapukākī. We have brought together leading climate scientists to discuss the implications of climate change on Hawai'i's water resources: Dr. Thomas W. Giambelluca, one of Hawai'i's premier hydrologists, provides a gripping analysis of the future of Hawai'i's rainfall scenarios and the implications for the freshwater cycle in our islands. Esteemed climate scientist Dr. Charles H. Fletcher III provides a concise yet robust view of the impacts of climate change on Hawai'i's coastal areas and water systems. And the O'ahu Board of Water Supply offers a chilling narrative focused on the future of the Red Hill aquifer system and the at times leaky underground storage tanks of the U.S. Navy (this essay was initially written in 2019, prior to the more recent spill in 2021). This section closes with the powerful essay "Uē ka Lani, Pouli ka Hōnua: The Community Fight for Water and Life" by O'ahu Water Protectors member Wayne Tanaka and Sharde Mersberg Freitas, which features the voices of several community advocates organizing on the Red Hill crisis.

In part 3, "Fluid Relationships: Management and Stewardship," we have gathered a number of thoughtful essays challenging the reader to consider the complex relationships required to mālama our wai. It begins with Dr. T. Kā'eo Duarte's reflective piece discussing his lessons learned from working in Hawai'i's streams for the last twenty-plus years.

Dr. Ayron M. Strauch, the primary stream ecologist for the Hawai'i State Commission on Water Resource Management, offers his perspective on the integrated approach needed for stream restoration in the islands. The section ends with a brief chapter on the hierarchy of water management in Hawai'i that will be useful for readers to understand some of the relevant regulators across our islands.

We close our book with two short pieces from the editor of *Waiwai: Water and the Future of Hawai'i,* Dr. Kamanamaikalani Beamer. The final part, "Let Wai Flow," includes Dr. Beamer's final remarks as a commissioner on the CWRM, and the book ends with an epilogue titled "What Will We Do? Seven Actions for the Future of Water in Hawai'i."

So what will we do for the *future* of wai in our islands? This book is just one resource that can help us to understand our sacred relationship to wai. None of us create water, but we all have a kuleana—a duty to ensure that all the lives that come after us, the trees, the fish, the ferns, the shrimp, the babies, and keiki we bring into our world, will have the fresh, clean, and life-giving waters of our islands. For the future of all Hawai'i, E OLA I KA WAI!

## Notes

1. Kamanamaikalani Beamer, Te Maire Tau, and Peter M. Vitousek, *Islands and Cultures: How Pacific Islands Provide Paths toward Sustainability* (Yale University Press, 2022).
2. Kate Raworth, "Doughnut Economics," keynote speech, University of Hawai'i 2023 Pi'o Summit: Advancing a Circular Economy, https://www.youtube.com/watch?v=fVp-ZtTwqU4.
3. Kamanamaikalani Beamer, "Quote of the Month," *Environment Hawaii* 25, no. 5 (November 2014): 2.
4. L. Gonschor and K. Beamer, "Towards an Inventory of Ahupua'a in the Hawaiian Kingdom: A Survey of Nineteenth- and Early Twentieth-Century Cartographic and Archival Records of the Island of Hawai'i," *Hawai'i Journal of History* 48 (2014): 53–67.
5. K. Beamer, K. Elkington, P. Souza, A. Tuma, A. Thorenz, S. Köhler, K. Kukea-Shultz, K. Kotubetey, and K. B. Winter, "Island and Indigenous Systems of Circularity: How Hawai'i Can Inform the Development of Universal Circular Economy Policy Goals," *Ecology and Society* 28, no. 1 (2023): 9.

## Bibliography

Beamer, Kamanamaikalani. "Quote of the Month." *Environment Hawaii* 25, no. 5 (November 2014): 2.

Beamer, K., K. Elkington, P. Souza, A. Tuma, A. Thorenz, S. Köhler, K. Kukea-Shultz, K. Kotubetey, and K. B. Winter. "Island and Indigenous Systems of Circularity: How Hawai'i Can Inform the Development of Universal Circular Economy Policy Goals." *Ecology and Society* 28, no. 1 (2023): 9.

Beamer, Kamanamaikalani, Te Maire Tau, and Peter M. Vitousek. *Islands and Cultures: How Pacific Islands Provide Paths toward Sustainability.* Yale University Press, 2022.

Gonschor, L., and K. Beamer. "Towards an Inventory of Ahupua'a in the Hawaiian Kingdom: A Survey of Nineteenth- and Early Twentieth-Century Cartographic and Archival Records of the Island of Hawai'i." *Hawai'i Journal of History* 48 (2014): 53–67.

Raworth, Kate. "Doughnut Economics." Keynote speech, University of Hawai'i 2023 Pi'o Summit: Advancing a Circular Economy. https://www.youtube.com/watch?v=fVp-ZtTwqU4.

PART I

# OLA I KA WAI

## Foundations of Water Law and Governance

"Ola i ka Wai: Foundations of Water Law and Governance" explores key historical and contemporary components necessary for understanding Hawai'i's water resources. Relationships between 'Ōiwi and the natural environment were foundational in the establishment and management of systems that governed wai in ancestral Hawaiian society. These systems resulted in balance and productivity, and were perpetuated by ali'i for hundreds of generations. Leading up to and after the illegal overthrow of the Hawaiian Kingdom, wai was forcefully transitioned from being an abundant public resource to a commodity benefiting private sectors. Understanding what followed this transition through legal actions, codes, or rights held within the public trust doctrine is essential to address the current and future state of Hawai'i's water resources.

CHAPTER 1

# Wai Governance and the Ancestral Circular Economy

*Kamanamaikalani Beamer, Pua Souza, and Kawena Elkington*

Our earliest moments of life begin in water. The fluids that hold us in our mothers' wombs provide space for life to grow, protecting and feeding us until we find ourselves ready to take our journey into the ao honua (day, light, time, the earthly and visible realm). After we leave our mothers' waters, the rest of our lives are spent being nourished, cleansed, and cared for by the waters that flow from the earth. Our ancestors formed relationships with and understood these waters as an essential part of sustaining life and maintaining balance in health and well-being. From an 'Ōiwi (Native Hawaiian) worldview, wai (freshwater) is recognized and cared for as a physical manifestation of the akua[1] Kāne. Kāneikawaiola (Kāne that produces life-giving waters) is one of many forms of Kāne revered for its sacred natural elements and cycles that support life through the movement and regeneration of freshwater.[2] Relationships between 'Ōiwi, akua, and the natural environment served as a foundational pillar in the establishment and management of systems that governed our freshwater resources and organized fundamental social structures.

The function and flow of water as the essential element for growth, along with the intricate processes of the water cycle, were intimately understood by our ancestors. Wai is the tethering force between all life, connecting the peaks of our highest mountains to the smallest of kāhuli (endemic Hawaiian land snail) in our forests. Numerous mo'olelo (stories and histories), oli (chants), and hula also identify akua Laka, Lono, and Kanaloa throughout the stages of the water cycle, making clear the vital roles of several akua within processes of evaporation, condensation, and precipitation.[3] Kinship between these akua formulate interactive cycles of water based on reciprocity and redistribution. Wai moves and transforms as it rotates through each phase of the cycle, creating with it new opportunities and spaces for regrowth and revitalization. These natural cycles and their regenerative properties guided and shaped social values within the use and allocation of resources, along with the inheritance of responsibilities among governing roles. A water cycle–informed systematic redistribution of resources created holistic and well-balanced structures of governance within ancestral societies.

Wai is found in many forms throughout ʻōlelo Hawaiʻi (Hawaiian language) and is often attached to things that hold essential value in both health and daily society. Its reduplication in the word "waiwai" can be interpreted as wealth, or primary importance. Attaining waiwai was often connected to one's knowledge of place, along with their relationships to their surroundings. Waiwai reflected an abundance of resources and an ability to reproduce life. Maintaining healthy water cycles was about survival, a responsibility to protect Kāne so that he will give us life in return.[4] These intimate relationships allowed for a way of living and being in which you yourself were an embodiment of the waiwai of your surroundings. Understandings of the circularity of the water cycle even contributed to the design of structures of ancestral governance such as Kālaiʻāina. Kālaiʻāina was modeled after the water cycle that "carved out the lands," while continuously redistributing and moving resources from mauka to makai. Early law revolving around the use and management of freshwater has also been associated with the word "kānāwai," now used as a general term for law and regulation.[5] In ancestral times, legal title and ownership to water in relation to land or domestic use was nonexistent. Aliʻi and konohiki acted as trustees[6] and oversaw circular systems of freshwater management for the benefit and well-being of all who depended on the resource. Water used for ceremony, crop production, drinking, or other purposes was attributed to Kāneikawaiola, whose health is interconnected with the quality and longevity of all other forms of life.[7] As we begin to better understand how environmental cycles were related to governing systems in ancestral society, we better inform ourselves about how and why to manage this precious resource in today's communities.

## Wai, Waiwai, and Kālaiʻāina—Water's Influence on Ancestral Economy and Governance

Wai was seen as the most essential and primary organizer of life in ancestral Hawaiʻi. It was the element that made life possible, and the sanctity of this resource was deeply respected. Discussed in the introduction of this book, "He Mele no Kāne" offers us a glimpse into the interconnectivity of water resources as documented through the words of our ancestors. The changes in the water cycle are easily observable in high island systems where streams, springs, and the ocean are geographically close to one another. Over the generations, our ancestors studied these interactions and the ways in which the flow and return of water could alter the landscape while also giving life to all that was in its path. They named and categorized thousands of place- and attribute-specific kinds of rain,[8] cloud formations, the ways the mist formed in the mountains, and dew on the blades of grass. They observed, named, and consumed species of fish, crustaceans, and mollusks that were dependent on the flow of freshwater and its interactions with the ocean. They even gave particular names to different regions of the stream in relation to the flow of water.

For instance, our ancestors designated an area a muliwai, to specifically reference the latter portions of a stream and its interface with the ocean—an area that provides essential nutrients and environments for aquatic organisms. While "muli" has the connotation of being the last in a procession, it can also carry the meaning of being a source of something, as in "ma muli o," often translated as "because of." Naming the region where freshwater mixes with saltwater as "muliwai" or "because of freshwater" connotes the essential life-giving function of freshwater to this ecosystem. It identifies this region as being freshwater dependent and existing only "because of" its freshwater interface; it thus makes clear that wai brought life and essential nutrients to the biggest and smallest of organisms within this space.

Wai's flow in the forms of streams or in the mist on leaves in our forests was what enabled both people and species to thrive across all of Hawaiʻi's diverse ecosystems. The ancestral society's drive to thrive and function like wai can be seen in its use of the term "waiwai" or "water, water" to describe the condition and state of wealth. A wealthy place was one that had access to water resources, while a wealthy person was one who could feed and provide for those around them. Dr. George Hueu Kanahele describes waiwai as follows:

> Waiwai is the Hawaiian word for wealth. It means water-water. In the book of Hawaiian symbols, water is the life-giving element. It is the perfect symbol for the process of giving that was the Way every true seeker of wealth must follow. This was the Way of Kāne, who gave the breath of life to man; of Lono, who gave the cloud-bearing rains; of Kū who gave the canoe and so many other crafted things; of Kanaloa, who gave the mighty sea with all its riches; of the innumerable gods for whom generous giving was the fullest measure of their divinity, of their mana. The gods are the great givers, for in that is their purpose and their greatness; an ungiving god was, to the Hawaiian believer, a contradiction beyond imagining. Giving, the Way to Wealth, was preordained from primeval times by the divine exemplars, who were honored in the sacred moʻolelo.[9]

When our water cycle is allowed to flow uninterrupted, so too will our cycles of life. Kanahele sheds light on giving as a generator of wealth, an action that mimicked those of our akua. Poorly managed and stagnant wai reflected a lack of commitment to caring for one's resources, thus resulting in the inability to give generously. In this sense, "wealth was not an economic quantum but a social ideal."[10] In a contemporary context, wealth is commonly understood within its relationship to capital and product. A survey of current and popular social media platforms will reveal words like "bling" or "bougie" commonly used to highlight forms of richness or abundance, illustrating contemporary societies' obsession with not just having but ostentatiously flaunting flashy products or materials as symbols of wealth. Social media is one of the most influential and powerful tools of this decade, making it the perfect space for massive market-driven corporations to exploit and perpetuate these misguided ideas of wealth.[11]

"Waiwai" being utilized as both the term and symbol of wealth essentially grounded the goal of Hawaiʻi's ancestral society's economic mode of production to a state where water is healthy, shared, and abundant for all. The goals of our ancestral leadership often focused on creating circular systems that functioned like the water cycle, along with producing a state of flow where things are not allowed to stagnate.

Systems worked to flush out the bad while continuously feeding, redistributing, and providing for the needs of society—from the smallest to the largest forms of life. How might the images from social media change if and when someone were to see wealth as intimately bound and idyllically symbolized by our planet's primary life-giving resource?

Wai is found throughout the origins and births of various elements and foundations of life within one of the most significant sources for understanding ancestral perspectives on the creation of our universe—the cosmogonic creation chant, Kumulipo. The Kumulipo is composed of more than 2,000 lines and broken up into sixteen wā (period of time, era), each of which is divided into two major sections: Pō (establishes the conception and creation of the universe and earth; animals, plants, marine life, and akua are born) and Ao (humans are born, genealogies are further established). It was composed for the birth of ali'i Kalaninui'īamamo, who was born centuries prior to the arrival of Cook in Hawai'i.[12] A variation of the phrase "O kane iā Wai'ololī, o ka wahine iā Wai'ololā" (the male essence in the narrow waters, the female essence in the broad waters) appears in the first, second, third, fourth, and eighth wā of the famous creation chant.[13] Additionally, water is a central element in Wā 'Ekahi, in which life emerges from darkness and primordial slime. In each successive birth from the first coral polyps to the various algae and nearshore succulents, each species is born with the phrase, "He wai ka 'ai a ka lā'au" (water is the source of food for trees, plants, forests, etc.), making clear that the forests and plants are not just connected to but also play roles in ensuring continued flow within the water cycle.[14]

Ancestral understandings of the myriad forms of Kāne, intimate knowledge of place, and water's essential life-giving functions for society inspired ali'i to establish ancestral systems of governance that emulated the water cycle. One of the most important structures and processes of governance was the Kālai'āina. This process was named by a combination of the words "kālai" (to carve) and "'āina" (land or that which feeds). The Kālai'āina mimicked circular and regenerative aspects of the water cycle, its goal being to create a systematic redistribution of resources to achieve balance and productivity between the ali'i, 'āina, and all others.[15] Dr. Lilikalā Kame'eleihiwa succinctly describes Kālai'āina thusly: "The mechanics of the Kālai'āina were such that upon the death of a Mō'ī, all 'Āina would automatically revert to the new Mō'ī. He or she then would redistribute these ʻĀina according to the advice of his Kālaimoku (divider of the island), keeping in mind the aid certain Aliʻi Nui had proffered to the Mōʻī on his rise to power."[16] Kameʻeleihiwa's descriptions highlight the cyclical nature of the Kālaʻiʻāina process. The death of a reigning mōʻī (ruler, monarch) and the subsequent passing of land awards to the incoming mōʻī was like water evaporating and returning to the sky and clouds. The mōʻī, as a representation of the akua, would metaphorically foster the new precipitation of the future waiwai over the island. The new mōʻī would now have the responsibility of redistributing the lands amongst the chiefly circle. In this way, he or she would be metaphorically carving out the ʻāina to increase abundance, much like water does when it rains and transports nutrients from upland regions into oceans. This process of redistributing resources was a key development that enabled a sustainable and intergenerational approach toward resource management, consumption, and the inheritance of resources. Rather than a system that rewarded the accumulation of resources and the passing onto the firstborn heir (as was the case in primogeniture), this ancestral Hawaiian system sought to provide a mechanism to cycle resources in alignment with their observations of the water cycle.

Similarly to the Kālaiʻāina, the Makahiki procession and ceremony was intimately tied to observations in the water cycle, specifically the return of the wet season over the islands.[17] Unlike the Kālaiʻāina, which was a generational redistribution of resources, the Makahiki was an annual period of renewal, celebration, and ceremony as related to the season of the akua Lono[18] during winter months. Storms brought heavy rains and overflowing streams, while animals, wildlife, and the earth came alive with activity. During the Makahiki, different parts of our islands experienced an overabundance of rainfall, often resulting in the ceasing of farming activities. In other, more commonly dry areas of our islands, the abundance of rainfall and water was utilized to plant food crops that could be redistributed throughout the year. People were immersed in ceremony and play, while the aliʻi nui traveled in processions across the islands, stopping at various moku (large land division or district) and ahupuaʻa (land division) to conduct ceremonies and collect waiwai (goods or provisions produced by makaʻāinana and gathered by konohiki).[19]

These processions honored Lono through ceremonies and rituals, and were often strategic in that aliʻi were able to reestablish relationships with maka'āinana, konohiki, and those who cared for 'āina as means to better understand what was required of them as leaders. It allowed for an annual review of the productivity of 'āina and the diversity of resources available across districts, while also providing an exchange of goods in the celebration of the akua and arrival of a new year.

### Wai in the Ancestral Circular Economy

In ancestral Hawai'i, wai, waiwai, Kālai'āina, and the Makahiki acted as pillars of Hawai'i's economy. Systems of resource management were based on circular models that encouraged balance and regeneration between people and nature. Current circular economy (CE) models adopted by countries and cities around the world have created similar structures in an effort to "close the loops" within their own markets.[20] Unlike our current linear economy, a closed-loop system repurposes and recycles materials and resources within the consumption stream for as long as possible, both preserving product value and minimizing waste. CE models reflect the regenerative nature and organic cycles of our environment in an effort to decouple economic growth with environmental degradation.[21]

Early forms of CE models are seen within Indigenous economies and systems of resource management, such as those that once flourished in Hawai'i. Indigenous Ancestral Circular Economies (ACE) achieved equity and balance through circular resource management models.[22] Present-day CE policy and implementation frameworks lack a considerable amount of focus on social equity and institutional reform.[23] One analysis of 114 definitions of the circular economy found that modern CE models are frequently depicted as "a combination of reduce, reuse and recycle activities" with little to no discussion on the impact of CE on social equality or community stakeholders.[24] The Indigenous ACE, however, created robust institutions that emphasized social and relational well-being over individualism.

As CE models continue to gain traction across the world, we are seeing more discussions and calls to action for Indigenous-led economic reform. The United Nations Development Programme recently addressed CE within Indigenous communities as "economies centered around holistic approaches, where one process or action feeds into another, fostering resilience, reciprocity and respect between people and nature."[25] In 2019, the Ellen MacArthur Foundation[26] held its very first

CE Pacific Summit in Rotorua, Aotearoa, where CE proponents, Māori experts, and community leaders explored the connections between Indigenous epistemologies and the CE. Reflections on the summit highlighted connections between the two, in that "the term circular economy may not come naturally to te ao Māori, but in terms of what that means and in practice—we do it."[27]

The ACE allowed for a natural ebb and flow between people and nature that resulted in social, cultural, and economic stability. In the Hawaiian ACE, nearly every resource and material held a purpose and function within a broader cyclical system that, like Kālai'āīna, mirrored the structure of our water cycle. Resource and material management within the Hawaiian ACE varied from place to place. Communities worked to ensure that resources were continuously cycled through a process of gathering, distributing, and recycling back into the environment. These practices ensured continuity over decades of management.

Structures designed using circular models were also embedded within smaller ACE management systems. For example, 'auwai (ditches; canals) were created to divert small quantities of water from mainstreams and imitated the natural organization of wai. 'Auwai systems connected three primary parts that together created a circular water management system: po'owai (the headwater; dam), makawai (small waterways), and ho'iwai (outlet at which water is returned to the mainstream). This was a popular practice in the cultivation of wetland lo'i kalo, where water flowed throughout an 'auwai system, providing and receiving nutrients within each lo'i before returning back through the ho'iwai.

Those who were tasked to oversee the division of water were known as kahuwai, appointed by the konohiki to inform the distribution and irrigation of freshwater sources. The amounts of water distributed were determined based on need and the number of people who depended on stream flow for crop production. Kahuwai were responsible for the maintenance and care of the 'auwai, inspecting the po'owai daily to determine its condition. This was especially important during times of extreme drought or extreme flooding. Any blockages or damages to the 'auwai that might harm the natural flow of water were quickly fixed in order to prevent any further obstructions to others who relied on the water after its ho'i to the mainstream. If the 'auwai was poorly managed and negatively impacting water flow, any persons tasked with its care would lose rights to their water until it was shown to be tended to properly.

Since water is one of the most precious resources in an island society, water laws were strictly obeyed and adhered to. If one was found disobeying any law related to water use or care, they became subject to harsh consequences. If anyone were to be caught tampering with or even breaking a poʻowai, they were to be "slain by the shoulders of that dam," their bodies used "as a temporary stopgap, thus serving as a warning to others who might be inclined to act similarly."[28]

With wai as an informant and guiding model, Kālaiʻāina as the implementing institution, Makahiki as an annual assessment, and waiwai as the intended outcome, the Hawaiian ACE maintained the ability to sustain a population upwards of one million people (figure 1.1).[29] Grounded in familial relationships between ʻŌiwi and the surrounding environment, it enabled equity and balance in ancestral societies.

**Figure 1.1** Four components and functions of the Hawaiian Ancestral Circular Economy.
*Source:* Pōʻai ke Aloha ʻĀina Research Lab.

The key principles of the Hawaiian Ancestral Circular Economy are: 1) prioritization of relationship building; 2) balanced governance structures; 3) systematic and regular redistributions of wealth and power; 4) promoting regenerative socioecological processes; and 5) environmental kinship.[30] Through a *give, take, regenerate* model,[31] the ACE worked to ensure that resources and materials were filtered through systems of management designed to promote ʻāina momona[32] (sustainable resource abundance with perpetual surplus) and support the health of socioecological systems. The Hawaiian ACE was developed from profound ancestral understandings of wai and ʻāina. It strengthened the ability to create regenerative social systems that harmoniously interacted with our natural ecosystems, creating governance that could be renewed with each new generation or leader without destabilizing the balance between ʻāina and people.

Wai was also included in the establishment of land and resource boundaries. Our ancestors developed a complex system of palena (boundary, limit) of which demarcated the boundaries of ahupuaʻa and moku divisions. An ahupuaʻa was organized using palena and can be interpreted as a "culturally appropriate, ecologically aligned, and place specific unit with access to diverse resources."[33] Palena were often informed by water resources and instrumental in balancing the management of resources between the inhabitants of places and the aliʻi. Inoa ʻāina or place-names were also often classified by their water resources. There are at least 121 ahupuaʻa that begin with the word "wai" across all of Hawaiʻi.[34] Some are very specific, such as Waiāhole or Waiʻanae on Oʻahu, which reference the place's resource relationships between freshwater and a marine species of fish. There are even common names, such as Waimea, that occur on Hawaiʻi, Oʻahu, and Kauaʻi islands that reference the condition of water as it interacts with the specific kinds of soil present in these places.

Additionally, the palena of larger districts or moku have held important roles in management systems. Recent scholarship has suggested using the term the "moku system" to "describe the Hawaiian biocultural resource management system, which divided large islands into social-ecological regions and further into interrelated social-ecological communities."[35] These ecological zones "created a mosaic that contained forested landscapes, cultural landscapes, and seascapes, which synergistically harnessed a diversity of ecosystem services to facilitate an abundance of biocultural resources."[36] The care and flow of wai were a principal part of the larger moku system. Regions such as Halele'a on Kaua'i even used a unit of management called a kalana (division of land smaller than a moku) to best make use of several streams in the region that crossed ahupua'a boundaries. The traditional palena of moku are also of significance for wai as they often correspond with contemporary understandings of the boundaries between hydrological units for groundwater management. Figures 1.2 and 1.3 showcase the alignments between ancestral moku boundaries on Hawai'i Island and present-day groundwater hydrologic unit boundaries.

Groundwater, aquifers, springs, and sources of drinking water were known to our ancestors all across the islands. They were named and noted as life-giving resources, frequented and cared for by those who relied on them. One example of this can be found in Honomaka'u, Kohala, where a specific submarine source of freshwater is recorded within an article in the Hawaiian language newspaper *Ka Makaainana*. The article provides invaluable insight into the ways deep-sea wai was accessed by the people of Kohala, who recognized the significance of this water source through naming and skillfully collecting its contents.

**Wai Momona Malalo o ke Kai**

O Wai Puapuao, aia maluna aku o Haea, he 24 kapuai ka hohonu o ke kai. He wai huihui keia mai ke kuahiwi mai, eia i ka hohonu o ke kai. O na aina kokoke i keia wai, o ia ko lakou wai e inu mau ai. I keia wa, aole kiina no ka mea, ua lako i na luawai a ua makauia i ka i'a huhu o ka moana, a kakaikahi no hoi na poe ike i ka luu.

Ma ka makahiki 1843, ua hele au me ka'u aikane ae kona kaikoeke i ka luu wai i keia Wai Puapuao, a he ku no i ka makau ke nana aku. Na ke kaikoeke i luu ko makou wai, na ka mea ahaloa i ka luu, a he kamaaina i ka luu ia wai.

Aole nae e pono i ka huewai wahanui, o komo ke kai; he huewai loloa a uuku o ka waha, i ole ai e komo ke kai. Ina e luu, pani ka waha o ke huewai, a hiki ilalao o ka hohonu, wehe ka lima a piha, pani hou ka lima a ea hou ae iluna, pela a pau na huewai i ka piha. O ka inoa o ia huewai he ole. He huewai lolo, i pili i ka poo, i pono ka luu ilalo, i ole ai e pau ke aho a hoolanaia iluna. Oia na mea hai aku ia oe e like me kau noi. Ina papau a hohonu, ua like no. Eia ka wai luu i Kohala nei, he eha anana ka hohonu iloko o ka papaku o ka moana.

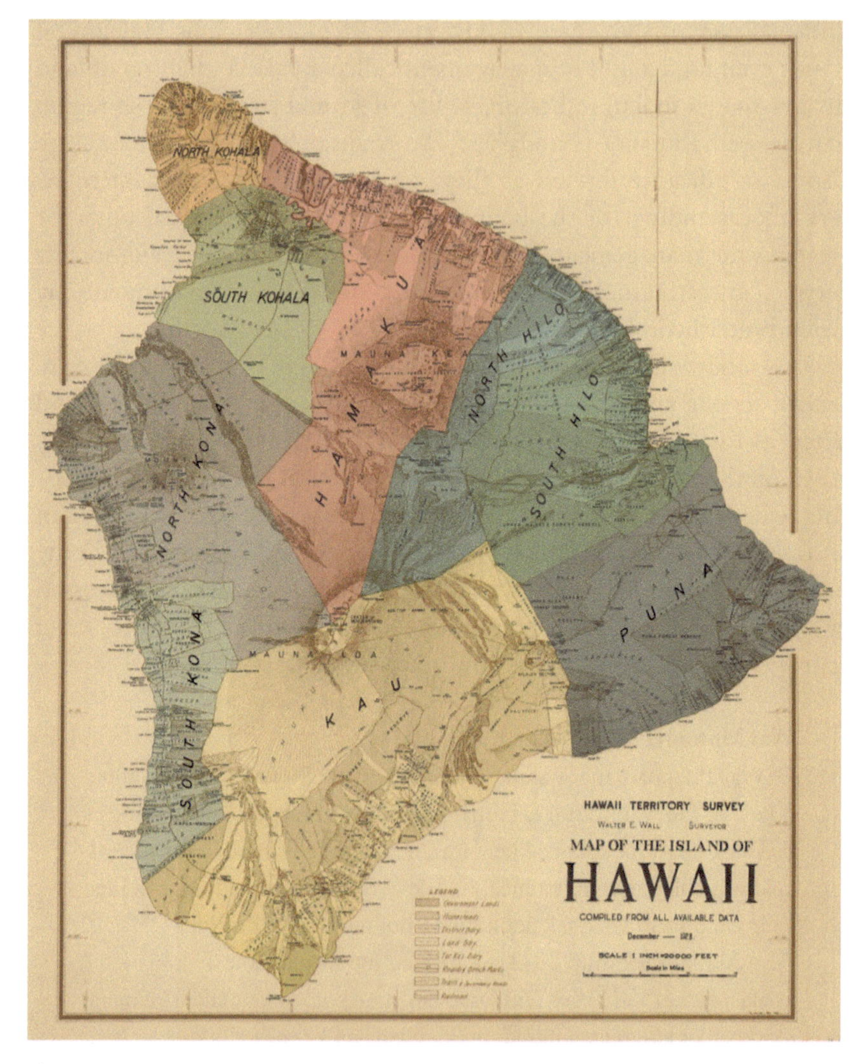


**Figure 1.2** 1928 map of Hawai'i Island with moku names and boundaries. Map authors: Joseph Iao and Walter E. Wall (surveyor).

*Source:* University of Hawai'i, Hawai'i Government and Territory Survey Maps Collection, https://evols.library.manoa.hawaii.edu/items/a66acb20-d476-4d26-80c3-a6fb6acd2649.

## Abundant Freshwater Source beneath the Ocean

Wai Puapuao, located upland from Haea, is twenty-four-feet deep within the ocean. The chilly water is brought down from the mountains into the depths of the ocean. It is from this area that the people of this 'āina would get their water to drink. At present, it is not

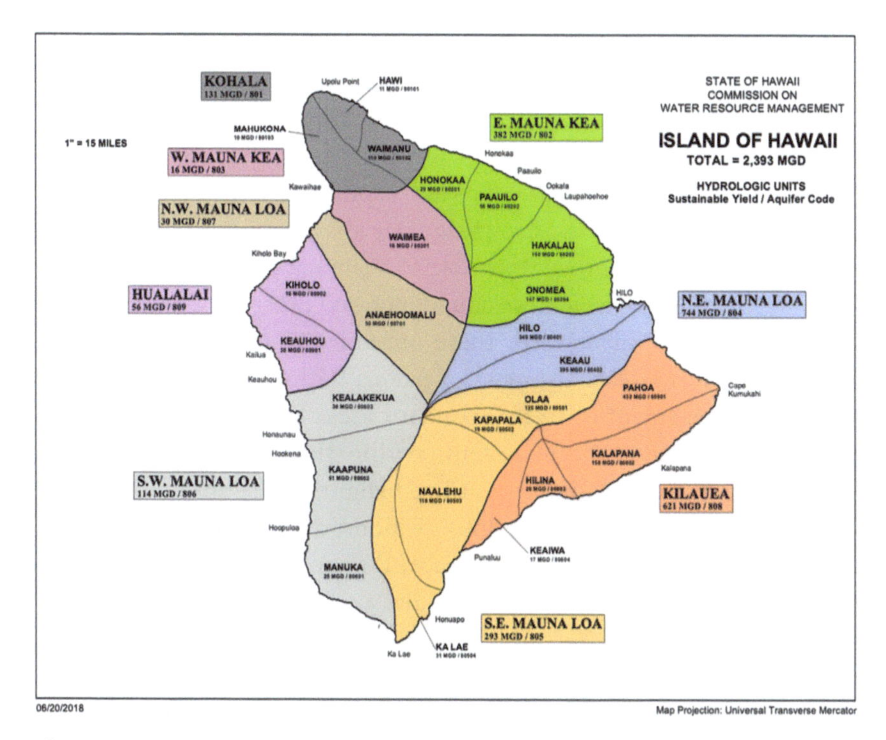


**Figure 1.3** Commission on Water Resource Management's map of Hawai'i Island groundwater hydrologic units and boundaries, https://dlnr.hawaii.gov/cwrm/groundwater/hydrounits/.

fetched because [they] are provided for by a water well and are afraid of provokable fish of the sea, and those who know how to dive for the water are rare.

In 1843, I went with my companion and his brother-in-law to dive for water at Wai Puapuao, and was halted in fear while looking at it. It was the brother-in-law that dove for our water, because he was the one who had the longest breath for diving, and was well acquainted with the waters. You don't need a water container with a large mouth lest sea water will go inside. A long container with a small mouth is used so that the salt water will not go inside. If you dive, shut the mouth of the container until you reach the deepest depths, then cover, take your hand all the way off, then cover the mouth again with your hand and go up for air, repeat this process until the container is full. The name of this container is an 'olo; it's a long container all the way to the top so that you can dive with it, and won't waste your breath from its flotation. These are the things I say to you, like you had requested, if it's shallow or deep it's the same. Here is the diving water of Kohala, its depth is four anana[37] (double arm spans) inside the seafloor of the ocean. (S. C. Luhiau, Kohala, September 3, 1896)[38]

As illustrated in the case of Wai Puapuao, ancestral society relied upon unique, place-specific knowledge sets about ground and freshwater resources. With these, they were able to sustainably access and care for these waters as a way to achieve abundance and prosperity.

As previously mentioned, these ground and freshwater resources were critical in the establishment of palena within a moku, leading to strategic management of water-abundant areas that supported broader agriculture and aquaculture systems within the ACE. Across the islands, every moku and ahupua'a utilized slightly different management techniques and relied upon kilo (observations, data collection) to inform them of the unique needs of their place. Palena informed the allocation of resources during Kālai'āina, and Kālai'āina informed cycles of management within the ACE—which are still in existence today. Kawainui fishpond is a clear case in point.

Located in Kailua, O'ahu, Kawainui fishpond sits at the piko of the ahupua'a and was once surrounded by 'auwai that carried freshwater into lo'i kalo and fed into other dryland crops. Life-sustaining provisions such as the delicious lepo'ai produced by Kawainui were sought after by ali'i and maka'āinana alike, and it is said that this single fishpond could grow up to 500,000 pounds of fish per year.[39] Present-day restoration efforts of Kawainui have been growing with every new generation. Using ACE frameworks, 'Ōiwi, community, and cultural practitioners gather at and around the fishpond, working to revitalize and reclaim the waiwai of this revered 'āina.[40]

Across the island in Waikīkī are Kānewai and 'Āpuakehau freshwater springs. These are two of many that fed acres upon acres of lo'i kalo in O'ahu's ancestral society, making the ahupua'a one of the most productive and abundant on the island. In the midst of mass social and economic changes happening in the era of the Hawaiian Kingdom, these springs provided the local population with means to grow enough food to feed not only their own ahupua'a, but neighboring ones as well. In the early 1900s, dredging and draining of Waikīkī's wetlands by the hands of Hawai'i's Territorial government drastically reduced the natural mauka-to-makai flow of these waters, forcing farmers to abandon centuries-old land management practices.[41]

Despite the tremendous loss, Kānewai and 'Āpuakehau continue to flow and nourish the people of Waikīkī today. Their waters circulate through 'auwai systems that feed lo'i throughout places like Mānoa Valley, where farmers and educators currently use the springs as sources to teach ancestral stewardship to students and community.

Standing up against decades of environmental pollution, resource mismanagement, and political injustices, the Hawaiian ACE has withstood the test of time. Today, Kailua and Waikīkī are widely known to the world as popular centers of tourism and highly sought-after vacation attractions (also often used as locations to flaunt an excess of material wealth on social media). Their once-thriving fishponds and freshwater springs have been dangerously polluted or covered up by cement, yet despite all odds, ancestral principles of circularity and regeneration remain within cultural revitalization and water management efforts within their communities. This is a testament to the power of ancestral knowledge and the vital role wai has in creating cyclical and regenerative stewardship in 'Ōiwi society.

## Wai Today

The last century of Anglo-American imperialism in the islands has made it impossible to return to the pristine island nation and the ecologically balanced governance system of our ancestors. And let us be clear and say that is not our intent. The ancestral knowledge about wai provided throughout this chapter is about our future. While wai might not be a predominant organizer within our current state of governance and affairs, there are many community groups and leaders who have been at the forefront of advocacy for wai and water justice in our islands for over a century.[42] Kāneikawaiola and other akua continue to be honored both within cultural practices and in decision-making surrounding water use or misuse today.[43] Make no mistake, wai has never lost the respect and sanctity it deserves as a primary and life-giving element within Hawaiian culture and community.

In one of his lessons during a class at the Kamakakūokalani Center for Hawaiian Studies, Hawaiian kupuna and longtime water rights activist Uncle Calvin Hoe shared with a group of University of Hawai'i students that "whoever controls the water controls the future."[44] He made clear that wai (particularly clean and well-managed wai) is a source of power. To control this life-giving and life-sustaining resource can grant us the ability to control and make decisions for the future state of our islands and communities. Uncle Calvin's words are echoed in communities like Waiāhole-Waikāne, East Maui, Nā Wai ʻEhā,[45] and other areas across Hawaiʻi that have fought for generations for water justice and control. We wonder, what would our islands look like if we created governance and community structures like our ancestors that attempted to mimic the water cycle? What if we had systems that placed such a value on water that wasting or overuse would be an egregious act of violence against society? What if we redesigned our systems of managing water to reflect the interconnectedness of this resource with the overall health and wealth of the environment and society?

For instance, in our current paradigms of management, state and county governments have multiple agencies responsible for the management of wai. Terms such as *stormwater, wastewater, groundwater,* and *surface water* not only influence our perception of how water flows, they often determine agency authority and decision-making over this precious resource. As an example, think of the early goals of managing "stormwater." At a high level, these goals are to get water out and mitigate potential flooding as quickly as possible to protect people and property. Separating "stormwater" from the other aspects of the water cycle and viewing it as its own particular branch of management led to the hardening of stream beds and the construction of fully cemented channelized drainage basins. These structures are effective in getting large quantities of water from one place to the next, while providing some protection from flood damages. There are also negative impacts.

On islands such as Hawaiʻi, pulse rain events and flashy streams that are channelized into canal systems too quickly transport harmful and even toxic runoff into the precious muliwai regions and onto reefs. Stormwater runoff can be as toxic as untreated sewage, especially in the first twenty-four hours of an intense period of rain. The harmful impacts of such systems, many of which had been built in the islands over the past one hundred years, have been tremendous. It is not hyperbole to suggest they have been slowly killing our reefs for decades—all the while accomplishing their goals within a paradigm that managed stormwater as distinct from other parts and functions of the water cycle. One can see the harsh contrast between this paradigm and framework and the ancestral system that recognized the muliwai as a source of life for the ocean. A recent report done by the American Society of Civil Engineers graded Hawaiʻi's infrastructure, marking systems for "stormwater" at a D–, while offering slightly higher marks for both "drinking water" and "wastewater" at a D+.[46] We must do better for Hawai'i's wai and embrace regimes of care and management that acknowledge the entire water cycle and its importance for life in our islands.

There are also contemporary large-scale movements toward a more unified approach to water management. The One Water movement recognizes that "all water has value and should be managed in a sustainable, inclusive, integrated way."[47] It also is somewhat critical of previous paradigms of water resource management, noting that many of the previous "assumptions and approaches used over the past 200 years to design, build, and operate our existing systems are insufficient to meet the 21st-century challenges we face."[48] This is a ripe moment for Hawai'i to adopt a more holistic One Water approach to water management in the islands, but it would be even more progressive for the islands to value and place 'Ōiwi frameworks and peoples at the forefront of this change. As our world looks for solutions to the complex set of problems posed by the 200 years of building systems and uplifting values that caused the modern climate crisis, 'Ōiwi water management practices and ancestral knowledge systems need to take the forefront in conversations around future policy for our islands.

Wai is the foundational element of Hawaiian society. It provides potential for health and abundance, along with the ability to feed and care for ourselves and others. Our ancestors intimately understood the intricacies of Kāneikawaiola, and how the health of our communities was deeply dependent on the health of this akua. Life-sustaining cycles of Kāneikawaiola are mimicked and honored in cultural practices, protocols, and in the systems that once governed our communities. Accessibility and abundance of wai were indicators of wealth and well-being in ancestral society, while the knowledge about the many forms and manifestations of water aided in forming holistic structures of resource management. Structures like the Kālai'āina mirrored the natural organization of our water, and allowed for a continuous flow of resource redistribution and regeneration within a broader Ancestral Circular Economy. Contemporary systems and leadership that now oversee the majority of our island's water have created linearity within the management of wai. We are now faced with untangling years of misuse and mistreatment that have led to excessive amounts of pollution, waste, and resource scarcity. The Kālaiʻāina and the Hawaiian Ancestral Circular Economy show us that the pono systems created by our ʻŌiwi leaders were able to produce economic surplus through modeling the circular cycles of our environment.

How might relearning the role of wai in processes like the Kālaiʻāina carve a different path for Hawaiʻi's future economy? What might a contemporary Ancestral Circular Economy look like? And what would it mean for our systems to embody a state of waiwai? Like a new mōʻī inheriting the responsibility to redistribute ʻāina amongst their people, every new generation carries with them an opportunity for new, more imaginative cycles of governance. Much like the wai that flows through an ʻauwai, our communities continue to move and adapt with changes in our ʻāina, innovating and pushing through their own challenges to keep our people and places well nourished. When it comes time for each of us to "hoʻi i ka wai," to return to the waters and soil of our honua, how will we ensure that the generations that flow after us remain cared for?

## Notes

1. Natural elements, cycles, and life-giving energies of our environment, commonly referenced as godly deities or ancestors. While we are unable to give akua the substantial amount of discussion required within this one chapter, we believe that knowing and building relationships with akua (particularly those associated with wai), is integral to understanding water's spiritual and cultural significance to ʻŌiwi. These relationships ultimately set the foundations in which our ancestral water management systems were built upon.
2. E. S. Craighill Handy and Elizabeth Green Handy, with the collaboration of Mary Kawena Pukui, *Native Planters in Old Hawaiʻi: Their Life, Lore and Environment* (Honolulu: Bishop Museum Press, 1972); Pualani Kanakaʻole Kanahele, *Ka Honua Ola: Eliʻeli Kau Mai* (Honolulu: Kamehameha Publishing, 2011).
3. For more information and discussion on akua, along with the different roles they hold within the water cycle, see Davida Malo, *The Moʻolelo Hawaiʻi of Davida Malo*, vol. 2, *Hawaiian Text and Translation*, edited and translated by Charles Langlas and Jeffrey Lyon (Honolulu: University of Hawaiʻi Press, 2020); Kanahele, *Ka Honua Ola;* Kalei Nuʻuhiwa, "Makahiki—Nā Maka o Lono Utilizing the Papakū Makawalu Method to Analyze Mele and Pule of Lono and the Makahiki" (PhD diss.,

University of Waikato, 2020, 193–198); Marie Alohalani Brown, *Ka Poʻe Moʻo Akua: Hawaiian Reptilian Water Deities* (Honolulu: University of Hawaiʻi Press, 2022).

4. Puhipau and Joan Lander, *Stolen Waters* (Nā Maka o ka ʻĀina, 2005, DVD).
5. D. Kapuaʻala Sprout, "Wai through Kānāwai: Water for Hawaiʻi's Streams and Justice for Hawaiian Communities," *Marq. L. Rev.* 127, no. 95 (2011–2012).
6. Handy, Handy, and Pukui, *Native Planters in Old Hawaiʻi.*
7. Kanakaʻole Kanahele, *Ka Honua Ola,* 98.
8. Collette Leimomi Akana with Kiele Gonzalez, *Hānau Ka Ua: Hawaiian Rain Names* (Honolulu: Kamehameha Publishing, 2015).
9. George Huʻeu Sanford Kanahele, *Kū Kanaka: Stand Tall* (Honolulu: University of Hawaiʻi Press, 1986), 364.
10. Ibid.
11. Shosana Zuboff, *The Age of Surveillance Capitalism: The Fight for a Human Future at the New Frontier of Power* (New York: Public Affairs, 2020).
12. For more information and discussion on the Kumulipo, see Martha Warren Beckwith, *The Kumulipo: A Hawaiian Creation Chant* (Honolulu: University of Hawaiʻi Press, 1972); Rubellite Kawena Johnson, *The Kumulipo Mind: A Global Heritage: In the Polynesian Creation Myth* (n.p., 2000); Noenoe K. Silva, *The Power of the Steel-Tipped Pen: Reconstructing Native Hawaiian Intellectual History* (Durham, NC: Duke University Press, 2017), 193–196.
13. The use of "male" and "female" in this translation does not conform to Anglo-American conceptions of gender. The essences of both kāne and wahine correlate to the roles they play in the creation of life forms throughout the Kumulipo.
14. Beckwith, *The Kumulipo,* 50–54.
15. "Mimic" is used throughout this chapter in reference to biomimicry practices, the emulation of models, designs, and structures found in nature for the purpose of creating more sustainable systems. For more information, see Janine M. Benyus, *Biomimicry: Innovation Inspired by Nature* (New York: HarperCollins, 2002).
16. Lilikalā Kameʻeleihiwa, *Native Lands and Foreign Desires: Pehea Lā E Pono Ai?* (Honolulu: Bishop Museum Press, 1992), 51.
17. Nuʻuhiwa, "Makahiki," 193–198.
18. Ibid. Akua that brings fertility, healing, and abundance to ʻāina and ʻŌiwi.
19. Ibid.

20. Modern circular economy models are built upon closed-loop systems in which little to no waste is generated from material and product consumption, and everything is recycled and reused in an effort to preserve product value. For more discussion, see Yuliya Kalmykova, Madumita Sadagopan, and Leonardo Rosado, "Circular Economy—From Review of Theories and Practices to Development of Implementation Tools," in "Sustainable Resource Management and the Circular Economy," edited by Callie Babbitt et al., special issue, *Resources, Conservation and Recycling* 135 (August 2018): 190–201.
21. Kamanamaikalani Beamer, "An Aloha ʻĀina Economy," in *The Value of Hawaiʻi 3: Hulihia, the Turning,* edited by Noelani Goodyear-Kaʻōpua et al. (Honolulu: University of Hawaiʻi Press, 2020), 83.
22. K. Beamer et al., "Island and Indigenous Systems of Circularity: How Hawaiʻi Can Inform the Development of Universal Circular Economy Policy Goals," *Ecology and Society* 28, no. 1 (2023): 9.
23. Vincent Moreau et al., "Coming Full Circle: Why Social and Institutional Dimensions Matter for the Circular Economy," *Journal of Industrial Ecology* 21, no. 3 (June 2017): 497–506; Sébastien Sauvé, Sophie Bernard, and Pamela Sloan, "Environmental Sciences, Sustainable Development and Circular Economy; Alternative Concepts for Trans-Disciplinary Research," *Environmental Development* 17 (January 2016): 48–56; Alan Murrary, Keith Skene, and Kathryn Haynes, "The Circular Economy: An Interdisciplinary Exploration of the Concept and Application in Global Context," *Journal of Business Ethics* 140 (May 2015): 369–380.
24. Julian Kirchherr, Denise Reike, and Marko Hekker, "Conceptualizing the Circular Economy: An Analysis of 114 Definitions," *Resources, Conservation and Recycling* 127 (September 2017): 221–232.
25. Alana Craigen, "For a Truly Circular Economy, We Need to Listen to Indigenous Voices," *United Nations Development Programme* (blog), November 11, 2021, https://www.undp.org/blog/truly-circular-economy-we-need-listen-indigenous-voices.
26. The Ellen Macarthur Foundation is a nonprofit charity focusing on designing and accelerating the transition toward a circular economy. For more information, see https://ellenmacarthurfoundation.org.
27. "Indigenous Worldviews and the Circular Economy," *Scion Connections* 32 (June 2019), https://www.scionresearch.com/about-us/about-scion/corporate-publications/scion-connections/past-issues-list/scion-connections-issue-32,-june-2019/indigenous-worldviews-and-the-circular-economy; Manatū Mō Te Taiao

(Ministry for the Environment), "Ōhanga Āmiomio Pacific Summit—Day 2 Highlights," video from Circular Economy Pacific Summit 2019, Rotorua, North Island, New Zealand, updated April 2021, https://environment.govt.nz/what-you-can-do/stories/ohanga-amiomio-circular-economy-pacific-summit-2019/.

28. Emma Metcalf Nakuina, "Ancient Hawaiian Water Rights: And Some of the Customs Pertaining to Them," *Organization and the Environment* 20 (2007): 506–509.
29. Natalie Kurashima, Denise Reike, and Marko Hekkert, "The Potential of Indigenous Agricultural Food Production under Climate Change in Hawai'i," *Nature Sustainability* 2 (February 2019): 191–199.
30. Beamer et al., "Island and Indigenous Systems of Circularity."
31. Beamer, "An Aloha 'Āina Economy."
32. Fertile, rich, fruitful.
33. Lorenz Gonschor and Kamanamaikalani Beamer, "Toward an Inventory of Ahupua'a in the Hawaiian Kingdom: A Survey of Nineteenth- and Early Twentieth-Century Cartographic and Archival Records of the Island of Hawai'i," *Hawaiian Journal of History* 48 (2014): 71.
34. Mary Kawena Pukui, Samuel H. Elbert, and Esther T. Mookini, *Place Names of Hawai'i* (Honolulu: University of Hawai'i Press, 1974).
35. Kawika B. Winter et al., "The *Moku* System: Managing Biocultural Resources for Abundance within Social-Ecological Regions in Hawai'i," *Sustainability Journal* 10, no. 10 (2018).
36. Ibid.
37. Fathom, measurement from fingertip to fingertip. It is important to note here that anana vary by height and are different for each person. Anana are measured with arms outstretched to the side.
38. "Wai momona ma lalo o ke kai," *Ka Makaainana,* November 2, 1896, 2.
39. Maya Saffery, "Mai ka piko a ke mole: Clearing Paths and Inspiring Journeys to Fulfill Kuleana through 'Āina Education" (PhD diss., University of Hawai'i at Mānoa, 2019).
40. Ibid., 239.
41. Ellen-Rae Cachola, "Beneath the Touristic Sheen of Waikīkī," in *DeTours: A Decolonial Guide to Hawai'i,* edited by Hokulani K. Aikau and Vernadette Vicuna Gonzalez (Durham, NC: Duke University Press, 2019), 283–292.
42. For more discussion on fights for water justice, along with the culpability of water-greedy companies throughout Hawai'i's history, see chapter 2, "Law and Water Management in the Hawaiian Kingdom," and chapter 3, "Reframing Wai as Waiwai: The Public Trust Paradigm in Hawai'i Nei."

43. One recent example of how cultural practices and relationships to Kāneikawaiola and other akua have guided actions and decision-making in areas of water management can be found in the fight for water justice at Kapukakī (Red Hill). For more information, see Oʻahu Water Protectors, "Learn More," https://oahuwaterprotectors.org/learn-more/, and "Aloha ʻĀina: Kapūkakī (Red Hill Bulk Fuel Storage Facility)," University of Hawaiʻi—West Oʻahu Library Resource Guide, https://guides.westoahu.hawaii.edu/alohaaina/redhillfueltanks.
44. Firsthand knowledge shared with Kamana Beamer.
45. For more information and discussion on fights for water justice in Waiāhole-Waikāne, East Maui, and Nā Wai ʻEhā, see resources shared at "Hawaiʻi—EA Hawaiian Activism Movements 1960–2010: Wai," University of Hawaiʻi at Mānoa Library, Hawaiian and Pacific Collections Resource Guide, https://guides.library.manoa.hawaii.edu/c.php?g=1126349&p=8217305; Hui O Nā Wai ʻEhā, "Historical and Cultural Background," https://www.huionawaieha.org/nawaiehainformation.
46. American Society of Civil Engineers, "2019 Hawaiʻi Infrastructure Report Card Executive Summary," https://infrastructurereportcard.org/state-item/hawaii/.
47. US Water Alliance, "One Water Roadmap, the Sustainable Management of Life's Most Essential Resource (2016)," https://uswateralliance.org/resources/one-water-roadmap-the-sustainable-management-of-lifes-most-essential-resource/ (accessed June 19, 2023).
48. Ibid.

## Bibliography

Akana, Collette Leimomi, with Kiele Gonzalez. *Hānau Ka Ua: Hawaiian Rain Names.* Honolulu: Kamehameha Publishing, 2015.

"Aloha ʻĀina: Kapūkakī (Red Hill Bulk Fuel Storage Facility)." University of Hawaiʻi—West Oʻahu Library Resource Guide. https://guides.westoahu.hawaii.edu/alohaaina/redhillfueltanks. Accessed June 19, 2023.

American Society of Civil Engineers. "2019 Hawaiʻi Infrastructure Report Card Executive Summary." https://infrastructurereportcard.org/state-item/hawaii/. Accessed June 19, 2023.

Babbitt, Callie, Gabrielle Gaustad, Angela Fisher, Gang Liu, and Weiqiang Chen, eds. "Sustainable Resource Management and the Circular Economy." Special issue, *Resources, Conservation and Recycling* 135 (August 2018).

Beamer, Kamanamaikalani. "An Aloha ʻĀina Economy." In *The Value of Hawaiʻi 3: Hulihia, the Turning,* edited by Noelani Goodyear-Kaʻōpua, Craig Howes, Jonathan Kay Kamakawiwoʻole Osorio, and Aiko Yamashiro, 83–90. Honolulu: University of Hawaiʻi Press, 2020.

Beamer, K., K. Elkington, P. Souza, A. Tuma, A. Thorenz, S. Köhler, K. Kukea-Shultz, K. Kotubetey, and K. B. Winter. "Island and Indigenous Systems of Circularity: How Hawaiʻi Can Inform the Development of Universal Circular Economy Policy Goals." *Ecology and Society* 28, no. 1 (2023): 9.

Beckwith, Martha Warren. *The Kumulipo: A Hawaiian Creation Chant.* Honolulu: University of Hawaiʻi Press, 1972.

Benyus, Janine M. *Biomimicry: Innovation Inspired by Nature.* New York: HarperCollins, 2002.

Brown, Marie Alohalani. *Ka Poʻe Moʻo Akua: Hawaiian Reptilian Water Deities*. Honolulu: University of Hawaiʻi Press, 2022.

Cachola, Ellen-Rae. "Beneath the Touristic Sheen of Waikīkī." In *Detours: A Decolonial Guide to Hawaiʻi,* edited by Hokulani K. Aikau and Vernadette Vicuna Gonzalez, 283–292. Durham, NC: Duke University Press, 2019.

Craigen, Alana. "For a Truly Circular Economy, We Need to Listen to Indigenous Voices." *United Nations Development Programme* (blog), November 11, 2021. https://www.undp.org/blog/truly-circular-economy-we-need-listen-indigenous-voices.

Gonschor, Lorenz, and Kamanamaikalani Beamer. "Toward an Inventory of Ahupuaʻa in the Hawaiian Kingdom: A Survey of Nineteenth- and Early Twentieth-Century Cartographic and Archival Records of the Island of Hawaiʻi." *Hawaiian Journal of History* 48 (2014): 71.

Handy, E. S. Craighill, and Elizabeth Green Handy, with the collaboration of Mary Kawena Pukui. *Native Planters in Old Hawaii: Their Life, Lore and Environment.* Honolulu: Bishop Museum Press, 1972.

"Hawaiʻi—EA Hawaiian Activism Movements 1960–2010: Wai." University of Hawaiʻi at Mānoa Library, Hawaiian and Pacific Collections Resource Guide. https://guides.library.manoa.hawaii.edu/c.php?g=1126349&p=8217305. Accessed June 19, 2023.

Hui O Nā Wai ʻEhā. "Historical and Cultural Background." https://www.huionawaieha.org/nawaiehainformation. Accessed June 19, 2023.

"Indigenous Worldviews and the Circular Economy." *Scion Connections* 32 (June 2019). https://www.scionresearch.com/about-us/about-scion/corporate-publications/scion-connections/ast-issues-list/scion-connections-issue-32,-june-2019/indigenous-worldviews-and-the-circular-economy. Accessed June 19, 2023.

Johnson, Rubellite Kawena. *The Kumulipo Mind: A Global Heritage: In the Polynesian Creation Myth.* N.p., 2000.

Ka Makaainana. "Wai Momona Malalo o ke Kai," November 2, 1896, sec. 2. https://www.papakilodatabase.com/. Accessed June 19, 2023.

Kameʻeleihiwa, Lilikalā. *Native Land and Foreign Desires: Pehea Lā E Pono Ai?* Honolulu: Bishop Museum Press, 1992.

Kanahele, George Huʻeu Sanford. *Kū Kanaka: Stand Tall.* Honolulu: University of Hawaiʻi Press, 1986.

Kanakaʻole Kanahele, Pualani. *Ka Honua Ola: Eliʻeli Kau Mai.* Honolulu: Kamehameha Publishing, 2011.

Kirchherr, Julian, Denise Reike, and Marko Hekkert. "Conceptualizing the Circular Economy: An Analysis of 114 Definitions." *Resources, Conservation and Recycling* 127 (September 2017): 221–232.

Kurashima, Natalie, Lucas Fortini, and Tamara Ticktin. "The Potential of Indigenous Agricultural Food Production under Climate Change in Hawaiʻi." *Nature Sustainability* 2 (February 2019): 191–199.

Malo, Davida. *The Moʻolelo Hawaiʻi of Davida Malo.* Vol. 2, *Hawaiian Text and Translation.* Edited and translated by Charles Langlas and Jeffrey Lyon. Honolulu: University of Hawaiʻi Press, 2020.

Manatū Mō Te Taiao (Ministry for the Environment). "Ōhanga Āmiomio Pacific Summit—Day 2 Highlights." Video from Circular Economy Pacific Summit 2019, Rotorua, North Island, New Zealand, updated April 2021. https://environment.govt.nz/what-you-can-do/stories/ohanga-amiomio-circular-economy-pacific-summit-2019/. Accessed June 19, 2023.

Moreau, Vincent, Marlyne Sahakian, Pascal van Griethuysen, and François Vuille. "Coming Full Circle: Why Social and Institutional Dimensions Matter for the Circular Economy." *Journal of Industrial Ecology* 21, no. 3 (June 2017): 497–506.

Murrary, Alan, Keith Skene, and Kathryn Haynes. "The Circular Economy: An Interdisciplinary Exploration of the Concept and Application in Global Context." *Journal of Business Ethics* 140 (May 2015): 369–380.

Nakuina, Emma Metcalf. "Ancient Hawaiian Water Rights: And Some of the Customs Pertaining to Them." *Organization and the Environment* 20 (2007): 506–509.

Nuʻuhiwa, Kalei. "Makahiki—Nā Maka o Lono Utilizing the Papakū Makawalu Method to Analyze Mele and Pule of Lono and the Makahiki." PhD diss., University of Waikato, 2020.

Oʻahu Water Protectors. "Learn More." https://oahuwaterprotectors.org/learn-more/. Accessed June 19, 2023.

Puhipau and Joan Lander. *Stolen Waters.* Nā Maka o ka ʻĀina, 2005. DVD.

Pukui, Mary Kawena, and Samuel H. Elbert. *Hawaiian Dictionary*. Honolulu: University of Hawaiʻi Press, 1986.
Pukui, Mary Kawena, Samuel H. Elbert, and Esther T. Mookini. *Place Names of Hawaiʻi*. Honolulu: University of Hawaii Press, 1974.
Saffery, Maya. "Mai ka piko a ke mole: Clearing Paths and Inspiring Journeys to Fulfill Kuleana through ʻĀina Education." PhD diss., University of Hawaiʻi at Mānoa, 2019.
Sauvé, Sébastien, Sophie Bernard, and Pamela Sloan. "Environmental Sciences, Sustainable Development and Circular Economy; Alternative Concepts for Trans-Disciplinary Research." *Environmental Development* 17 (January 2016): 48–56.
Silva, Noenoe K. *The Power of the Steel-Tipped Pen: Reconstructing Native Hawaiian Intellectual History*. Durham, NC: Duke University Press, 2017.
Sproat, D. Kapuaʻala. "Wai through Kānāwai: Water for Hawaiʻi's Streams and Justice for Hawaiian Communities." *Marq. L. Rev.* 127, no. 95 (2011–2012).
US Water Alliance. "One Water Roadmap, the Sustainable Management of Life's Most Essential Resource (2016)." https://uswateralliance.org/resources/one-water-roadmap-the-sustainable-management-of-lifes-most-essential-resource/. Accessed June 19, 2023.
Winter, Kawika B., Kamanamaikalani Beamer, Mehana Blaich Vaughan, Alan M. Friedlander, Mike H. Kido, A. Nāmaka Whitehead, Malia K. H. Akutagawa, Natalie Kurashima, Matthew Paul Lucas, and Ben Nyberg. "The *Moku* System: Managing Biocultural Resources for Abundance within Social-Ecological Regions in Hawaiʻi." *Sustainability Journal* 10, no. 10 (2018).
Zuboff, Shosana. *The Age of Surveillance Capitalism: The Fight for a Human Future at the New Frontier of Power*. New York: Public Affairs, 2020.

CHAPTER 2

# Law and Water Management in the Hawaiian Kingdom

*Kawena Elkington, Pua Souza, Ikaika Lowe, and Kamanamaikalani Beamer*

"Kānāwai," the Hawaiian word for "law," may be literally translated to "belonging to the waters," and is based on ancient ideas regarding equal sharing of water.[1] Furthermore, some scholars credit the basis of the word "kānāwai" in a historical context to the regulation of water, since the earliest laws are said to have been in relation to the resource.[2] In the emergence of the Hawaiian state, the concept would evolve as Hawaiian ali'i (chiefs) embraced European concepts of law as a tool of nonviolent manipulation to codify Hawaiian ideas of governance and resource management.[3] Considering the fundamental importance of the resource to Hawaiian society, water resource management efforts in particular were expansive. Ancestral decision-making trends were perpetuated by ali'i during the kingdom era through law in order to influence water issues with place-based knowledge, and balance the demands of kupa'āina (citizens) and foreign entities. However, water would endure a forced transition from being managed as a public resource to benefit private sectors. Leading up to and after the overthrow, planters in the sugar industry imposing their priorities to make profits drastically altered the environment, government, and socioeconomics in Hawai'i–and it all began when they started moving water.

## Water Laws in the Hawaiian Kingdom

### *1840 Kumu Kānāwai (Constitution)*

Three decades after King Kamehameha I had united the islands under one government, Kamehameha III and his advisors authored the

Hawaiian Kingdom's first written constitution in 1840.[4] It established a new governance system by combining Hawaiian values concerning the environment with European ideas of law and property and explicitly maintained traditional ideas regarding resource management when it stated:

> Eia ke ano o ka noho ana o na'lii a me ka hooponopono ana i ka aina. O Kamehameha I., oia ke poo o keia aupuni, a nona na aina a pau mai Hawaii a Niihau, aole nae nona ponoi, no na kanaka no, a me na'lii, a o Kamehameha no ko lakou poo nana e olelo i ka aina.[5]
>
> The origin of the present government, and system of polity, is as follows. Kamehameha I, was the founder of the kingdom, and to him belonged all the land from one end of the Islands to the other, though it was not his own private property. It belonged to the chiefs and people in common, of whom Kamehameha I was the head, and had the management of the landed property.[6]

This section posits strong public trust conditions since it determined that 1) resources, including water, were not owned by anyone, not even the king, and 2) the management of water was assigned to the king, who acted as a trustee by ensuring that water was shared amongst the citizens of the kingdom. The 1840 Constitution was one of the first official government documents to codify ancient Hawaiian values that were already embedded in Hawaiian society for centuries, ensuring the perpetuation of management practices that took into account both environmental health and community needs.[7] Arguably, the 1840 Kumu Kānāwai marks a significant point in the history of Hawaiian governance and the evolution of laws regarding water regulation. Hawaiian ali'i ensured the continuance of ancient values by commanding foreign technologies to install safeguards to protect water resources.

### *The Act for the Protection and Preservation of Woods and Forests (1876) and the Act to Prevent the Waste of Artesian Water on the Island of Oahu (1884)*

The 'ōlelo no'eau (proverb) "Hahai no ka ua i ka ululā'au" means "rains always follow the forest."[8] Mosses, ferns, shrubs, and trees of Hawai'i's forests are water-gathering engines that capture water from passing moist clouds and rain. The watershed's geological design collected water that either remained on the surface in flowing streams or rivers, or moved down into subterranean geological structures that purify and store water. Therefore, the proverb encodes the lesson that the health of native forests was directly related to environmental and community health as important water sources.

Drought caused by deforestation from lumbering and ranching in the nineteenth century caused a significant water crisis, motivating the Hawai'i legislature to pass the Act for the Protection and Preservation of Woods and Forests (Forestry Act) in 1876. It authorized the protection of woods and forest lands that were "best suited for the protection of water sources,"[9] and installed a superintendent position in charge of managing these protected areas. The superintendent would be responsible for constructing fences and barriers, and they would have the power to arrest trespassers or anyone violating the statute.

Considering the political and economic context of the time, it is important to note societal complexity and different parties' investment in water issues such as watershed protection. Because of the relationship between the destruction of forests and diminishing water quantities in streams, plantation owners and sugar interests vehemently supported watershed protection through reforestation and controlling destructive ungulate species. Reforestation would also meet their needs for wood to fulfill construction efforts.[10] Ultimately though, by ensuring the protection of forested areas, the Forestry Act was arguably another method in which the Hawaiian Kingdom worked to codify protections to sustain watershed function and ensure they continue to fulfill their crucial role in capturing and collecting water.

James Campbell's successful artesian well in 1879 brought to light a previously untapped water supply: subterranean water sources. Campbell's artesian well made clear the potential for water to be available in much larger quantities, and to be transported anywhere.[11] The extent of the water supply artesian wells could provide was uncertain, but general consensus was that it was limited. In an effort to prevent the waste of water as much as possible, a law regulating artesian wells was passed: the Act to Prevent the Waste of Artesian Water on the Island of Oahu. It dictated that every flowing artesian well must be capped by the owner in order to control water flow, and condemned the waste of water.[12] In the midst of the expanding sugar industry, however, the act's allowance of water use for "irrigation, domestic and other useful purposes" did not restrict the *amount* of water drawn and wells bored, which proved the law to be inadequate in accomplishing the intended results of preventing exploitation of the resource. Nevertheless, the purpose of the act to preliminarily prevent the waste of water based on inherent beliefs in protecting water sources lends support to the kingdom's intention of extending those same beliefs through law.

## Hawaiian Kingdom Water Commissioners

In 1860, the Hawaiian legislature formed the Commission of Private Ways and Water Rights to address water rights disputes that arose from land titles established during the 1848 Māhele. Three representatives were appointed to act as commissioners in each election district to "hear and determine all controversies respecting rights of way and rights of water."[13] Former water commissioners, including Emma Nakuina (Kona: 1892–1909), Joseph Nawahī (Hilo: 1882—1887), and John Richardson (Wailuku: 1882–1887), would be appointed by the minister of the interior, and their occupations varied, to include mahi'ai kalo (taro farmers), lawyers, district judges, or members of the Privy Council.[14] In early disputes, oral testimony was delivered, and commissioners determined water irrigation by day and time, rather than defining exclusive ownership (see figure 2.1).

With a few exceptions, Hawaiian Kingdom election districts were based on traditional moku (district) designations, which allowed for decisions to be rendered by representatives with place-specific, local familiarity at the district level.

For example, two appeals made to the Kona, O'ahu, district commissioners in October 1870 concerned rights to access water in Kunawai. S. P. Kalama, who was elected as luna (supervisor; headman) of this water source, advocated for residents' appeals by providing testimony based on his place-specific knowledge of lands, water flows, and when these residents have received water in the past. His testimonies resulted in many of their appeals being granted.[15] In 1867, a dispute between a local farmer and a plantation and mill company diverting water from the Wailuku River was appealed to the Supreme Court, which ruled that neither party had exceptional rights: "Kalaniauwai, Kamaauwai and the mill water course are entitled to all the water which have flowed in them from time immemorial . . . The original purpose of these water courses was to supply *kalo* [taro] patches, and the intention of the

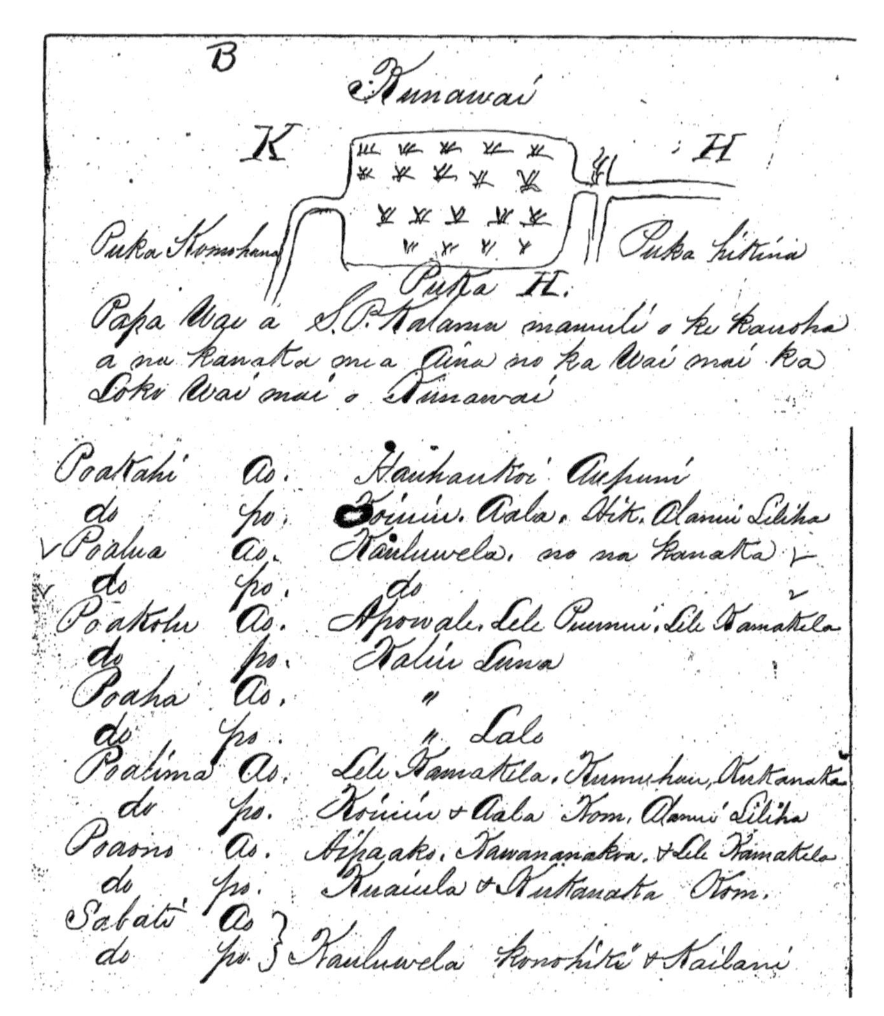


**Figure 2.1** An example of a time schedule assigning residents access to water from a freshwater pond in Kunawai, Kona, Oʻahu. Access was determined by water commissioners, and granted according to days of the week and time of day (daytime or nighttime).

*Source:* Record of the Commission of Private Ways and Water Rights 1859 to 1872, p. 116.

*konohiki* [headman of the ahupuaʻa] must have been to give all the *kalo* lands on this *Ahupuaa* rights of water at all times when needed."[16] This ruling is one example of the commission's dedication to upholding ancestral rights to water and traditional systems of water divisions.[17] These commissions have been compared to ancestral konohiki systems due to awareness of local issues and accountability.[18]

Over the lifetime of the Commission of Private Ways and Water Rights—from 1860 to 1900—more than half of the commissioners were ʻŌiwi (Native Hawaiian).[19] These place-specific appointments of commissioners, the acknowledgment of customary water rights laws and residents' relationships to place in their decisions, and a majority ʻŌiwi representation on the commission demonstrated a commitment to address issues and engage with community members in place-based contexts. This was bolstered by active community members, who more often than not provided multiple testimonies in support of generational water access rights.

## Ditch Systems and Water Licenses

Water diversion structures engineered by the sugar industry are often referred to as "ditches," which is misleading. The term connotes narrow channels, relatively small in size and scale, when in reality, these structures were a widespread network of flumes, siphons, and tunnels that invaded the four main islands to serve the needs of private plantations and water companies. Modern sugar irrigation in Hawaiʻi began with the construction of the Rice Ditch on Kauaʻi in 1856 by William Harrison Rice, founder of the Lihue Plantation. This initial ditch had its shortcomings, but was nevertheless significant since it proved to planters that perfect conditions for their industrial agriculture efforts could be *created* virtually anywhere in Hawaiʻi—by transporting water.

In 1876, the Act to Aid the Development of the Resources of the Kingdom was passed, which allowed watercourses to be leased for up to thirty years. In addition, the passing of the Act to Regulate the Passage of Water over the Lands not Benefited Thereby allowed requesters to petition for a right-of-way to transport water over another's land.[20] Samuel Alexander and Henry Perrine Baldwin, descendants of Maui missionary families and owners of Haiku Sugar Co., submitted the first request for a water lease in 1876, on the heels of the Reciprocity Treaty. Largely supported by sugar planters, the treaty's duty-free exportation of products enabled planters to invest in their own development.[21] Inspired by Rice's ditch on Kauaʻi, Alexander and Baldwin were granted a license for a major irrigation project, and they constructed the Hamakua Ditch on Maui in 1878—Hawaii's first concrete-lined ditch. Thereafter, licenses to divert water from streams for plantation irrigation were regularly granted,[22] and Hawaiʻi would be subject to dramatic investments in industrialized agriculture to accommodate a rapid expansion of the sugar industry.[23] These leases were one of the first efforts to regulate major, concrete-lined irrigation projects across the islands. The Hawaiian government ensured that leases were subject to the traditional and customary rights of the people and their access to water, which was necessary for farming. For example, Alexander and Baldwin's lease included a provision within it stating that "existing rights of present tenants of said lands, or occupiers along said streams shall in no ways be lessened or affected injuriously by reason of anything herein before granted or covenanted." Of course, diversion projects and diversion structures were not without problems. For example, in September 1891, Wailuku residents issued a petition challenging the plantation's monopolization of water, and demanded the settlement of water rights.[24] The government expanded the district of Lahaina's water commissioners' authority because of continued problems with water distribution.[25]

The Hawaiian Kingdom, within the context of the socioeconomic environment at the time, took advantage of the sugar industry as a source of revenue to pay for new roads, schools, and hospitals through taxation, while at the same time making a conscious effort to protect the rights of its people.[26] The government maintained some control by limiting the length of time planters could divert water to up to thirty years. Doing so allowed the government to retain the power to repurpose water upon the expiration of licenses.

## The Transition of Water to Commodity after the Overthrow

In the midst of the sugar industry's rapid expansion, the business community began making demands of Kalākaua and the Hawaiian government for political changes to strengthen their power and protect their sugar economy. In an effort to ensure their continued prosperity, a coalition of government opponents consisting of missionary descendants, members of the Hawaiian legislature, and stakeholders in the sugar economy rebelled on July 1, 1887, and presented a hastily written constitution for the kingdom to the king on July 6. An armed militia club supporting the rebellion, made up mostly of U.S. citizens, was employed to threaten Kalākaua into signing, forcing his approval of the new constitution under duress; as a result, the illegal 1887 Constitution would come to be known as the "Bayonet Constitution."[27] The new constitution limited Hawaiian suffrage while expanding the political power of white foreigners, causing a radical change in the government and legislature. Businessmen poured into the legislature, crafting legal and policy institutions that would become crucial for the development of the sugar industry in the future.[28] Liliʻuokalani's attempt to enact a new constitution in response to overwhelming Hawaiian demands for one illustrated her political mastery of the situation. It also demonstrated her unwavering commitment to her people's demands to control underhanded American influence, reinstall the Hawaiian nation, and strengthen the native government by restoring the monarch's power.[29] Threatened by the possibility of their political gains being reversed, emboldened sugar planters and business interests believed that an overthrow of the monarchy was justified. In January 1893, with the aid of U.S. Navy forces, a small annexationist group engaged in extralegal methods to stage a coup against the autonomy of the Hawaiian Kingdom, and violently deposed Liliʻuokalani.[30]

In the Hawaiian Kingdom, official statutes like the laws mentioned earlier in this chapter and the 1848 Māhele[31] preserved traditional ideas of resource use and management, while preserving the rights of community members. The overthrow is one of the most crucial—if not *the* most crucial—event in the history of sugar development in Hawaiʻi, since it marked U.S. intervention into land and water management. The United States wrested control of the government and crown lands from the Hawaiian Kingdom, therefore also gaining control of the rights to grant licenses for water irrigation projects. This significantly extended the environmental reach of the sugar industry and fundamentally affected water resources across the four main islands. In the two decades following the overthrow, more than twenty new major water irrigation projects diverted surface water to plantations through miles of tunnels and concrete-lined waterways.[32]

In 1895, the Republic of Hawaiʻi passed what would come to be commonly referred to as the Land Act,[33] which prompted drastic change in Hawaiʻi by overwriting Hawaiian Kingdom land laws. The Land Act combined the crown lands with government lands and renamed the resulting land bank "public lands,"[34] which effectively 1) eliminated the protections for crown lands established by the 1848 Māhele, 2) severed crown lands from Hawaiian rulers, and 3) allowed the estimated 1,772,640 acres of the now "public lands" to be leased or sold.[35] MacLennan credits the Land Act as "the mechanism for achieving land security for the sugar planters."[36] As a result of new land policies, sugar and ranching industries significantly increased their total acreage through government leases. By 1898, 1,384,903 acres of land were being leased from the government—an 83 percent increase since the beginning of the decade—mostly by sugar plantations and ranchers.[37] This would not have been possible without the overthrow.

In 1907, through Act 56, the powers and duties of the Commission of Private Ways and Water Rights were transferred to circuit courts.[38] Previously, the regionally based commissions perpetuated traditional water management similar to the konohiki system in many ways. However, the centralization of authority from the locally based structure of three commissioners appointed to each election district to the distant, impersonal circuit court judges was a drastic, detrimental change in determining water rights controversies.[39]

It was during Hawaii's territorial period between the overthrow and annexation that the Big Five emerged. A result of organizing missionary family wealth, the Big Five was a powerful system of corporate consolidation that commanded political and economic power in the islands. The symbol of this relationship between government and big business toward water management was embodied in Act 124—passed in 1911—which gave powers of eminent domain to corporations to transport water, therefore giving corporations the right to "condemn rights of way over lands and property for ditches, tunnels, flumes and pipe-lines necessary or proper for the construction and maintenance of a system for conveying, distributing and transmitting water for irrigation."[40] Act 124 indicated corporate seizure of the power of the state, ensuring corporations' authority to arrange water transport infrastructure to benefit private institutions while paralyzing the state's capacity to confront the problems, rights, and needs of the citizenry. The centralization of wealth in the agricultural sector under the control of a handful of families and their companies contributed to an expedited accumulation of both wealth and property on a monumental scale: in 1889, four companies controlled 56 percent of the sugar crop in Hawai'i, and by 1920, five companies controlled 94 percent of the sugar crop, a steep incline of growth that was the product of consolidation of plantation holdings, scientific advances, and irrigation projects (figure 2.2).[41]

Between the building of the Rice Ditch on Kaua'i in 1856 and the overthrow, eight ditches were built on Kaua'i and Maui that collectively transported an average of 49 million gallons per day (mgd) of water

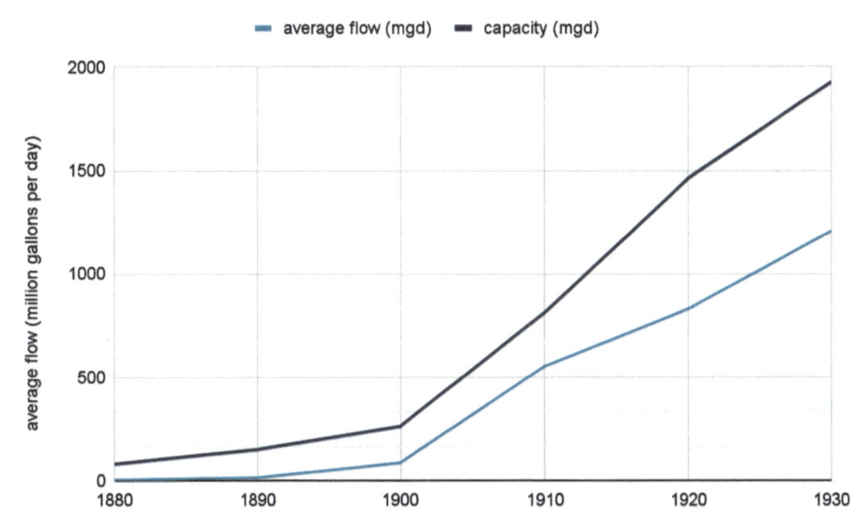


**Figure 2.2** Water diversion from 1880 to 1930. The graph indicates the exponential increase of water being transported away from its natural sources as more irrigation projects were completed in the three decades after the overthrow of the Hawaiian Kingdom in 1893. Data collected from *Sugar Water* by Carol Wilcox.

by 1893. This amount does not include the water being pumped by the ʻEwa Plantation Company on Oʻahu, who dug 22 artesian wells in 1890 with the capacity to transport up to 50 mgd. After 1893, because of the shifted power dynamic instituted by the overthrow, more than 20 major ditch systems were built over three decades, collectively moving an average of 1,205,000,000 gallons of water per day by 1926.[42]

Prominent Windward Oʻahu community activist Calvin Hoe has said, "Whoever controls the water controls the future." This sentiment has rung true throughout the ages. During the precolonial Hawaiian civilization, water was considered a resource that could not be owned by anyone, but was instead managed to ensure the health and well-being of the environment and society. In the Hawaiian Kingdom, aliʻi enacted laws that continued to protect water as a public resource by codifying ancient Hawaiian values in legislation like the 1840 Kumu Kānāwai, the Forestry Act, the Act to Prevent Artesian Well Water Waste, and enabling the influence of local knowledge on water rights controversies through Hawaiian Kingdom water commissioners. However, the series of events stimulated by the sugar industry to influence law to benefit their corporations and consolidate authority would alter the power dynamics of water and the future of Hawaiʻi.

When the Reciprocity Treaty enabled planters to invest in the development of their own infrastructure, water licenses prompted the construction of major irrigation systems that would transport monumental amounts of water away from their original sources. Planters would influence reducing the monarch's power through the 1887 Bayonet Constitution. The 1893 overthrow of the Hawaiian Kingdom and U.S. seizure of government and crown lands, along with the 1895 Land Act, empowered sugar planters to exponentially increase the acreage of their plantations. Furthermore, Act 56 would eliminate local influence on water issues by transferring the powers and duties of the Hawaiian Kingdom water commissioners to circuit court judges. As the authority over water was wrested from ʻŌiwi, it would seem that so too was the future.

Despite the series of events in which sugar growers arranged the socioeconomic environment of Hawaiʻi to support the development of their own industry, defiance on the part of the ʻŌiwi community continuously challenged and criticized the sugar industry's exploitation of Hawaiʻi. In December of 1911, Jonah Kūhiō Kalanianaʻole penned a complaint to Washington, DC, voicing his grievances about then-governor Walter Frear's relationship with the Big Five:

> The domination of Hawaii by the sugar plantations, which are in turn directly controlled by the sugar agencies in Honolulu, has been progressing and extending throughout the Governor's administration, and this fact has been winked at, certainly not challenged by Governor Frear . . . The Vital trouble is that the people who control the industrial life of Hawaii have become so blinded by long continued prosperity and the habit of controlling everything from their own standpoint that they, themselves, do not realize how deadly that policy is to the ultimate welfare of the Territory . . . He [Governor Walter Frear] is too strongly affiliated with the plantation interests to be able to see through them and to deal with them as an outsider.[43]

Prince Kūhiō also embedded within the Hawaiian Homes Commission Act language necessary to secure water access to homesteaders—one of the purposes of the act includes "providing adequate amounts of water and supporting infrastructure, so that homestead lands will always be usable and accessible."[44] Despite the drastic turn of events in the previous two decades, Kūhio remained steadfast in his dedication to provide the resources necessary for the well-being of his people, continuing the legacy of ali'i before him throughout the kingdom era.

With the impending climate crisis and the possibility of less rainfall to recharge the aquifers, hard decisions will have to be made for the next fifty to a hundred years, yet current efforts in water restoration still face difficult challenges. What can we learn from our past to inform future decisions? Hawaiian Kingdom water governance was influenced by ancestral values for wai. While the kingdom struggled in some ways, it also empowered community governance and collective "ownership" over wai. The Hawaiian Kingdom Commission of Private Ways and Water Rights appointed three local commissioners to each election district, and was majority 'Ōiwi over the lifetime of the commission. Today, a total of seven members serve on the Hawai'i State Water Resource Management Commission to address water issues across the islands. In the last thirty years, there have been four 'Ōiwi commissioners: Michael Chun, Doug Ing, Kamana Beamer, and Neil Hannahs. A fifth 'Ōiwi commissioner, Aurora Kagawa-Viviani, was recently appointed in 2021. The Water Commission has restored fewer than fifty streams in the past three decades. While the earliest ditch and irrigation systems of the sugar industry were constructed before 1893, the overthrow and the United States' commandeering of a substantial base of Hawai'i land became the impetus for sugar development to increase exponentially on Hawai'i landscapes and gain power and control in the political environment.

The overthrow was more than a century ago, yet much of Hawai'i's streams remain diverted as a result of American occupation. This chapter has demonstrated that it is necessary to understand the processes that occurred as a result of the overthrow in order to overcome the challenges we face today. The Hawai'i State Water Commission has the authority and duty to address these lingering issues of historical injustice. The commission continues to grapple with rectifying water issues that are a direct result of the overthrow and sugar infrastructure left in communities across the islands. While there is hope in increasing 'Ōiwi governance over water, the harsh reality is that the legacy of the overthrow permeates landscapes and communities across Hawai‘i today. The continued diversion of streams through massive irrigation projects and the power that enables this to remain as the status quo is one example of this legacy. While representation is crucial, outcomes are necessary in stimulating the kind of change needed to re-institutionalize the values that prioritize protecting life-giving resources and ensuring community health. We must restore Hawai‘i's streams and let the wai flow.

## Notes

1. Elizabeth A. H. K. P. Martin et al., "Cultures in Conflict in Hawai‘i: The Law and Politics of Native Hawaiian Water Rights," *University of Hawai‘i Law Review* 18, no. 1 (Winter/Spring 1996): 87–88.
2. Shaundra A. K. Liu, "Native Hawaiian Homestead Water Reservation Rights: Providing Good Living Conditions for Native Hawaiian Homesteaders," *University of Hawai‘i Law Review* 85, no. 7 (Winter 2002): 86.
3. Kamanamaikalani Beamer, *No Mākou Ka Mana: Liberating the Nation* (Honolulu: Kamehameha Publishing, 2014), 105.
4. Ibid., 116–127. Kauikeaouli enacted Ke Kumu Kanawai (Source of Law, or Constitution), and Ke Kanawai Hooponopono Waiwai (Law Regulating Taxation, Property, and the Rights of Classes; also called Declaration of Rights) the year before, in 1839; they are the first formal body of written laws.
5. Hawaiian Kingdom Constitution, "He olelo hoakaka i ka pono o na kanaka a me na'lii," "Ka hoakaka i ke ano o ka noho o na'lii."
6. Hawaiian Kingdom Constitution of 1840, official government translation.
7. D. Kapua‘ala Sproat and Mahina Tuteur, "The Power and Potential of the Public Trust: Insight from Hawai‘i's Water Battles and Triumphs," in *ResponsAbility: Law and Governance for Living Well with the Earth,* edited by Betsan Martin, Linda Te Aho, and Maria Humphries-Kil (London: Routledge, 2018), 194.
8. Mary Kawena Pukui, *‘Ōlelo No‘eau: Hawaiian Proverbs and Poetical Sayings* (Honolulu: Bishop Museum Press, 1983), 50.
9. An Act for the Protection and Preservation of Woods and Forests, Statute laws of His Majesty Kamehameha III, 1876.
10. Planters' Labor and Supply Company, "The Destruction of Forests," *Planters' Monthly* 6, no. 10 (October 1887): 437–439.
11. Carol Wilcox, *Sugar Water: Hawaii's Plantation Ditches* (Honolulu: University of Hawai‘i Press, 1997), 16.

12. Superintendent of Public Works, *History of the Honolulu Water Works* (Honolulu: Hawaiian Gazette, 1914), 277.
13. An Act of the Settlement of Controversies Respecting Rights of Way and Rights of Water, Statute laws of His Majesty Kamehameha III (1860), 12; Carol A. MacLennan, *Sovereign Sugar: Industry and Environment in Hawaii* (Honolulu: University of Hawai'i Press, 2014), 256.
14. Interior Department Commissioners List, Hawaii State Archives, Honolulu.
15. Appeal of R. E. Wakeman to settle his rights to water, before the Committee of Private Ways and Water Rights, p. 73 (October 1870); Appeal of J. Robinson for water for his land in Koiuiu, before the Committee of Private Ways and Water Rights, p. 75 (October 1870).
16. Peck v. Bailey, 8 Haw. 658 (1867).
17. MacLennan, *Sovereign Sugar*, 256–257.
18. Wilcox, *Sugar Water*, 31.
19. Record of Interior Office Commissions, Aug. 1854–June 1889, Archives of Hawaii; Index to Int. Dept Commissions, Aug. 1859–July 1894, Archives of Hawaii; Commissions–Department of the Interior, Sept. 1894 to Apr. 1900, Hawaii State Archives.
20. S. A. K. Derrickson et al., "Watershed Management and Policy in Hawaii: Coming Full Circle," *Journal of the American Water Resources Association* 38, no. 2 (2007): 568–569.
21. MacLennan, *Sovereign Sugar*, 145–146.
22. Ibid., 257.
23. S. J. La Croix and C. Grandy, "The Political Instability of Reciprocal Trade and the Overthrow of the Hawaiian Kingdom," *Journal of Economic History* 57, no. 1 (1997): 161–189.
24. Interior Department Correspondence Maui, Hawaii State Archives, Sept. 21, 1891. "There is plenty of trouble of the water at Wailuku, and there is plenty of quarrelling of the kanakas, since the Plantation takes most of the water . . . the land owners are meeting to complain for the settlement of the Water Rights of all concerned."
25. An Act to Regulate the Awarding and Distributing of Water in the District of Lahaina, Island of Maui, p. 544, 1870. "It will be of great advantage to increase the jurisdiction of the Commissioners of Private Ways and Water Rights."
26. The Kuleana Act, "The people also shall have a right to drinking water, and running water"; The 1840 Constitution of the Hawaiian Kingdom, chapter 3, section 15: Of the Division of Water for irrigation, "in all places which are watered by irrigation, those farms which have not formerly received a division of water, shall when this

new regulation respecting lands is circulated, be supplied in accordance with this law."

27. Keanu D. Sai, *Ua Mau Ke Ea—Sovereignty Endures: An Overview of the Political and Legal History of the Hawaiian Islands* (Honolulu: Pūʻā Foundation, 2013), 59–65.
28. MacLennan, *Sovereign Sugar,* 74–75.
29. Tom Coffman, *Nation Within: The History of the American Occupation of Hawaiʻi* (Durham, NC: Duke University Press, 2016).
30. Liliuokalani, *Hawaii's Story by Hawaii's Queen* (Boston: Lee and Shepard, 1898); *Act of War—The Overthrow of the Hawaiian Nation*, directed by Puhipau Lander and Joan Lander (Honolulu: Na Maka o ka Aina, 1993); Noenoe Silva, *Aloha Betrayed: Native Hawaiian Resistance to American Colonialism* (Durham, NC: Duke University Press, 2004).
31. Kamanamaikalani Beamer and Wahineʻaipōhaku Tong, "The Māhele Did What? Native Interest Remains," *Hūlili: Multidisciplinary Research on Hawaiian Well-Being* 10 (2016): 125–145.
32. MacLennan, *Sovereign Sugar,* 157–158.
33. Act 26, An Act Relating to Public Lands, Statute laws relating to the Provisional Government of Hawaiʻi, 49.
34. Kawēlau Wright, "The Homesteads of Haʻikū, Maui: A Territorial Attempt at an American Colony" (master's thesis, University of Hawaiʻi, 2015), 9.
35. MacLennan, *Sovereign Sugar,* 261.
36. Ibid.
37. Ibid., 262.
38. Act 56, Relating to Private Ways and Water Rights Amending Sections 2199 to 2205 Statute laws relating to the Territory of Hawaii 1907, 66. "The word 'Commissioner' wherever used in this chapter shall refer to the judge of the circuit court. The word 'court' shall refer to and mean the supreme court."
39. Wilcox, *Sugar Water,* 31
40. Act 124 Statute laws relating to the Territory of Hawaii, 1911, 182.
41. MacLennan, *Sovereign Sugar,* 82.
42. Wilcox, *Sugar Water,* 62–67.
43. Jonah Kuhio Kalanianaole, *The Complaint of Hon. Jonah Kuhio Kalanianaole Delegate in Congress from Hawaii against the Administration of Hon. Walter F. Frear, Governor of Hawaii* (Washington, DC: Press of the Sudwarth Co., 1911), 2.
44. Hawaiian Homes Commission Act, section 101.

## Bibliography

Beamer, Kamanamaikalani. *No Mākou Ka Mana: Liberating the Nation.* Honolulu: Kamehameha Publishing, 2014.

Beamer, Kamanamaikalani, and Wahineʻaipōhaku Tong. "The Māhele Did What? Native Interest Remains." *Hūlili: Multidisciplinary Research on Hawaiian Well-Being* 10 (2016): 125–145.

Coffman, Tom. *Nation Within: The History of the American Occupation of Hawaiʻi.* Durham, NC: Duke University Press, 2016.

Derrickson, S. A. K., M. P. Robotham, S. G. Olive, and C. I. Evensen. "Watershed Management and Policy in Hawaii: Coming Full Circle." *Journal of the American Water Resources Association* 38, no. 2 (2007): 563–576.

Kalanianaole, Jonah Kuhio. *The Complaint of Hon. Jonah Kuhio Kalanianaole Delegate in Congress from Hawaii against the Administration of Hon. Walter F. Frear, Governor of Hawaii.* Washington, DC: Press of the Sudwarth Co., 1911.

Kauikeaouli. *Kumu Kanawai, a me Na Kanawai o ko Hawaiʻi Pae Aina (Hawaiian Kingdom Constitution).* Honolulu: 1841.

La Croix, S. J., and C. Grandy. "The Political Instability of Reciprocal Trade and the Overthrow of the Hawaiian Kingdom." *Journal of Economic History* 57, no. 1 (1997): 161–189.

Lander, Puhipau, and Joan Lander, dirs. *Act of War—The Overthrow of the Hawaiian Nation.* Honolulu: Na Maka o ka Aina, 1993.

Liliuokalani. *Hawaii's Story by Hawaii's Queen.* Boston: Lee and Shepard, 1898.

Liu, Shaunda A. K. "Native Hawaiian Homestead Water Reservation Rights: Providing Good Living Conditions for Native Hawaiian Homesteaders." *University of Hawaiʻi Law Review* 25, no. 1 (Winter 2002): 85–130.

MacLennan, Carol A. *Sovereign Sugar: Industry and Environment in Hawaiʻi.* Honolulu: University of Hawaiʻi Press, 2014.

Martin, Elizabeth A. H. K. P., David L. Martin, David C. Penn, and Joyce E. McCarty. "Cultures in Conflict in Hawaii: The Law and Politics of Native Hawaiian Water Rights." *University of Hawaiʻi Law Review* 18, no. 1 (Winter/Spring 1996): 71–200.

Planters' Labor and Supply Company. "The Destruction of the Forests." *Planters' Monthly* 6, no. 10 (October 1887): 437–439.

Pukui, Mary Kawena. *ʻŌlelo Noʻeau: Hawaiian Proverbs and Poetical Sayings.* Honolulu: Bishop Museum Press, 1983.

"Report of the Superintendent of Public Works to the Governor of the Territory of Hawaii for the Year Ending June 30, 1913." Honolulu: Hawaiian Gazette, 1914.

Sai, Keanu D. *Ua Mau Ke Ea—Sovereignty Endures: An Overview of the Political and Legal History of the Hawaiian Islands.* Honolulu: Pūʻā Foundation, 2013.

Silva, Noenoe. *Aloha Betrayed: Native Hawaiian Resistance to American Colonialism.* Durham, NC: Duke University Press, 2004.

Sproat, Kapua, and Mahina Tuteur. "The Power and Potential of the Public Trust: Insight from Hawaiʻi's Water Battles and Triumphs." In *ResponsAbility: Law and Governance for Living Well with the Earth,* edited by Betsan Martin, Linda Te Aho, and Maria Humphries-Kil, 193–215. London: Routledge, 2018.

Superintendent of Public Works. *History of the Honolulu Water Works.* Honolulu: Hawaiian Gazette, 1914.

Wilcox, Carol. *Sugar Water: Hawaii's Plantation Ditches.* Honolulu: University of Hawaiʻi Press, 1997.

Wright, Kawēlau. "The Homesteads of Haʻikū, Maui: A Territorial Attempt at an American Colony." Master's thesis, University of Hawaiʻi, 2015.

CHAPTER 3

# Reframing Wai as Waiwai

## The Public Trust Paradigm in Hawai'i Nei

*D. Kapua'ala Sproat and Mahina Tuteur*

Dr. Martin Luther King Jr.'s reflection that "[t]he arc of the moral universe is long, but it bends toward justice" aptly describes Hawai'i's revolution in water resource management over the last two centuries and the trajectory necessary to secure both fresh water and the future of our island homes.[1]

Both require justice. Initially, our Indigenous management regime was grounded in Maoli[2] values of aloha 'āina (love, respect) and mālama (to care for, preserve, protect), which yielded a profusion of both wai (fresh water) and waiwai (values, wealth, abundance).[3] Over time, however, increasing foreign demands and influences, and the rise of industrialized agriculture and sugar plantations in particular, redirected how and where water flowed and the principles enabling that. This change in course devastated Hawai'i's natural and cultural resources and the native people and practices dependent upon them. With the rise of the aloha 'āina movement in the 1960s and 1970s, political and legal action sought justice for Kānaka and our resources, including in the realm of wai. Legal developments, such as constitutional amendments and a Water Code, resurrected Maoli values and management principles—such as the public trust—to once again drive policy. Despite what appeared to be favorable changes, decision-makers resisted this progress, and protracted litigation was necessary to bend the moral universe toward justice, restoring streams and communities and holding trustees to their fiduciary duties. Even with these important steps forward, much remains to be done.

As we ponder our water future, especially in this age of global climate change, Hawai'i's public trust doctrine can and should play a pivotal role.[4] This idea that wai is a resource belonging to all and owned by none, but managed by the sovereign for present and future generations, is an inherently Maoli concept. Indeed, our ancient system of managing wai is the original public trust. With its foundation in Hawai'i's laws since the earliest days of our kingdom and enshrined in Hawai'i's constitution since 1978, today's public trust doctrine is a unique hybrid of Native Hawaiian custom and western law, which was developed largely in the context of wai. Utilized to its fullest extent, our public trust could be the most important tool for resource management both in Hawai'i and beyond our shores. It is our cultural heritage as Kānaka, our kuleana (responsibility and privilege), and our legacy. Yet, time and time again, decision-makers have largely failed to honor their responsibilities under the public trust doctrine, and that kuleana has rested with communities and the courts to uphold the law. This chapter explores the public trust as paradigm in Hawai'i nei and its potential to bend the moral universe toward justice for Kānaka Maoli and all who call these islands home. It also serves as a call to action for decision-makers to deploy the law as it was intended and to do their part in restoring waiwai for generations to come.

## Water's Role and Significance in Ancient Hawai'i[5]

### *Water as a Public Resource in Native Hawaiian Society*

Ola i ka wai: Water is life. Since time immemorial, fresh water has been the lifeblood of Hawai'i's Indigenous people, culture, and resources. Internalizing the innate spiritual connection that Kānaka Maoli share with natural resources is key to understanding how the public trust has evolved in Hawai'i as a legal concept. One creation story is in the Kumulipo, the great chant of the cosmos that traces the birth of Maoli to the beginning of time in Hawai'i.[6] The Kumulipo explains that Maoli descend from akua (ancestors or gods) and are physically related to all living things in the Hawaiian archipelago.[7] As younger siblings, Native Hawaiians are bound to our extended family and have a kuleana to care for Hawai'i's natural and cultural resources.[8] Given the familial relationship between Maoli and our Native environment, elder siblings support younger ones by providing the resources necessary to sustain human and other life.[9] In return, Kānaka Maoli care for our elder siblings by managing those resources as a trust for generations yet unborn.[10] The principle of aloha ʻāina[11] is therefore directly linked to conserving and protecting not only the land and its resources, but also humankind and the spiritual realm as well.[12]

As island people who rely on fresh water to survive, Kānaka Maoli developed an intimate and complex relationship with our resources. Fresh, free-flowing water was necessary for distributing flow sufficient to cultivate our staple crop kalo (*Colocasia esculenta* or taro). Water was also revered as a kinolau (physical manifestation) of Kāne, one of the Maoli pantheon's four principal akua (gods, ancestors). Kāne was the "embodiment of male procreative energy in fresh water, flowing on or under the earth in springs, in streams and rivers, and falling as rain (and also as sunshine), which gives life to plants."[13] Given the physical and spiritual nature of our relationship to these sacred life-giving waters, Kānaka Maoli held these resources in trust.

Laws and customs preceding western contact and continuing through Hawaiʻi's sovereign kingdom reflected these important principles, recognizing that water could not be owned in any sense, but instead must be proactively managed as a resource for generations to come.[14] For instance, the Kingdom of Hawaiʻi's first western-style constitution of 1840 included strong public trust provisions, declaring that the land along with its resources "was not [the king's] private property. It belonged to the Aliʻi and the people in common, of whom [the king] was the head, and had the management of the landed property."[15] These ancient values were embedded in Maoli society long before any written constitution and were strictly enforced by kahuwai (water stewards) who managed the flow of water within and between ahupuaʻa (loosely defined as watersheds) to ensure, for example, that if water was taken from a stream for kalo cultivation, it was returned to the same stream so that downstream users had enough water to satisfy agricultural or other needs. These management practices respected the environment while also taking into account the competing needs of the larger community.

### *Water as Private Property in Plantation Society*

The institution of private property via the Māhele,[16] subsequent consolidation of land ownership by foreign (largely U.S.) interests, and growing recognition that Hawaiʻi's climate and year-round growing season made plantation agriculture, and especially sugarcane, a lucrative business, drove the foreign takeover of Maoli ancestral land. Water resources seemed destined to follow this history of dispossession.

To establish and expand plantations, massive irrigation systems were constructed to transport and use water in ways that nature never intended. To satisfy thirsty crops, sugar planters constructed ditches that diverted streams from wet, Windward (or Koʻolau), predominantly Maoli communities, to the drier Central and Leeward plains where sugar was cultivated, and also drilled wells to siphon groundwater. All of this was done with no consideration of or consultation with affected Maoli communities. This rapid change altered the natural environment while also inflicting significant physical and cultural harm on Kānaka Maoli, much of which remains unaddressed to this day. Plantations and their irrigation systems took root on each of the major Hawaiian islands, fundamentally changing how and where water was used. Sugar's rise to dominance rewrote the social contract. Despite kingdom laws (such as the 1840 Constitution) that formalized and reduced some of our Native Hawaiian custom and tradition to writing, large agricultural plantations increased their influence and soon controlled a vast portion of Hawaiʻi's resources. The law was no exception, and cases during Hawaiʻi's kingdom and territorial periods also began to reflect increasingly western approaches to water use and management.

Soon, conflicts over water ensued—first, between plantation interests and Kānaka Maoli, and later, between competing sugar plantations.[17] The case of *McBryde Sugar Company v. Robinson* (1973) brought the tensions over water as a public resource or private property to a head.[18] Two sugar companies litigated their respective rights to take water from the Hanapēpē River on Kauaʻi. The Hawaiʻi Supreme Court, led by the late, great Chief Justice William S. Richardson (who was Maoli), addressed the larger issue of water management in Hawaiʻi and clarified that although the parties in that case may have had rights to use water, they had no ownership interest in the water itself.[19] Relying on a legal analysis of the historical treatment of water rights under the kingdom, the court affirmed that such ownership rights were never included when a hybridized form of private property was instituted in Hawaiʻi via the Māhele. Instead, the court ruled that the state, as successor to the mōʻī (sovereign), holds all water in trust for the benefit of the larger community.[20] Other cases followed, including *Robinson v. Ariyoshi* and *Reppun v. Board of Water Supply* (both decided in 1982), which respectively considered the public nature of Hawai'i's water resources and the rights of downstream kalo growers to maintain Maoli agricultural practices.[21]

Maoli tradition and custom, along with cases and laws from the Kingdom of Hawai'i and consistent rulings from the Hawai'i Supreme Court upholding the public trust over Hawai'i's water resources, firmly established the principle that natural resources, including water, were not private property but were held in trust by the government for the benefit of the people. Still, opposition by entrenched powers persisted. The black letter of the law carried moral and legal authority, which collided with the political power wielded by plantation and other aligned interests. Thus, more needed to be done to bring legal protections to life on the ground and in our communities.

## Hawai'i's Legal Regime for Water Resource Management

Around the time that the *McBryde* litigation was unfolding, sugar plantations began to lose their economic dominance to tourism and the military. Concerned communities took this opportunity to reexamine the legal regime and manage Hawai'i's water resources more proactively for the benefit of the larger society. The 1978 Constitutional Convention ("ConCon") proved critical in this regard. Thanks to the efforts of young Maoli and environmentally conscious delegates and staff, the ConCon crafted amendments that were later ratified by Hawai'i's voters to enshrine resource protection and Native Hawaiian rights as constitutional mandates. Article 11, section 1 of the Hawai'i Constitution now declares: "For the benefit of present and future generations, the State and its political subdivisions shall conserve and protect Hawaii's natural beauty and all natural resources, including land, water, air, minerals and energy sources, and shall promote the development and utilization of these resources in a manner consistent with their conservation and in furtherance of the self-sufficiency of the State."[22] The provision articulates the general public trust principles and the responsibilities of the state and counties to uphold the public trust. Article 11, section 7 makes specific reference to water and affirms that the state "has an obligation to protect, control and regulate the use of Hawaii's water resources for the benefit of its people."[23] Hawai'i's highest court has confirmed that these two provisions, "article XI, section 1 and article XI, section 7[,] adopt the public trust doctrine as a fundamental principle of constitutional law in Hawai'i."[24] In addition, article 11, section 9 proclaims: "Each person has the right to a clean and healthful environment, as defined by laws relating to environmental quality, including control of pollution and conservation, protection and enhancement of natural resources."[25] Moreover, delegates independently safeguarded traditional and customary Maoli rights and practices. Article 12, section 7 "reaffirms and shall protect all rights, customarily and traditionally exercised for subsistence, cultural and religious purposes and possessed by ahupua'a tenants who are descendants of native Hawaiians who inhabited the Hawaiian Islands prior to 1778."[26]

In 1987, the state legislature enacted Hawai'i's Water Code (Hawai'i Revised Statutes chapter 174C), establishing a comprehensive water resource management regime that sought to balance resource protection with "reasonable and beneficial use"; specific provisions also protected Maoli rights and practices.[27] The legislature gave the Commission on Water Resource Management (the "Commission") primary authority over water use and management, deeming it the primary public trustee of water, although the constitutional nature of the public trust doctrine in Hawai'i reserved an important role for the Hawai'i Supreme Court.[28]

Today, under Hawai'i's Constitution, the Water Code, and common law, the "water resources trust" applies to "all water resources without exception or distinction."[29] The public trust establishes "a dual mandate of (1) protection and (2) maximum reasonable and beneficial use."[30] The Water Commission, therefore, has an "affirmative duty to take the public trust into account in the planning and allocation of water resources, and to protect public trust uses whenever feasible."[31]

Organizers and Maoli communities wasted no time utilizing these new legal tools to bend the arc of the moral universe toward justice and redress the more than century-long theft of these life-giving waters. The first major case seeking stream restoration under the new legal regime arose in Waiāhole, O'ahu.

## The Arc of Hawai'i's Moral Universe Begins to Bend toward Justice

### *Ke Kalo Pa'a o Waiāhole: The Hard Taro of Waiāhole*

Waiāhole Valley is nestled on the Windward side of the soaring Ko'olau mountain range. The largest streams on O'ahu flowed through Waiāhole and its neighboring valleys, enabling cultivation of the most extensive lo'i kalo (wetland taro terraces) on the island and sustaining a thriving

Maoli community and culture. At the turn of the twentieth century, plantation owners sought water for their fields on Oʻahu's dry Central plain. In 1916, a twenty-five-mile tunnel and ditch system was completed that drained the majority of surface and groundwater from the Windward valleys to subsidize Oʻahu Sugar's operations on the other side of the island. Massive diversions of Windward water began without any thought for the communities or resources that depended on that wai. These diversions, which continued unchecked for almost one hundred years, devastated streams and destroyed fishponds and estuaries, forcing many to abandon their subsistence lifestyles.

In the early 1990s, a small but dedicated coalition of Kānaka Maoli, family farmers, and community activists took on some of the most powerful economic and political forces in Hawaiʻi in a protracted legal battle over Windward stream flows, which resulted in the first distillation of what the constitutional public trust actually means in Hawaiʻi. The case led to the first-ever restoration of streams in Hawaiʻi's history, fundamentally correcting the course of water law across our archipelago. Traditional moʻolelo (stories or history) about this area speak of Ke Kalo Paʻa o Waiāhole, the hard taro of the region, which also alludes to the character of its people, who are steadfast in the face of adversity.

In 1992, in response to a community petition, the Commission designated Windward Oʻahu a groundwater management area under the Water Code, bringing wai in the area under the Commission's regulatory control and requiring water users to apply for permits. Soon after, Oʻahu Sugar announced that it would be closing. This set the stage for a pitched battle over the future allocation of this life-giving resource. The Windward Parties sought the return of wai to their streams and communities via the establishment of Interim Instream Flow Standards ("IIFSs") as mandated by the code. At the same time, a dozen users (the "Leeward Parties"), which included some of the largest landowners and powerful political interests in Hawaiʻi, coveted that water for large-scale agribusiness, private golf course development, and landscaping, among other things. Several county, state, and federal agencies also joined the fray, and all were in favor of maximizing stream diversions. In fact, the collective water demands of the Leeward Parties exceeded the entire flow of the Waiāhole Ditch System.

Years of proceedings followed, including a ten-month administrative trial, which involved hundreds of exhibits, dozens of witnesses, and a dizzying number of motions and other filings. In 1994, a mediated agreement between the parties partially restored stream flows on an interim basis, and then a 1997 Commission decision split the water between the Windward streams and Leeward users; three separate appeals were subsequently filed over the course of more than a decade. Ultimately, the Hawai'i Supreme Court played a crucial role in rising above the political wrangling that had tainted the Commission's decisions and the plantation mentality that had subverted the public trust as a management principle for more than a century.

### *The Reemergence of the Public Trust Doctrine*

In August 2000, the Hawai'i Supreme Court issued its opinion, which was recognized globally for its pronouncements on the public trust doctrine and also returned to Windward streams and communities roughly half of the diverted flows that had been appropriated for nearly a century.[32] This landmark decision affirmed the public nature of Hawai'i's water resources, clarified the burden of proof for those seeking stream restoration and diversions, and explicitly recognized the value of free-flowing streams and the perpetuation of Indigenous culture. Both legally and metaphorically, *Waiāhole* created significant opportunities for Maoli and other communities throughout Hawai'i to seek justice, and illuminated both the promise and potential of the public trust doctrine in particular, which has largely failed to achieve the stated purpose of protecting and restoring Hawai'i's freshwater resources or the cultural practices dependent upon them. There have been successes, most often achieved through protracted litigation, but also too many setbacks. Time and time again, on Moloka'i, Maui, Kaua'i, and beyond, the Water Commission has wholly failed to embrace its duties under the public trust and proactively protect water resources, in large part due to lack of political will, but also given its uncritical and even disturbing acceptance of the status quo "plantation" approach to resource extraction (as opposed to management). One case on Kaua'i, however, demonstrates that at least some government officials have been taking their constitutional obligations seriously, embracing the public trust paradigm and applying it in their own contexts—precisely as article 11, section 1 of the constitution contemplates. Indeed, one of the most important developments since *Waiāhole* has been the analytic framework proffered by the Hawai'i Supreme Court in the *Kaua'i Springs* case.

### Kaua'i Springs

#### *The Public Trust as Paradigm Prevails*

Kaua'i Springs operates a private water bottling business in Kōloa, Kaua'i, where it collects spring water supplied from a former plantation ditch system and distributes five-gallon containers to island residents and businesses.[33] For some, Kaua'i Springs' basic business model raised serious concerns—Can a commercial operation bottle and sell public trust resources for private profit?—and challenged the very notion of water as a public trust. The company operated for several years before being informed that it could not continue its industrial processing on agricultural land without securing permits from the County of Kaua'i.[34]

In 2007, the County Planning Commission denied Kaua'i Springs' applications for the necessary permits, concluding that the company had failed to demonstrate that it was entitled to extract and sell water on a commercial basis.[35] After several appeals, the case made its way to the Hawai'i Supreme Court, which in 2014 upheld the Planning Commission's decision and sent the case back to the county for more hearings.

With respect to the public trust, the high court strongly reaffirmed the doctrine as a fundamental principle of constitutional law in Hawai'i.[36] The court emphasized that applicants bear the burden of justifying proposed water uses in light of the protected trust purposes and that the public trust provides independent authority to guide decision-makers in fulfilling their mandates.[37] The court distilled six principles that agency officials and other decision-makers must apply to uphold the public trust in rendering determinations:

1. "The agency's duty and authority is to maintain the purity and flow of our waters for future generations and to assure that the waters of our land are put to reasonable and beneficial use";
2. Officials "must determine whether the proposed use is consistent with the trust purposes";
3. Decision-makers need to "apply a presumption in favor of public use, access, enjoyment, and resource protection";
4. Authorities must "evaluate each proposal for use on a case-by-case basis, recognizing that there can be no vested rights in the use of public water";

5. "If the requested use is private or commercial, the agency should apply a high level of scrutiny"; and
6. Administrators must apply "a 'reasonable and beneficial use' standard, which requires examination of the proposed use in relation to other public and private uses."[38]

These principles establish "duties under the public trust independent of [any] permit requirements"[39] and provide more specific direction about what the heightened standard of review actually requires under Hawai'i's public trust paradigm.[40] Moreover, these requirements apply to all state and county officials making determinations that affect public trust resources, not simply water.[41] Hopefully, this further elucidation will guide and inspire decision-makers, who have floundered in the more than twenty years since the initial *Waiāhole* decision, to embrace the public trust as Hawai'i's paradigm for comprehensive resource management.[42]

In the aftermath of *Waiāhole* and *Kaua'i Springs,* various state and county agencies have struggled with how to properly discharge their fiduciary duties under the public trust.[43] Although the doctrine has been well developed in the context of Hawai'i's freshwater resources, it has not been applied and litigated as extensively in other arenas, and decision-makers have grappled with how and when to apply it.[44] More recently, however, the Hawai'i Supreme Court has recognized the doctrine's applicability beyond water, including land zoned conservation,[45] and 'āina within the "Public Land Trust," which refers to the roughly 1,200,000 acres now held in trust by the State of Hawai'i for Kānaka and the general public.[46] The Supreme Court also upheld a duty to aloha 'āina in 2019.[47] Yet even in the context of water, where agencies and courts have been deploying the public trust for decades, the Commission has netted mixed results. In some areas, such as Moloka'i and Hawai'i Island, entrenched interests have resisted calls to operationalize the public trust, and litigation has persisted. In other 'āina, especially where the Water Commission has partnered with the community, it has realized greater success uplifting biocultural knowledge and fulfilling its fiduciary duties.

As these questions of wai and waiwai make their way through our political and legal systems, the public bears witness to how administrative agencies and even the courts reach different results when interpreting the same provisions depending on each decision-maker's understanding of injustice and how the law contributes to it. One recent struggle in Wai'oli, Kaua'i, provides a glimmer of hope and a compelling example of how embracing the public trust as paradigm, rooted in Maoli custom and tradition, can produce substantive results.

## Charting a Path Forward

### *Juggling the Promise and Potential of the Public Trust in Wai'oli, Kaua'i*

> Aia i Wai'oli ke aloha 'āina, ia 'āina momona no ka hui kalo
> (There at Wai'oli is aloha 'āina, that fertile land for the hui kalo)[48]

Literally translated as "joyous waters," Wai'oli has epitomized aloha 'āina and long sustained 'āina momona—the community's ability to live in balance with and cultivate an abundance of natural and cultural resources. Wai'oli's traditional lo'i kalo system has been utilized since time immemorial to sustain the people of that place, and today, it continues to nurture folks across Hawai'i pae 'āina (the archipelago), largely through its kalo and the poi that is made from it. For example, of the forty-six identified kuleana awarded in Wai'oli during the Māhele process (starting in about 1845), forty-one parcels had at least one portion used for kalo cultivation. The Water Commission also identified Wai'oli Valley as one of only six areas throughout Hawai'i that historically supported more than fifty acres of kalo.[49]

In April of 2018, Kaua'i's north shore experienced record-breaking rains, which were recorded in Wai'oli's neighboring ahupua'a of Waipā at nearly fifty inches within a twenty-four-hour period. This severe weather event triggered cataclysmic flooding that devastated the island's infrastructure, generating massive landslides and destroying homes and businesses. The flooding also severely damaged the lo'i kalo irrigation system that the farmers of the Wai'oli Valley Taro Hui (the "Hui") used to transport water from Wai'oli Stream to their fields.

Disaster recovery efforts revealed that the farmers' mānowai, po'owai, and some 'auwai—crucial components of the irrigation system they have stewarded for generations—were on state conservation land, which sparked a maze of permitting, exemptions, and other requirements. Much of this work had to be completed before restoration efforts could begin. So, over the next four years, the Hui navigated a gauntlet of administrative and legislative processes with the assistance of the

Environmental Law and Native Hawaiian Rights Clinics at the William S. Richardson School of Law,[50] simply to continue traditional farming practices that had thrived in Waiʻoli for centuries. In the process, the Hui epitomized the promise and potential of the public trust, including how everyone benefits when Maoli values and resource management principles are the basis for agency decision-making. It also exposed the complexity of fitting the traditional practices of a living Indigenous culture into western legal frameworks where decisions are often made by nonnatives who are usually not cultural practitioners.

At the outset, it is critical to recognize that the Hui's place-based biocultural knowledge played a critical role in persuading decision-makers every step of the way. First, though the ʻohana of Waiʻoli and Haleleʻa more broadly have comprehensively managed this ʻāina for generations, it was not until the 2018 flooding that they formally established a nonprofit organization with a mission to educate and engage the community, and empower future generations with traditional knowledge. The next step was to secure from Hawaiʻi's Board of Land and Natural Resources ("BLNR") a right of entry, which later became a perpetual easement for the mānowai, poʻowai, and ʻauwai, and a revocable permit for the Hui's water use. The Hui's easement for its Native Hawaiian irrigation system was BLNR's first comanagement agreement for terrestrial resources, and has been hailed as an exemplar that will hopefully guide both resource management and the way that agencies interact with Native practitioners and kalo farmers in particular.

While this ensured immediate access to the irrigation system and use of Waiʻoli Stream water, the permitting was just getting started. The Hui embarked upon the complex and highly contentious long-term water leasing process under Hawaiʻi Revised Statutes ("HRS") chapter 171, which regulates the disposition of water leases on state land. This required completing the environmental review process under HRS chapter 343, establishing numeric Interim Instream Flow Standards ("IIFSs") with the Water Commission, and consulting with the Department of Hawaiian Home Lands on a water reservation. The Hui also completed a draft watershed management plan and a comprehensive cultural impact assessment and environmental assessment as part of this cumbersome process.

Because of the highly politicized nature of water licenses, not one has been issued for many years. Due to community pressure, the Hawaiʻi state legislature took up the hot-button issue in the 2021 and 2022 sessions. In 2021, the legislature passed a concurrent resolution authorizing BLNR to enter into a long-term water lease with the Hui by direct negotiation, avoiding the public auction process. In doing so, the legislature recognized that the Hui's tireless work to perpetuate traditional lo'i kalo cultivation had succeeded in maintaining "cultural lifeways, and community identity" for centuries.[51] Despite the plain language of HRS 171–58(c) and the concurrent resolution's findings that the Hui's kalo cultivation was nonpolluting and nonconsumptive, the Department of Land and Natural Resources ("DLNR") would not directly negotiate with the Hui because some staff believed that kalo cultivation was consumptive, as a tiny amount of water is lost to evaporation. In addition, the same staff were concerned that this lo'i kalo irrigation system returned some water from Wai'oli Stream to the Hanalei River. In 2022, the legislature again intervened on the Hui's behalf, adding a new section, HRS 171–58(h), which exempted the instream use of water for traditional and customary kalo cultivation from the water leasing statute altogether.[52] This benefited both the Hui and other traditional kalo cultivars across the islands and highlighted the importance of Indigenous knowledge and practices as a foundation for Hawai'i's public trust. Though Hawai'i's Constitution already embraced strong protections for traditional and customary rights, narrow interpretations of "nonconsumptive" use and the challenge of forcing traditional and customary Native Hawaiian practices into the contemporary legal framework of water leasing nearly ended centuries of kalo cultivation in Wai'oli. Fortunately, in December 2022, Water Commission and BLNR Chair Suzanne Case informed the Hui that, given the legislature's 2022 amendment to the water leasing process, HRS 171–58 "shall not apply" to the Hui "because it is an instream use of water for traditional and customary kalo cultivation practices."[53] This conclusion was made possible because of the Hui and the Clinics' tireless efforts, including amending Wai'oli Stream's IIFS, completing a Final Environmental Assessment ("FEA"), and securing a Finding of No Significant Impact in the FEA, all of which are described in more detail below. As we shape the future of wai in Hawai'i, the mo'olelo of Wai'oli's farmers reveals much about the power and potential of the public trust in uplifting a community-centered and place-based model of abundance by respecting and actually utilizing biocultural resource management in decision-making. It also illustrates the incredible challenges inherent in actualizing existing legal protections when—as then BLNR Chair Suzanne Case explained: "What we're trying to do here is fit an old system into a new legal system."[54]

As one example, the FEA evaluating the environmental impacts of the Hui's lo'i kalo cultivation revealed that their work had broad and significant positive benefits on water, cultural, and socioeconomic resources, just one example being that the farmers provide important flood mitigation for the surrounding and densely populated Hanalei area. Because the vast majority of projects subject to HRS chapter 343 have adverse impacts on the environment and the Hui's work had beneficial impacts, these benefits were difficult to measure and capture within the standard chapter 343 process.

Another salient example is the effort to support the Water Commission in amending the IIFS for Wai'oli Stream in 2021. The Commission manages water in two ways: by delineating boundaries between surface and ground water, and along hydrologic lines. These boundaries can seem artificial, and even contrary to comprehensive water resource management, especially to Kānaka who understand water as a kinolau of Kāne and recognize the inherent connection of these resources through the water cycle. Under the Water Code, HRS 174C, IIFSs dictate the minimum amount of streamflow "necessary to protect the public interest in the particular stream." For the first time in modern history, the Commission recognized "the traditional land and water management system of kalana" in the Halele'a moku (district), in that Wai'oli was an important part of the larger watershed and shared water resources with the neighboring ahupua'a of Waipā and Hanalei. A kalana is a land division that is smaller than a moku, but contains several smaller ahupua'a or was a "distinct area within a large ahupua'a."[55] Importantly, contrary to broader understandings of ahupua'a as distinct and independent resource management systems operating from ma uka to ma kai, kalana are "associated more with systematic biocultural resource management and community identity rather than governance."[56] This recognition was significant because for many centuries the Hui's lo'i kalo irrigation system proactively managed across the watershed boundaries reflected on the Commission's maps. For example, two of the Hui's seven ho'i return water flowing out of lo'i to the Hanalei River, and the remaining five ho'i return water to Wai'oli Stream, which flows into the ocean in the Waipā ahupua'a. Absent this crucial historical and cultural context from the Hui, the Commission may have based its decision solely on its maps, disregarding traditional knowledge and relying only on western cartography. Instead, by working collaboratively with the Hui and respecting its traditional place-based understandings and practices, the Commission was empowered to fulfill its duties under the public trust and encouraged to advance an adaptive management model that was more inclusive of community input and flexible in terms of responding to climate change impacts. Indeed, embracing Wai'oli as part of the larger integrated Hanalei Kalana also aligned with scientific and historical research characterizing the Hanalei Kalana as the collective lands that discharge water into Hanalei Bay. This is a welcome departure from DLNR staff's narrow interpretation of nonconsumptive use and rejection of the Hui's ancient practices as not fitting within its interpretation of 171–58(c) because some water from Wai'oli Stream was returned to Hanalei River, for example.

The Hui's journey—from destruction to recovery, not simply to sustain their community, but to thrive in abundance—can hopefully encourage others to raise their voices and share their stories in these complex administrative processes. Importantly, while Indigenous precepts such as aloha 'āina are actually the foundation of Hawai'i's legal regime, especially regarding the public trust, decision-makers must begin to better understand and then explicitly uplift and incorporate them into decision-making—particularly in controversial cases where communities' livelihoods are at stake, and especially as we adapt to our new climate reality with more frequent and severe weather events.

The Hui's comanagement agreement with BLNR was hailed as an exemplar to guide agency interaction with Native practitioners and kalo farmers in particular. But what lessons should Kānaka take from this? Why did the Water Commission embrace the cultural precept of kalana while amending the IIFS after the Land Division rejected the idea of managing water resources across lines on a map as not fitting within a narrow definition of instream use? One fundamental difference is the training and cultural competency of the Water Commissioners and their deputy director in 2021, when Wai'oli Stream's IIFS was amended. At that time, some commissioners and their deputy were Native Hawaiian cultural practitioners (including kalo farmers) who are fluent in both Hawai'i's Indigenous language and culture. Dr. Kamanamaikalani Beamer, for example, is a leading scholar in a range of Kānaka Maoli disciplines, including ʻāina and wai management and aloha ʻāina in particular. This expertise helped the Commission to process and contextualize the biocultural and other knowledge that the Waiʻoli farmers brought to the Commission—after four to six generations of stewarding this ʻāina and wai—and embrace it as part of their decision-making process. Importantly, this collaborative decision-making process incorporating biocultural knowledge respected Indigenous precepts and practices while also embracing the public trust as paradigm.

## The Power and Potential of Mālama as a Public Trust Duty

Aloha ʻāina is a piko at the very core of Maoli identity and spirituality. It is a philosophy, relationship, worldview, cultural obligation, and more. Indeed, renowned scholar-activist Lilikalā Kameʻeleihiwa described this cultural value as the "first lesson of Wākea,"[57] referencing the Kumulipo and Native Hawaiians' inherent responsibility to respect and care for our elder sibling, the kalo plant, and in turn, all natural and cultural resources. According to this Indigenous worldview, rights and responsibilities are inextricably intertwined, as embodied in the concept of kuleana, which "grows from reciprocity: regular return, cultivation of relationship, and active work to nurture abundance."[58] In the same way, Maoli culture and identity are bound to and inseparable from Native resources, especially ʻāina and wai.

Historically, these understandings of the reciprocal nature of kuleana and aloha ʻāina created a system of "checks and balances" between resource managers and public trust beneficiaries in their collective reverence for the spiritual and physical worlds upon which they depended. Makaʻāinana (people of the land) worked the ʻāina, while kahuwai and konohiki (water and land stewards) who were Aliʻi (leaders) collected tributes and directed the day-to-day activities of cultivation and fishing, playing a pivotal role in ensuring the ahupuaʻa functioned properly.[59] If any party (makaʻāinana, konohiki, or Aliʻi) failed in their respective duties, they were ousted.[60] This resource management regime also underpinned "a social system rooted in aloha ʻāina that was hybridized during the . . . Hawaiian Kingdom."[61]

Despite colonization's influences, including rapid population decline and the appropriation of land and water, this resource management regime—or the public trust as paradigm—endured. For example, the concept of ʻāina as the relationship between land and Kānaka, not simply the land itself,[62] was reinforced by the 1839 Declaration of Rights, in which Kauikeaouli (Kamehameha III) specifically requested that konohiki reflect on the tasks that would enrich not only the ʻāina but also the people who live on and work the ʻāina: "e noonoo i ka hana e waiwai ai ka aina, a me na makaainana malalo iho o oukou maluna hoi o ka aina."[63] Later, in 1847, Kauikeaouli declared that he was the konohiki of all of Hawaiʻi's ʻāina, embracing his duty to aloha ʻāina, to care for all the ʻāina in his kingdom on behalf of his people, precisely as a trustee of public trust resources can and must. In this way, konohiki and kahuwai are synonymous with the concept, value, and fiduciary duty of aloha ʻāina.[64]

Just as aloha ʻāina endured, so too has the trust relationship between the government and the people, with the expectation that decision-makers will ʻauamo (shoulder) this kuleana. The sad reality is that although these tools have long been a vital foundation of both the sovereign Hawaiian Kingdom's and the State of Hawaiʻi's legal regimes, government agencies—as our modern konohiki and kahuwai—have regularly failed to deploy them, especially within the last forty years.[65] Even after victories, such as *Waiāhole* and *Kauaʻi Springs,* decision-makers and entrenched interests have continued to push back against the public trust as paradigm by opposing preemptive measures to conserve land and water, such as the precautionary principle.

In *Waiāhole I,* the Hawaiʻi Supreme Court recognized that the constitutional mandate to conserve and protect our natural resources is grounded in the historical understanding of the state's trusteeship and that the constitutional framers expressly confirmed that protection of natural resources is for the benefit of present and future generations "because it affirms the ethical obligations of this generation toward the next and is entirely consistent with the concept that the Constitution should provide for the future."[66] In the face of dwindling resources, political upheaval, and climate change impacts, the time is now for decision-makers to heed the call to proactively care for and protect our wai and other resources in a way that honors the past, considers present needs, and restores waiwai—both abundance and values that enable it—for future generations.[67]

As we return to the origin of Hawaiʻi's public trust doctrine—aloha ʻāina—we are hopeful that our decision-makers will turn toward the light and be inspired to reclaim their roles as true stewards, cultivating a kahuwai mindset for the future. E ʻauamo i ke kuleana kūkulu waiwai, it is time for our leaders and decision-makers to "shoulder the responsibility and privilege of building prosperity through pono management of our resources."

## Reframing Wai as Waiwai

### *The Public Trust as Paradigm in Hawaiʻi Nei*

For Kānaka Maoli, the history of water and the law in Hawaiʻi is a painful one. The arc of this moral universe feels incredibly long, and although we have been successful in bending it toward justice, the end result is by no means guaranteed. We are still composing that story. In doing so, we must move forward firmly grounded in both our unique laws and the cultural waiwai or values that underpin them—such as the duty to aloha ʻāina. Decision-makers must ʻauamo this kuleana with the courage to make the choices necessary to ensure abundance for generations yet unborn, and must move from managing wai as a commodity to a life-giving source of spiritual, physical, and cultural mana. Utilized to its fullest extent, our public trust doctrine could be the most important tool for water resource management—and resource management period—in Hawaiʻi nei. It has already been developed; it need only be deployed. To do so effectively, our decision-makers must embrace the public trust as a paradigm—especially its Indigenous roots and values; only then will aloha ʻāina, the first lesson of Wākea, endure and will "justice roll down like water and righteousness like a mighty stream."[68]

## Notes

1. Martin Luther King Jr., "Out of the Long Night," *Gospel Messenger,* Feb. 8, 1958 (quoting Theodore Parker), available at https://archive.org/details/gospelmessengerv107mors.
2. Native Hawaiian, native Hawaiian, Hawaiian, Kānaka Maoli, Maoli, Kānaka ʻŌiwi, and ʻŌiwi are used interchangeably and without reference to blood quantum. Kānaka Maoli or Maoli is the Indigenous Hawaiian name for the population inhabiting Hawaiʻi at the time of the first western contact. Mary Kawena Pukui and Samuel H. Elbert, *Hawaiian Dictionary* (Honolulu: University of Hawaiʻi Press, 1986), 127 (noting that Kanaka Maoli historically referred to a full-blooded "Hawaiian person").

3. Ibid. at 21, 232, 377, 380.
4. For more on the role of the public trust doctrine in recent climate change litigation, see Michael C. Blumm and Mary Christina Wood, "'No Ordinary Lawsuit': Climate Change, Due Process, and the Public Trust Doctrine," *American University Law Review* 1, no. 67 (2017): 1–87.
5. Some text from this chapter previously appeared in D. Kapua'ala Sproat, "A Question of Wai: Seeking Justice through Law for Hawai'i's Streams and Communities," in *Nation Rising: Hawaiian Movements for Life, Land, and Sovereignty,* edited by Noelani Goodyear-Ka'ōpua, Ikaika Hussey, and Erin Kahunawaika'ala Wright (Durham, NC: Duke University Press, 2014), 199–219; and D. Kapua'ala Sproat and Isaac H. Moriwake, "Ke Kalo Pa'a o Waiāhole: Use of the Public Trust as a Tool for Environmental Advocacy," in *Creative Common Law Strategies for Protecting the Environment,* edited by Clifford Rechtschaffen and Denise Antolini (Washington, DC: Environmental Law Institute, 2007), 247–284.
6. See Martha Warren Beckwith, *The Kumulipo: A Hawaiian Creation Chant* (Honolulu: University of Hawai'i Press, 1972). The Kumulipo is detailed and complex, with sixteen wā (intervals) and over 2,000 lines. The Kumulipo explains that in the beginning there was pō or darkness, and from this darkness came life. Pō gave birth to two children: a son named Kumulipo and a daughter named Pō'ele. Through their union, Kumulipo and Pō'ele created the natural world. The first child born to them was the coral polyp, which created the foundation for all life in the sea. Born in continuing sequential order were all of the plants and animals in Hawai'i nei, which became 'aumakua or guardians that continue to watch over Kānaka Maoli. Pō had many children that comprised all aspects of Hawai'i's natural world. After all of the Hawaiian Islands were born, Wākea (Sky Father) had a child with Ho'ohōkūkalani, who was stillborn. They buried it outside of their home, and a kalo plant grew from its grave. Wākea and Ho'ohōkūkalani had a second child, named Hāloa in honor of its elder sibling, which was the first Kānaka Maoli—the first human child born in Hawai'i. Ibid. at 37–117.
7. Ibid. at 117.
8. See Melody Kapilialoha MacKenzie, Susan K. Serrano, and Koalani Laura Kaulukukui, "Environmental Justice for Indigenous Hawaiians: Reclaiming Land and Resources," *Natural Resources and Environment* 21, no. 3 (2007): 37. Pukui and Elbert, *Hawaiian Dictionary,* 179.
9. See MacKenzie, Serrano, and Kaulukukui, "Environmental Justice," 37.
10. Ibid.

11. At its etymological roots, "'āina" means "that which feeds." Here, we use the term to refer to all natural and cultural resources that sustain communities, including wai.
12. MacKenzie, Serrano, and Kaulukukui, "Environmental Justice," 37.
13. E. S. Craighill Handy and Elizabeth Green Handy, with the collaboration of Mary Kawena Pukui, *Native Planters in Old Hawai'i: Their Life, Lore, and Environment* (Honolulu: Bishop Museum Press, 1972), 64.
14. *McBryde Sugar Co. v. Robinson,* 54 Haw. 174, 185–187, 504 P.2d 1330, 1338–1339 (1973).
15. Kingdom of Hawai'i Constitution of 1840, reprinted in *Translation of the Constitution and Laws of the Hawaiian Islands, Established in the Reign of Kamehameha II* (1842), 11–12, http://punawaiola.org. See also D. Kapua'ala Sproat, "Where Justice Flows like Water: The Moon Court's Role in Illuminating Hawai'i Water Law," *University of Hawai'i Law Review* 33, no. 537 (2011): 537.
16. In brief, the Māhele refers to the process of instituting a hybridized private property regime in Hawai'i over the course of many years beginning in about 1845; it resulted in stripping too many Maoli of ancestral land and resources. For a detailed explanation of this process, see Lilikalā Kame'eleihiwa, *Native Land and Foreign Desires: Pehea Lā E Pono Ai?* (Honolulu: Bishop Museum Press, 1992); Kamanamaikalani Beamer, *No Mākou ka Mana: Liberating the Nation* (Honolulu: Kamehameha Publishing, 2014).
17. See, e.g., John E. Bush, "Echoes," *Ka Leo o ka Lahui* (Sept. 18, 1891) ("With astonishing boldness, [Cabinet members] have offered for lease huge tracts of valuable government lands at ridiculously low figures, with the apparent purpose of furnishing a land grabbing picnic for the wealthy sugar barons, to the exclusion of the homesteader or small farmer. The announcement of the result of the sale of these leases . . . would seem to indicate that at least one minister does not believe that 'public office is a public trust.'"); John E. Bush, "Wailuku Water," *Ka Leo o ka Lahui* (Sept. 7, 1891) ("And to-day this family, like others who came here ostensibly as Christian teachers, claim the right to deprive the people of the water of the hills and valleys in which they were born. Verily, the missionary enterprise of these islands has been characterized by every species of hypocrisy and fraud: but a day of reckoning is at hand and please remember it most ignoble Marquis of lao.").
18. *McBryde,* 54 Haw. 174, 504 P.2d 1330.
19. *McBryde,* at 186–187, 504 P.2d at 1338–1339.
20. *McBryde,* at 186–187, 504 P.2d at 1339.

21. *Robinson v. Ariyoshi*, 65 Haw. 641, 658 P.2d 287 (1982); *Reppun v. Bd. of Water Supply*, 65 Haw. 531, 656 P.2d 57 (1982).
22. Hawai'i Constitution, article 11, section 1. This amendment was much broader than its predecessor and significantly strengthened the trustee's duties.
23. Hawai'i Constitution, article 11, section 7.
24. *In re Waiāhole Combined Contested Case* ("*Waiāhole I*"), 94 Hawai'i 97, 132, 9 P.3d 409, 444 (2000).
25. See *In re Application of Maui Electric Co.*, 141 Hawai'i 249, 408 P.3d 1 (2017) (recognizing that due process protections apply to the "right to a clean and healthful environment" as defined by the Hawai'i constitution and laws related to environmental quality).
26. Hawai'i Constitution, article 12, section 7 ('okina added).
27. The code declares that the "traditional and customary rights of ahupua'a tenants who are descendants of native Hawaiians who inhabited the Hawaiian Islands prior to 1778 shall not be abridged or denied by this chapter." Hawai'i Revised Statutes (HRS) § 174C-101(c). The code makes clear that such rights include but are not limited to the cultivation of kalo on one's own kuleana, as well as the ability to gather various resources for subsistence, cultural, and religious purposes, including hīhīwai (or wī), 'ōpae, 'o'opu, limu, thatch, kī, aho cord, and medicinal plants. HRS § 174C-101(c).
28. Hawai'i Constitution, article 11, section 7 ("The Commission shall set overall water conservation, quality and use policies; define beneficial and reasonable uses; protect ground and surface water resources, watershed and natural stream environments; establish criteria for water use priorities while assuring appurtenant rights and existing correlative and riparian uses and establish procedures for regulating all uses of Hawai'i's water resources."). See also *Waiāhole I*, 94 Hawai'i at 143, 9 P.3d at 455 ("As with other state constitutional guarantees, the ultimate authority to interpret and defend the public trust in Hawai'i rests with the courts of this State.").
29. *Waiāhole I*, 94 Hawai'i at 133, 9 P.3d at 445.
30. *Waiāhole I*, at 139, 9 P.3d at 451.
31. *Waiāhole I*, at 141, 9 P.3d at 453 (emphasis removed) (quoting *Nat'l Audubon Soc'y v. Superior Ct. of Alpine Cnty.* ["*Mono Lake*"], 658 P.2d 709, 728 [Cal. 1983]).
32. Sproat, "A Question of Wai," 207.
33. *Kaua'i Springs v. Planning Comm'n of the Cnty. of Kaua'i*, 133 Hawai'i 141, 146–147, 324 P.3d 951, 956–957 (2014).
34. *Kaua'i Springs*, at 147, 324 P.3d at 957.
35. *Kaua'i Springs*, at 152–153, 324 P.3d at 962–963.

36. *Kaua'i Springs*, at 171–175, 324 P.3d at 981–985.
37. *Kaua'i Springs*, at 171–175, 324 P.3d at 981–985.
38. *Kaua'i Springs*, at 174, 324 P.3d at 984 (citations omitted).
39. *Kaua'i Springs*, at 177, 324 P.3d at 987.
40. The court also underscored four affirmative showings that permit and other applicants must make to fulfill their responsibilities under the public trust:
    - (1) "[T]heir actual needs and the propriety of draining water from public streams to satisfy those needs";
    - (2) The absence of practicable alternatives, including alternate sources of water or making the proposed use more efficient;
    - (3) "[N]o harm in fact" to public trust purposes "or that the requested use is nevertheless reasonable and beneficial"; and
    - (4) "If the impact is found to be reasonable and beneficial, the applicant must implement reasonable measures to mitigate the cumulative impact of existing and proposed diversions on trust purposes, if the proposed use is to be approved."
    - *Kaua'i Springs*, at 174–175, 324 P.3d at 984–985 (citations omitted). Absent these requirements, "a lack of information from the applicant is exactly the reason an agency is empowered to deny a proposed use of a public trust resource." *Kaua'i Springs*, at 174, 324 P.3d at 984. Because the applicant here was unable to demonstrate the lack of impact on public trust resources, the Planning Commission proactively utilized the precautionary principle, correctly erring on the side of protecting wai.
41. Hawai'i Constitution, article 11, section 1. This pronouncement reaffirmed the holding in *Kelly v. 1250 Oceanside Partners* that counties are political subdivisions of the state and are therefore subject to the public trust. 111 Hawai'i 205, 140 P.3d 985 (2006).
42. On remand in 2018, the Kaua'i Planning Commission voted again to deny Kaua'i Springs permit applications, rejecting the hearing officer's proposed findings of fact, conclusions of law, and decision and order. The company shut down in April 2019, but asked the Circuit Court to reverse the Planning Commission's 2018 decision. In 2020, the Circuit Court ruled in favor of Kaua'i Springs, and the Planning Commission timely appealed to Hawai'i's Intermediate Court of Appeals ("ICA") In September 2024, the ICA vacated the Circuit Court's Judgment and remanded the case. *Kaua'i Springs v. Planning Dep't of the Cnty. of Kaua'i*, No. CAAP-20-0000011 (App. Sept. 23, 2024) (SDO).
43. See, e.g., *Missler v. Bd. of Appeals of the Cnty. of Hawai'i*, 140 Hawai'i 13, 396 P.3d 1151 (2017) (holding that a county planning department must review a planned unit development permit pursuant to its public trust duties, which are independent from permit requirements under the county code); *The Hilo Project, LLC v. Cnty. of Hawai'i Windward*

*Planning Comm'n,* 141 Hawai'i 380, 409 P.3d 784 (2018) (holding that a county planning commission must consider special management area permits in view of the public trust doctrine, implicitly ruling that land owned by private parties is still subject to the doctrine).

44. In *Waiāhole I,* the court purposefully declined to decide whether affirming public trust protection for a particular resource requires a separate constitutional provision explicitly protecting the resource (noting "[w]e need not define the full extent of article XI, section l's reference to 'all public resources' at this juncture"). *Waiāhole I,* 94 Hawai'i at 133, 9 P.3d at 445. Too frequently, constrained interpretations persist (for example, that the doctrine applies exclusively to wai and not land or other natural resources, despite the obvious failure to comport such a parable with the plain language and legislative history of article 11, section 1). This misperception maintains the status quo by turning a blind eye to the doctrine's grounding in Maoli custom, which encompasses everything from the depths of Kanaloa's (akua of the sea and voyaging) ocean to the expanses of Wākea's (Sky Father) sky.
45. *In re Contested Case Hearing re Conservation Dist. Use Application HA-3568 for the Thirty Meter Telescope at the Mauna Kea Sci. Reserve* (*In re TMT*), 431 P.3d 752, 773 (Haw. 2018).
46. See Hawai'i Constitution, article 12, section 4; *Ching v. Case,* 449 P.3d 1146, 1160 (Haw. 2019). The Public Land Trust is distinct from the Hawaiian Home Lands Trust and has a painful and complex history. In addition to water resources, the Hawai'i Supreme Court had already extended the public trust to navigable waters (*King v. Oahu Ry. & Land Co.,* 11 Haw. 717, 723–725 [1899]), shoreline areas (*County of Hawaii v. Sotomura,* 517 P.2d 57, 63 [Haw. 1973]), and lava extensions (*State v. Zimring,* 566 P.2s 725, 735 [Haw. 1977]).
47. In *Ching,* the Hawai'i Supreme Court recognized a "duty to 'malama 'aina,' which the court translated as 'to care for the land.'" *Ching,* 449 P.3d at 1160. This chapter will use "aloha 'āina" instead, as that better reflects the Maoli cultural practice of caring for the land as articulated by Kānaka. For more on the duty to aloha 'āina, see D. Kapua'ala Sproat and M. J. McDonald, "The Duty to Aloha 'Āina: Indigenous Values as a Legal Foundation for Hawai'i's Public Trust," *Harvard Civil Rights–Civil Liberties Law Review* 57 (2022): 525–576.
48. "Aia i Wai'oli ke Aloha 'Āina," written by U'ilani Tanigawa Lum, 2019.
49. Office of Hawaiian Affairs, *Ka Wai Ola* 36, no. 12 (Dec. 2019): 5. In contemporary times, Hanalei and Wai'oli's farmers grow about 80 percent of all locally produced kalo in Hawai'i.
50. This work involved four attorneys, over thirty students, and countless other supporters.

51. H.C.R. 163, 31st Leg. Reg. Sess. (Haw. 2021).
52. H.B. 1768, 31st Leg. Reg. Sess. (Haw. 2022).
53. Letter from Suzanne Case, former BLNR Chair, to the Wai'oli Valley Taro Hui, Inc. (Dec. 22, 2022).
54. Minutes of the Board of Land and Natural Resources (Feb. 2020).
55. Kawika B. Winter et al., "The Moku System: Managing Biocultural Resources for Abundance within Social-Ecological Regions in Hawaii," *Sustainability* 10, no. 10 (2018): 4.
56. Ibid. at 4.
57. Kame'eleihiwa, *Native Land,* 33. For a more detailed discussion of the importance of land in Maoli culture, and the practice of mālama 'āina, see ibid. at 25–33. See also Sproat and McDonald, "The Duty to Aloha 'Āina" (deconstructing aloha 'āina as a principle and legal duty).
58. Mehana Blaich Vaughan, *Kaiāulu: Gathering Tides* (Corvallis: Oregon State University Press, 2018), 48.
59. Ibid. at 29.
60. Ibid. at 30–31.
61. Kamanamaikalani Beamer, Axel Tuma, Andrea Thorenz, Sandra Boldoczki, Keli'iahonui Kotubetey, Kanekoa Kukea-Shultz, and Kawena Elkington, "Reflections on Sustainability Concepts: Aloha 'Āina and the Circular Economy," *Sustainability* 13, no. 5 (2021): 5.
62. See David Malo, *Hawaiian Antiquities,* translated by Nathaniel B. Emerson (Honolulu: Bishop Museum, 1951), 16.
63. He Kumu Kanawai a me Ke Kanawai Hooponopono Waiwai (1840), 21, available at http://punawaiola.org/. The word "ho'oponopono" is also used to describe the actions of resource managers, meaning "to put in order or shape, correct, revise, regulate, administer, supervise, manage." Pukui and Elbert, *Hawaiian Dictionary,* 341.
64. Kame'eleihiwa, *Native Land,* 33.
65. Only recently has the Water Commission taken the initiative to proactively protect water resources. For example, it set interim instream flow standards on its own, without the threat of litigation—in West Maui in March and November 2018 and attempted to do the same in Wailua, Kaua'i, in August 2018 (though a contested case hearing was requested). The Water Commission also ordered the full restoration of flows to ten East Maui streams in June 2018, which was considered unprecedented after a nearly two-decade battle pitting kalo farmers and practitioners against landowner Alexander & Baldwin.
66. *Waiāhole I,* 94 Hawai'i at 142, 9 P.3d at 453; Standing Comm. Rep. No. 77, reprinted in *1 Proceedings of the Const. Convention of Hawaii of 1978* (1980), 686.

67. Our previous works proffered an analytic framework based on Hawai'i's expressed commitment to restorative justice for Kānaka Maoli that provides one potential starting point for more fully realizing the public trust as paradigm. See D. Kapua'ala Sproat, "Wai through Kānāwai: Water for Hawai'i's Streams and Justice for Hawaiian Communities," *Marquette Law Review* 95, no. 1 (2011): 127–211; D. Kapua'ala Sproat and N. Mahina Tuteur, "Water and Climate Change in Hawai'i: A Restorative Justice Perspective," in *Climate Change: Hawai'i and the Pacific* (forthcoming). This approach to guide, and possibly compel, local decision-makers to proactively protect resources in the public interest requires focusing on four realms of self-determination for Native peoples: (1) cultural integrity; (2) lands and other natural resources; (3) social determinants and well-being; and (4) self-government.
68. Martin Luther King Jr., "I've Been to the Mountaintop," speech, Mason Temple, Memphis, TN, Apr. 3, 1968 (quoting Amos 5:24), https://www.americanrhetoric.com/speeches/mlkivebeentothemountaintop.htm.

## Bibliography


Beamer, Kamanamaikalani. *No Mākou ka Mana: Liberating the Nation.* Honolulu: Kamehameha Publishing, 2014.

Beamer, Kamanamaikalani, Axel Tuma, Andrea Thorenz, Sandra Boldoczki, Keli'iahonui Kotubetey, Kanekoa Kukea-Shultz, and Kawena Elkington. "Reflections on Sustainability Concepts: Aloha 'Āina and the Circular Economy." *Sustainability* 13, no. 5 (2021): 1–16.

Beckwith, Martha Warren. *The Kumulipo: A Hawaiian Creation Chant.* Honolulu: University of Hawai'i Press, 1972.

Blumm, Michael C., and Mary Christina Wood. "'No Ordinary Lawsuit': Climate Change, Due Process, and the Public Trust Doctrine." *American University Law Review* 1, no. 67 (2017): 1–87.

Bush, John E. "Echoes." *Ka Leo o ka Lahui* (Sept. 18, 1891).

Bush, John E. "Wailuku Water." *Ka Leo o ka Lahui* (Sept. 7, 1891).

Handy, E. S. Craighill, and Elizabeth Green Handy, with the collaboration of Mary Kawena Pukui. *Native Planters in Old Hawai'i: Their Life, Lore, and Environment.* Honolulu: Bishop Museum Press, 1972.

Kame'eleihiwa, Lilikalā. *Native Land and Foreign Desires: Pehea Lā E Pono Ai?* Honolulu: Bishop Museum Press, 1992.

King, Martin Luther, Jr. "I've Been to the Mountaintop." Speech, Mason Temple, Memphis, TN, Apr. 3, 1968 (quoting Amos 5:24). Available at https://www.americanrhetoric.com/speeches/mlkivebeentothemountaintop.htm.

King, Martin Luther, Jr. "Out of the Long Night." *Gospel Messenger,* Feb. 8, 1958 (quoting Theodore Parker). Available at https://archive.org/details/gospelmessengerv107mors.

MacKenzie, Melody Kapilialoha, Susan K. Serrano, and Koalani Laura Kaulukukui. "Environmental Justice for Indigenous Hawaiians: Reclaiming Land and Resources." *Natural Resources and Environment* 21, no. 3 (2007): 37–42.

Malo, David. *Hawaiian Antiquities.* Translated by Nathaniel B. Emerson. Honolulu: Bishop Museum, 1951.

Pukui, Mary Kawena, and Samuel H. Elbert. *Hawaiian Dictionary.* Honolulu: University of Hawai'i Press, 1986.

Sproat, D. Kapua'ala. "A Question of Wai: Seeking Justice through Law for Hawai'i's Streams and Communities." In *Nation Rising: Hawaiian Movements for Life, Land, and Sovereignty,* edited by Noelani Goodyear-Ka'ōpua, Ikaika Hussey, and Erin Kahunawaika'ala Wright, 199–219. Durham, NC: Duke University Press, 2014.

Sproat, D. Kapua'ala. "Where Justice Flows like Water: The Moon Court's Role in Illuminating Hawai'i Water Law." *University of Hawai'i Law Review* 33, no. 537 (2011): 537–579.

Sproat, D. Kapua'ala, and M. J. McDonald. "The Duty to Aloha 'Āina: Indigenous Values as a Legal Foundation for Hawai'i's Public Trust." *Harvard Civil Rights–Civil Liberties Law Review* 57 (2022): 525–576.

Sproat, D. Kapua'ala, and Isaac H. Moriwake. "Ke Kalo Pa'a o Waiāhole: Use of the Public Trust as a Tool for Environmental Advocacy." In *Creative Common Law Strategies for Protecting the Environment,* edited by Clifford Rechtschaffen and Denise Antolini, 247–284. Washington, DC: Environmental Law Institute, 2007.

Sproat, D. Kapua'ala, and N. Mahina Tuteur. "Water and Climate Change in Hawai'i: A Restorative Justice Perspective." In *Climate Change: Hawai'i and the Pacific.* Forthcoming.

Vaughan, Mehana Blaich. *Kaiāulu: Gathering Tides.* Corvallis: Oregon State University Press, 2018.

Winter, Kawika B., Kamanamaikalani Beamer, Mehana Blaich Vaughan, Alan M. Friedlander, Mike H. Kido, A. Nāmaka Whitehead, Malia K. H. Akutagawa, Natalie Kurashima, Matthew Paul Lucas, and Ben Nyberg. "The Moku System: Managing Biocultural Resources for Abundance within Social-Ecological Regions in Hawaii." *Sustainability* 10, no. 10 (2018): 1–19.

CHAPTER 4

# The Evolving Nature of Hawai'i Water Law

## From *McBryde* to *Waiahole* (1973–2000)

*William Tam*

### Introduction

Water is local, mobile, and fugitive. It changes form. Water moves, flows, floods, freezes and melts, infiltrates and condenses, passes through rock and earth, evaporates, and falls from the clouds. And often it is still. It cleans, purifies, and transforms people, animals, plants, and landscapes. Water is sometimes described as "property," a "commodity," an "asset" on a balance sheet, or a mineral, yet it has these characteristics and more. A gallon is heavy (8.34 pounds) and more expensive to move than electricity, yet less price sensitive. Everyone uses water and shares it in a variety of ways. Water crosses all boundaries and integrates everything it touches. It passes through every aspect of our environment and every living entity . . . only to transform itself and return again in the hydrologic cycle. Water makes people happy and land valuable. Water is critical to our health. It is most acutely felt when it is absent (dry periods) or in excess (flood). Water is essential to all life. There are no substitutes. We waste it unconscionably.

In Hawai'i, unlike the continental United States, there is no Colorado, Columbia, or Snake River to divert in order to relieve a drought or compensate for overgrowth. The Hawaiian Islands are geographically isolated, self-contained, and at the mercy of the trade winds. Hawai'i's topography is porous, volcanic rock. It is hydrologically dependent upon the prehistoric geology of each island and often steep.[1]

Water has a logic of its own. It is inexorable and asymmetric to other systems. Storms and droughts may be outlying events, but they dictate how infrastructure is sized and located. Intense rain, storm water runoff, and old reservoirs (like Lake Wilson, O'ahu) present special risks for both people and property. Locate a pipeline, reservoir, or pump station in the wrong place or design a system for too short a time frame and its failure will haunt you. All infrastructure has a lifetime and then fails.[2]

Water is a paradigm unlike any other. Energy and electricity usually move in one direction, but water circles back on itself. Water is not only used, but must be recycled, treated and/or disposed of. Understanding the full life cycle of water means not just learning where water comes from but where it goes. O'ahu's large, centralized wastewater treatment plants process and dispose of more than 100 million gallons/day into nearshore ocean waters. Yet fresh and wastewater systems are managed under separate jurisdictions by different departments with independent budgets, so there is little integration or coordination (except in billing). Moreover, the wastewater plants were built near coastlines only a few feet above sea level, making them increasingly vulnerable to rising seas and tsunamis.

Add chemicals, pollutants, and solids to water only if you are prepared to take them out again or face new environmental dislocations somewhere else. Go to the Environmental Protection Agency, State Department of Health, or United States Army Corps of Engineers websites and look at the enforcement actions to see how expensive and time-consuming cleanup is compared to prevention.

Only by living with water can we begin to understand it. Water is a truth serum that tests our philosophies, our relationship with the natural world . . . and our budgets. The mechanism societies have developed to memorialize these lessons is *law*. But water law is not simply a series of rules (like chess or the uniform commercial code) that prescribe a particular outcome. It is an evolving integration of values, needs, and circumstances that change over time and place, both in practice and conceptually.

The purpose of this chapter is to outline the chronological and conceptual evolution of Hawai'i water law from the 1973 Hawai'i Supreme Court *McBryde* decision, through the collateral attack on that decision in federal court in the *Robinson* litigation, the adoption of the 1987

Hawai'i Water Code, and then the Hawai'i Supreme Court's landmark response to decades of water conflict in the 2000 *Waiāhole* decision.[3]

To understand this epochal change, we must go back in time to the underlying assumptions that gave rise to these events. The two exhibits at the end of this chapter outline the evolving nature of Hawai'i water law *chronologically* (the history of problems and legal responses) and *conceptually* (toward an understanding of the integrated nature of water).[4] We begin with the 1848 Māhele (Division of Rights in Land)[5] and the historical forces at work during that time.

## The Sovereign's Duties and Ancient Prerogatives: 1846 Principles of the Māhele

Prior to the 1848 Māhele, customary Hawaiian practices governed water uses within the traditional ahupua'a land system. With the Māhele, a new set of historical forces arose to restructure and reallocate water resources. During the 1840s, Hawai'i experienced a rapid transition from a traditional land tenure and social system governed by chiefs to a "monarchy" and then a constitutional monarchy with a modified western private property system. The role of the king under the 1840 Constitution (later reflected in the 1846 Principles governing the division and distribution of land) was not that of an absolute monarch. Rather, the role of the sovereign in Hawai'i was more like a traditional guardian or trustee with reciprocal duties and responsibilities. The relationship reflected a sophisticated understanding going back hundreds of years in Hawai'i of both the natural world and human relationships. Each grew out of the islands' landscape ecology. As reflected in oral traditions, customs, and governance practices, water was at the center of food production, land tenure, and law giving. To that end, when the lands were divided during the Māhele, certain structural principles were recognized and established:

> What is the nature of and extent of that power which the King has Bestowed upon this board? It can be no other than his private or feudatory right as an individual participant in the ownership, not his sovereign prerogatives as head of the nation. Among those prerogatives which affect lands, are the following
>
> . . .
>
> *3rd. To encourage and even to enforce the usufructs of lands for the common good.*

> . . .
>
> These prerogatives, powers and duties, His Majesty ought not, and ergo, he cannot surrender. Hence, the following confirmations of the board, and the titles consequent upon them must be understood subject to these conditions.[6]

One of the most important "usufructs of lands for the common good" was (is) water, although this would not be fully articulated by the Hawai'i Supreme Court until the 1973 *McBryde v. Robinson* decision. The Court would describe the sovereign in terms of a trustee of the land with responsibilities, conditions, restrictions, and duties which could not be abrogated.

However, from the mid-1800s until statehood (1959), lawyers (and judges), trained in the concepts and traditions of western property law and eager to facilitate commercial enterprises, strained to define water as an incident of private property that could be transferred like a commodity. They were unsuccessful. It would take nearly 150 years for the older concept of stewardship over water and the role of the sovereign as trustee to be rediscovered and reaffirmed.

## Territorial Period 1900–1959: Confusion in *Territory v. Gay* (1930)

During the Territorial Period, the Hawai'i Supreme Court struggled to reconcile competing claims to surface water, mostly among sugar plantations. Between Cook's arrival (1778) and the mid-1800s, more than 80 percent of the Hawaiian population died or lost their lands. Taro cultivation declined in all but a few locations. Sugar plantations consolidated properties through acquisition, leases, quiet title actions, and "other" means. They built irrigation systems to transport stream water from wet to drier parts of the islands and argued about it later.

By 1920, the Native Hawaiian population had declined so fast that the federal government adopted a homestead program to restore the Hawaiian people to the land.[7] The effort failed. Hawaiian homestead "available lands" were selected from public properties but *excluded* "all cultivated sugar cane lands" (HHCA sec 203) (i.e., *productive* lands). The available Hawaiian home lands were by definition those areas with poor soil (often lava rock), little water, and no functioning infrastructure. Establishing a homesteading program on public lands where farming had not or could not succeed and which had inadequate water systems was a cruel and cynical promise that served only to highlight the importance of water to land.[8]

Hawai'i's streams are short, steep, and flashy. Flows vary greatly. In mathematical terms, the mean flow (average of a sum of measurements) and the median flow (midpoint in a series for measurements) often differ significantly. By comparison, rivers on the continental United States tend to be long, deep, and relatively consistent (within seasons). In Hawai'i, many windward streams are intermittent, both gaining *and* losing, and may at times have little measurable wet connection to the ocean. Conversely, high rain events deposit enormous volumes of water onto steep slopes in a brief time and effectively flood stream beds. The aquatic species (of 'o'opu, for example) that survive these conditions have adjusted to this variability.

Even where stream gauges accurately measure these large events, managing highly variable flows in any systematic way is difficult. Defining the boundaries of these flows for legal purposes became unworkable.

By 1930, the Hawai'i Supreme Court had tried a number of times without success to define a coherent theory of surface water law (groundwater would be addressed separately).[9] In *Territory v. Gay,* the Court again could not agree on a consistent rule to allocate storm versus normal water between 'ili's (small land divisions) based on a theory of konohiki rights. In fact, "normal" flows (let alone "storm" waters) rushing off Hawai'i's volcanic slopes (unlike voluminous, slow-moving rivers on the North American continent) defied any single measure. Consequently, in *Territory v. Gay,* a divided Court provided no single and controlling legal theory of surface water rights.

Consequently, while the Court continued to uphold appurtenant (taro) rights associated with kuleana parcels granted at the time of the Māhele, surface water law remained confused and uncertain through the Territorial Period, notwithstanding decisions resolving particular disputes that did not present fundamental structural problems.

Interestingly, in 1929, the Hawai'i Supreme Court decided its only groundwater case before *Waiāhole.*[10] In *City Mill v. Honolulu Board of Water Supply,* the Court considered multiple theories and adopted a "correlative rights" model of "shared use," as opposed to a "prior appropriation" ("first in time, first in right") or other exclusive rights rules. Under the correlative rights doctrine, every landowner has a right to drill a well on their property and withdraw a portion of the water *consistent with the rights of other overlying landowners to do the same.* It is similar to the *riparian* surface water doctrines (applicable in the eastern United States) in that it is a "shared use" approach to water allocation.

### Statehood (1959) and Rediscovering the Past: *McBryde v. Robinson* (1973)

The effort to rationalize water use for a new economic order based on an older land tenure system created major conceptual and practical problems. As a result of the overthrow of the monarchy (1893), the annexation of Hawai'i by the United States (1898),[11] and then the United States' adoption of the Organic Act (1900), the Hawai'i Supreme Court Justices *during the Territorial Period* were appointed and confirmed through a *federally* controlled process,[12] rather than through representatives chosen by the people of Hawai'i. This made the decisions by lawyers and judges representing large commercial interests all the more disconnected from the traditions and customs of Hawai'i's past.

Statehood in 1959 changed this. The governor of Hawai'i was now elected,[13] rather than appointed by the U.S. president, as had been the case during the Territorial Period.[14] And in turn, the governor nominated justices to the Hawai'i Supreme Court who were then confirmed by the Hawai'i Senate, rather than the United States Senate as before.[15]

The appointment of William S. Richardson (a Native Hawaiian) as chief justice following statehood laid the foundation for the Court to revisit and rediscover customs and traditions cast aside and ignored for more than a century. Water rights would be perhaps the most important (and certainly the most controversial) of the Court's new undertakings. Until statehood, the Hawai'i Supreme Court had given little consideration to Hawaiian customs and traditions.

In the early 1970s, the McBryde and the Gay & Robinson sugar companies both appealed a decision from the Fifth Circuit State Court (Kaua'i) involving a long-standing dispute over allocation of surface waters in the Hanapēpē River on the southwestern portion of Kaua'i. McBryde (a downstream plantation) complained that Gay & Robinson (an upstream plantation) had diverted too much water and deprived McBryde of water it needed.

To the surprise of some, the Court took the occasion to revisit Hawai'i's surface water law as it had evolved from the 1848 Māhele (in which private land titles were first created) through statehood.

First, the Court reconfirmed *appurtenant* rights for a duty of water to grow taro on kuleana parcels in the manner it was done at the time of the Māhele.

Second, the Court reviewed 125 years of surface water decisions and concluded there was no coherent and consistent rule. The matter was still open. The Court noted that previous decisions had failed to properly examine and incorporate Hawaiian customs and practices. To understand the Principles on which the Board of Commissioners to Quiet Land Titles established a new land tenure system in the 1848 Māhele, the Court revisited the relationships of the king as sovereign (as opposed to the king in his personal capacity), the chiefs, and the common people. The Court looked at the importance of water on all lands and for every person. The Court examined the role of the king in Hawai'i in his sovereign and governmental capacity and concluded that the king was a steward or guardian with an obligation to protect certain important "usufruct[s] of lands for the common good."[16] These important uses of the land the sovereign reserved and must preserve. These aspects (in Latin and civil law, "usufructs" or a limited *in rem* right to enjoy something of another without diminishing, altering, or destroying it) included public roads, the shoreline, gathering and access to the mountains and the sea for subsistence needs, and, of course, water.

Water has always been one of the important "usufruct[s] of lands for the common good." Roman, French, English, American, and Hawaiian society and laws have long recognized that water is a fundamental need shared by all people for a variety of common purposes and may not be held exclusively as "private property" as that term is conventionally understood.[17]

Over time, rules for using water have changed with changing circumstances (as they must). However, the Hawai'i sovereign's reserved and inherent power to manage and regulate this "usufruct of lands for the common good" remains. The original land granted at the Māhele was subject to this "condition" and remains so today. "These prerogatives, powers, and duties, [the sovereign] ought not, and ergo, cannot surrender."[18] The sovereign's ancient prerogatives have become, with the transfer of sovereignty to the "people," a public environmental right . . . and duty.[19]

The sovereign in Hawai'i not only reserved water from the grant of title to lands, but has regulatory police power authority over water.[20] Notwithstanding the state's authority, Hawai'i has also established important use rights to water by custom, common law, and statute.[21]

In the *McBryde* decision itself, the Court examined Hawaiian customs and concluded that long-established practices were based on *sharing* surface (stream) waters in some proportion. Water was not allocated on the basis of the "prior appropriation" rule, a "last in, first out" inventory principle common in the western United States. The Court also rejected earlier efforts to define normal versus surplus waters and a so-called konohiki rule that plantation agriculture claimed.

The old Hawaiian surface water practices were in fact remarkably similar to the "riparian" rule recognized in both England and the eastern United States by which landowners adjacent to a stream shared the waters of the stream in wet times and dry. In dry times, when it matters most, cutbacks were shared in some equitable manner among all landowners adjacent to a stream.[22] By comparison, in "prior appropriation" jurisdictions, the most recent (most junior) right holder gives up all their water before the next senior landowner gives up any water.

The riparian right is an incident of land ownership only on those lands through which the stream passed or on land next to a stream. Thus, lands that were *not* adjacent to a stream did not have any right to water in the stream. Moreover, under the "riparian" doctrine, a landowner may not divert water permanently away to nonadjacent lands because to do so would deprive downstream riparian landowners of their shared use. Importantly, under riparian doctrines, out-of-watershed transfers have no legal basis. This was a shock to the plantations that had diverted windward streams via elaborate ditch systems to the drier leeward and central plains on the larger islands for industrial-scale sugar and pineapple operations for more than a century. Yet despite the Hawai'i Supreme Court's 1973 decision, little changed on the ground until after the 1987 Water Code was adopted and the Hawai'i Supreme Court issued the *Waiāhole* decision in 2000.

In *McBryde,* the Court made an important statement about the very nature of water. The Court characterized the Māhele reservation as "very similar to the English common law rules that had evolved at that time that no one may acquire property to running water in a natural watercourse: that flowing water was *publici juris.*"[23]

This public trust concept *publici juris* originated in Roman law, migrated to France, and has long been recognized in English common law.[24] In 1892, the U.S. Supreme Court recognized and adopted the public trust doctrine in *Illinois Central Railroad v. Illinois.*[25] Seven years later, in 1899, and while still a republic (not recently, as many presume), the Hawai'i Supreme Court recognized the public trust doctrine in Hawai'i in *King v. Oahu Rail & Land Co.*[26]

In addition to the 1848 Māhele reservation of "usufructs for the common good," Hawai'i in 1892 adopted and incorporated "the common law of England as ascertained by English and American decisions" (including the public trust doctrine as the law of Hawaii).[27]

The public trust doctrine in this context (as distinguished from statutorily or affirmatively created "public trusts"), is a judicially recognized restraint on the permanent alienation or destruction of ocean and water resources. This is an important distinction. These are two similar but distinct ideas: 1) the historically evolving public trust doctrine, and 2) affirmatively created statutory public trusts.

In *McBryde,* the Court commented on the difficulties of reconciling old agricultural practices, traditional and customary rights, new urban demands, and the public interest with public duties. In a now-famous footnote number 15, the Court "invited" the legislature to reexamine Hawai'i's water law and establish a statutory code that could address water issues prospectively rather than through the rearview mirror of case-by-case litigation.[28]

### 1978 Hawai'i Constitutional Convention: Hawai'i Constitution art. XI, sec. 7

The political and cultural changes of the 1970s challenged land and water management practices that had dominated Hawai'i since the 1848 Māhele. When the citizens of Hawai'i voted to hold a Constitutional Convention in 1978,[29] many in the Hawaiian and environmental communities (especially those from Waiāhole in windward O'ahu who were engaged in legal battles to restore traditional taro cultivation) organized to take up the Supreme Court's 1973 invitation.[30] The *McBryde* decision was on appeal. O'ahu's Pearl Harbor aquifer was being overdrafted, which led to a statute to regulate groundwater.[31] The Constitutional Convention presented an opportunity to both memorialize the Court's legal determinations and direct the legislature to adopt a new water code.

Drawing on the experience in *McBryde* and the trial in what would later become *Reppun v. Honolulu Board of Water Supply,*[32] the Constitutional Convention articulated the state's affirmative obligation to regulate water for the public good.

> The State has an obligation to protect, control, and regulate the use of Hawai'i's water resources for the benefit of its people. The legislature shall provide for a water resources agency which, as provided by law, shall set overall water conservation, quality and use policies; define beneficial and reasonable uses; protect ground and surface water resources, watersheds and natural stream environments; establish criteria for water use priorities while assuring appurtenant rights and existing correlative and riparian uses and establish procedures for regulating all uses of Hawai'i's water resources. (Haw. Const. art. XI, sec. 7 [1978])

The amendment protected appurtenant *rights* (not just "uses") and *existing* correlative (groundwater) and riparian *uses,* and imposed a duty to protect watersheds and natural stream environments and establish criteria for water use priorities. The language did not speak of property and ownership (the "takings" claim in *McBryde* was on appeal), but more substantively of stewardship or guardianship deriving from the nature of trusts. Under long-established trust law (private or public), the title to property rests in the trustee (state), but the beneficial uses remain with water users. However, the terms of use are established under public criteria and subject to trust conditions. The nomenclature might change, but important uses would still be respected. The ability to use water is not absolute. It is subject to both more protective conditions and greater public review.

It would take nine more years (until 1987) before the legislature adopted a water code as mandated by the 1978 Constitutional Amendment. Critically, water decisions would now be made in public, under the administrative processes of the Water Code (subject to judicial review) and not, as before, by purely private actors for private purposes (subject to expensive case-by-case original actions in the courts). With public, constitutional, and statutory criteria established, water use decisions would receive closer and more public scrutiny, including administrative contested case hearings.[33]

### Rediscovering Balance: Robinson (1973–1989) and *Reppun* (1976–1982)

While the Constitutional Convention occupied public attention, two lawsuits proceeded through the courts.

#### *A.* Robinson v. Ariyoshi, *65 Haw. 641 (1982)*

After the Hawai'i Supreme Court's decision in *McBryde v. Robinson,* and before the petition of *certiorari* for review was denied in the United States Supreme Court, Robinson collaterally attacked the Hawai'i Supreme Court's *McBryde* decision by filing an original action in the U.S. District Court (Hawai'i). Robinson claimed the state's highest court had "taken" "property" in violation of the Fifth Amendment by deciding the dispute between Robinson and McBryde in a manner that upset "settled expectations." The challenge was strange in at least two ways: a) Appeals from a state supreme court decision may be taken only to the U.S. Supreme Court—not laterally to a parallel but separate federal district court; and b) courts do not "take" property by deciding a land dispute brought to them by parties (otherwise every losing party would claim in federal court that the state court "took" their property). Their sole remedy lies in a direct appeal from the state supreme court to the U.S. Supreme Court.

Nonetheless, the U.S. District Court of Hawai'i (Pence, J.) ruled against the state. On appeal, the U.S. Court of Appeals for the Ninth Circuit asked the Hawai'i Supreme Court to clarify six questions of Hawai'i law (as provided in Hawai'i's rules). In responding to the Ninth Circuit, the Hawai'i Court explained the uncertain nature of Hawai'i water law and the state's public trust obligations.[34] The Hawai'i Court reaffirmed the *McBryde* holdings as a matter of law, but did not enjoin the diversions. Injunctions are equitable remedies so that where there is no showing of harm, there is no basis for issuing an injunction. The Court concluded:

> The *McBryde* opinion was only the beginning of a necessary definition of the parameters of the State's authority and interests in Hawaii's waters. These parameters, we believe, should be developed on a case by case basis or by the legislature as the particular interests of the public are raised and defined. However, the opinion properly clarified the nature of respective rights to water, that is, it made clear that underlying every private diversion and application there is, as there always has been, a superior public interest in this natural bounty.[35]

Despite this explanation, the Ninth Circuit affirmed. The state appealed. The U.S. Supreme Court took the state's appeal and dismissed the case on Article III (constitutional judicial power) grounds because Robinson had a remedy for the alleged regulatory takings under Hawai'i law and because alternatives still existed that had not yet been fully tested.[36] The federal case was finally over, but only after Hawai'i adopted a new water code.

### *B.* Reppun v. Honolulu Board of Water Supply, *65 Haw. 531 (1982)*

Although the Hawai'i Supreme Court had always affirmed the appurtenant rights of kuleana parcels to a volume of water to grow taro in the same manner as it had been practiced on a parcel of land at the time of the Māhele, the time and costs of litigation (as a practical manner) meant farmers seldom used courts to resolve such disputes. In *Reppun,* the Honolulu Board of Water Supply had inserted a pipe into the side of the Ko'olau mountains (Windward O'ahu) just above where the Waihe'e stream emerged and diminished the stream flow such that taro farmers below suffered severe taro rot. On Windward O'ahu (as on the windward side of most Hawaiian islands), the groundwater and surface streams interact in direct and sometimes one-to-one relationship. The Supreme Court took note of this relationship in finding that all water resources are connected, integrated, and in effect "one water."

In *Reppun,* the Court described the traditional Hawaiian land and water system in detail and reiterated the priority of taro duties to water, based on the Māhele and the common law. The Court recognized a "public interest" exception to full stream restoration where a public entity withdrew water for public purposes and had come to rely upon the source for domestic consumption.

The Court also addressed the issue of severance where the landowner sold the land but attempted to reserve the water rights. This created a dilemma under contract law because the buyer had not paid "value" for the water. There appeared to be a lack of mutual consideration. However, water runs with the land. It is not personal and it cannot be severed from the land by contract. At most, the person who bought the land "without water rights" may be equitably estopped under the contract from using water for a period of time. But the *estoppel is personal* between the two contracting parties. It does not run with the land.

Otherwise, kuleana properties could be forever stripped of traditional and constitutionally protected water rights by the fiction of private contract agreements.

## 1987 Hawai'i Water Code: A New Consensus

Following the 1978 Constitutional Amendment, no consensus could be reached about how to enact a new water law. The governor and the legislature each established separate special commissions. For more than eight years, the debate continued. The legislature made multiple false starts. There was a stalemate. Three principal interests effectively blocked each other: corporate agriculture (old plantations); counties claiming "home rule"; and a collection of small farmers, governmental regulators, environmentalists, and Native Hawaiians.

In 1986, Peter Adler, Joann Yukimura, and a few individuals initiated a private Water Roundtable to see if a solution might be found outside of the limelight.[37] Participants were chosen because they were knowledgeable, they could act civilly, and they would participate in their personal capacity, not as a representative of any principal. After nearly eleven months of discussions, one-on-one sidebars, multiple redrafts, and many shared meals, a bill emerged. No one would claim it as their own. But when asked, no one would disown it. The critical accommodation involved a designation process to phase in the full regulatory force of the new law in areas that needed management and the postponement of regulating less critical areas until a later date. The cost was a bifurcated legal system and long delays. The reality was that the staff, time, and the cost to bring the law into effect statewide would be substantial and that it could not happen much faster in any case. Moreover, the water code required a long and steep learning curve by many people. Whole new processes had to be organized. Stream ecology and instream flow standards, Native Hawaiian rights, the integration of water and land uses, water duties for agriculture, the public trust, reasonable beneficial uses, conservation measures, water use permitting and due process, administrative rules, and a myriad other factors had to be developed and instituted.

In January 1987, the People's Water Conference under Martha Black's leadership invited Don Maughn, chair of the California Water Resources Control Board (and formerly the architect of the Arizona water law) to address the Hawai'i legislature and the public. His experience and gravitas conveyed one primary lesson: Hawai'i could not move forward without a water code; relying on past cases to resolve future problems would not work. Don Maughn's address and the roundtable's diligent work set the stage for the legislature to act. The bargaining began. But the roundtable participants and many others insisted the draft must be kept as a whole. Pull one strand out and the whole edifice would unravel. In the end, the roundtable draft remained almost as written. But it had to overcome still more hurdles.[38]

The 1987 Water Code passed in part because every interest secured something they thought they needed to fight in the future. Water fights are never over. They just change terrain, actors, issues, resources, and . . . lawyers. The Water Commission would be the new forum. It had a whole new set of tools, standards, obligations, and planning measures.

Fortunately, Bill Paty was chair of the Board of Land and Natural Resources and became chair of the Water Commission. He was respected and known by everyone. He was fair, always in good humor, and experienced.[39] It was a good start.

### Judicial Review, Midcourse Corrections, and the Public Trust: *In Re Waiāhole*

Organizing water law and practices after the 1987 legislation posed a number of new dilemmas. Should the administrative rules stake out the answers to every regulatory issue in advance? Or should the rules repeat the statute and evolve over time as experience and experiment dictate (the latter view prevailed)? How do you design instream flow standards when the prevailing models were created for continental stream ecologies (a new weighted factor scheme was devised)? How do you handle disputes in nondesignated areas where no water use permit process is in place (the Code has a tiered system of dispute resolution options based on the nature of the dispute and the parties)?

The Code called for interim and later final instream flows for all Hawai'i's streams. The reality is that streams change constantly. Determining a "permanent" standard based on statistical Q90 (or other) measurements proved elusive. A much more complex, qualitative ecology was required. In the meantime, interim instream flow standards were adopted based on the existing flows (as they may vary naturally from time to time) without any change in the diversion structures. Any change in the status quo required a petition to review. Streams would be addressed by petition, as part of adjudication, or in priority regions.

Permanent standards would require substantially more information, more work, and some procedural differences.

The Code created advance planning to integrate water and land use (state and county), water quality, conservation, state water projects, and agriculture.[40] By integrating water planning into the major natural resources that use or impact water, the Code began to build a more comprehensive and coherent natural resources management system. Little appreciated at the time, these planning requirements may prove to be the key to a unified public trust of natural resources in Hawai'i.[41] This was the beginning of a more inclusive and comprehensive approach. The creation of instream flow standards was among the most important provisions of the Code.[42]

Native Hawaiian rights occupy a special place in the Code.[43] Hawaiian traditions, practices, and customs permeate the law.[44] Yet efforts (and claims) to codify traditional and customary Hawaiian practices in administrative rules have largely failed because the full range of a people's history and customs can never be reduced to a code—nor can they be abrogated by the stroke of a pen. Hawaiian traditions and customs impact stream flows, permitting, growing taro, and the very philosophy of water management itself. Consequently, water and water "rights" are connected to land and land practices rather than to individuals themselves.

Hawaiian homes' rights to water (long neglected) under both the Hawaiian Homes Commission Act of 1921 (section 221) and the Code include not only a right to use water arising *on* Hawaiian homelands, but a priority to use water on public lands. Planning for this is long overdue.

The dispute resolution mechanisms in the Code are designed to address problems at the scale and specificity required. Many complain that the processes take too long, yet fail to recognize the historic nature of building new institutions. Public institutions (unlike private ones) must not only be concerned with precedent, but with public accountability. In many ways, the Commission is establishing a new common law in many fields at the same time (subject to judicial oversight).

In 1992, Waiāhole farmers petitioned to designate Windward O'ahu groundwater as a water management area and restore instream flows following the closure of O'ahu Sugar Co. ("OSC").[45] Beginning in the 1920s, OSC diverted major windward streams through a tunnel more than twenty miles to central O'ahu to irrigate sugar fields.

Although OSC began to close down sugar operations in the 1980s (due in part to the loss of the once-in-seven-years spike in world sugar prices), windward water still flowed to central Oʻahu. Much of it was simply dumped into Waikele Gulch and ran into Pearl Harbor. The *Waiāhole* contested case hearing became the first comprehensive test of modern Hawaiʻi water law.

Every new institution faces a test in which its very structure is measured. *Waiāhole* was that test. *Waiāhole* became the historic decision that reshaped not just water law, but natural resource law in Hawaiʻi.[46] The trigger was the restoration of Windward Oʻahu stream flows that had been diverted in the early twentieth century for central Oʻahu sugar plantations. But the historical moment was a paradigm shift that is still in motion.

Prior to the *McBryde* decision, the underlying and animating worldview of Hawaiʻi's unresolved (and confusing) water law was that water could be privately owned, taken, and sold without regard to public, cultural, or ecological consequences. It was viewed as a commodity and an incident of land like a crop or a mineral.

After the 1978 Constitutional Amendment, the 1973 *McBryde* and 1989 *Robinson* litigation, the 1987 Hawaiʻi Water Code, and the 1997 *Waiāhole* contested case and the 2000 Hawaiʻi Supreme Court decision, the historical, cultural, ecological, and legal values of water shifted to a public trust and stewardship model. The Court described the scope, purpose, and substance of the trust.[47] While private uses can be important, they are not "public trust uses," and must remain subservient to public needs. The Court recognized a procedural step.

> The Court discussed the "precautionary principle" in the context of water law: Where scientific evidence is preliminary and not yet conclusive regarding the management of freshwater resources which are part of the public trust, it is prudent to adopt "*precautionary principles*" in protecting the resource. That is, where there are present or potential threats of serious damage, lack of full scientific certainty should not be a basis for postponing effective measures to prevent environmental degradation. "Awaiting for certainty will often allow for only reactive, not preventive, regulatory action." *Ethyl Corp. v. EPA,* 541 F.2d 1, 25, 5–29 (D.C. Cir.), *cert. denied,* 426 U.S. 941 (1976). In addition, where uncertainty exists, a trustee's duty to

> protect the resource mitigates in favor of choosing presumptions that also protect the resource. *Lead Indus. Ass'n v. EPA,* 647 F.2d 1130, 1152–56 (D.C. Cir. 1976), *cert. denied,* 449 U.S. 1042 (1980).[48]

Judge Skelly Wright's seminal 1976 decision in *Ethyl Corp v. EPA,* 541 F.2d 1 (D.C. Cir.), adopted by the Hawai'i Supreme Court, has profound implications not just for water law, but for natural resource management across Hawai'i. A whole new worldview and decision-making model regarding natural resources was beginning to take shape.

The process came out of the shadows of private actors and into the sunlight. Water is no longer treated as a single, privately owned commodity for sale or trade, but is a shared natural resource, culturally significant, integrated with land and health matters, to be conserved, protected, and revered as if our very life depends upon it. The Waiāhole (and parallel Maui[49]) water struggles involved decades of perseverance and courage on the part of individual community members and public officials who faced extraordinary pressures. And their work could not be more timely.

For the existential threat of climate change is now upon us. The legal fights and evolution from statehood to *Waiāhole* are the precursor to the challenges presented by climate change. The institutions are now in place. They have met their preliminary tests.

## Decision-Making for the Future

The Water Code and the water law of Hawai'i are a toolbox to solve problems. Making decisions based on the limited basis of past cases alone was replaced with forward-looking concepts and instruments to shape the future. Above all, it is important to choose the appropriate means to solve problems.

### *Dispute Resolution*

The Water Code was adopted in part to look forward in settling disputes, to solve disagreements *prospectively* rather than with narrow rearview-mirror after-the-fact blinders. The Code establishes administrative processes by which communities and people without deep financial resources may obtain a fair hearing. Most importantly, the processes are open, and a record is created so that courts can correct mistakes. The multiple dispute resolution provisions in the Code are designed to create working tools appropriate for the problem at hand. In the beginning, contested cases required substantial time and resources. The Supreme Court is clarifying many procedures and the meaning of ambiguous statutes. Increasingly, parties will develop a better understanding of what to expect. This in turn will help narrow and focus disputes, making them more amenable to negotiated solutions. The water use permit process may be a matrix, but each of those steps is designed to examine a different aspect of water and recognize values that are often ignored.

### *Integrating Water and Land Use*

As Hawai'i's population grows and places more pressure on limited land and water resources, the need to integrate land and water use and management will accelerate. The Hawai'i Water Code anticipated this problem and established a set of water planning tools.[50] Each county is required to prepare a Water Use and Development Plan. The state must prepare an overall Water Resource Protection Plan, a State Water Projects Plan, and an Agricultural Plan, while addressing water conservation and alternative water resources.[51] These tools will be essential to address climate change and all the challenges it will bring.

### *Hawaiian Homes*

Hawaiian home lands have very strong water rights on paper. The HHCA, section 221 established a "first call" to use water on public land for DHHL needs.[52]

### *Taro and Appurtenant Rights*

The Hawai'i Supreme Court has always upheld the water rights of kuleana parcel owners to grow taro ("appurtenant rights") and did so in unmistakable language in the 1982 *Reppun v. Board of Water Supply* decision. The 1978 Constitutional Amendment cemented the right in constitutional terms.

### *Instream Flow Standards*

The Water Code establishes a mechanism to set instream flow standards for streams throughout the state.[53] Restoring streams to protect natural flora and fauna is also a means to assure that traditional and customary practices may continue. The process often takes a long time, but the very act of exploring a stream's natural ecology and history provides a foundation for later managing water and the stream. The burden of proof rests with the proposed user to demonstrate why an off-stream use is necessary and how it is a "reasonable-beneficial use" under the Code. These are important tests.

### *Wastewater Reuse*

Treated wastewater is a valuable alternative source of water that is literally thrown away. Approximately two-thirds of the groundwater pumped each day on O'ahu is used, processed, and dumped in the ocean. There are small-scale alternatives to the large, centralized wastewater treatment plants on the shoreline. Small R-1 units distributed throughout the island could deliver agricultural water to the community in which the wastewater is generated. Processing wastewater locally would reduce the high cost of energy to move water long distances.[54] This approach will require an integrated system and an appreciation of the full life cycle of water and energy.

Treated wastewater is often more reliable (and a more consistent supply of water) than old plantation surface systems and reduces agricultural demand for pumped groundwater or surface water diversions. The army is already doing this at Schofield Barracks.[55] Small distributed systems like this could reduce reliance on large treatment plants at risk of sea level rise, increase resilience to any single emergency, reuse "wasted" water for irrigation, and act as a reserve against drought.

This will require reconceiving water, reorganizing energy, and designing agriculture in new ways. Unfortunately, it may take a disaster to mobilize the political will to make such changes.

## Conclusion

Water is unlike any other substance. It circulates endlessly and connects us through every thread of life, every landscape, every culture, and across time. It is indispensable and inescapable. Hawai'i is discovering a new way to use, manage, and value water. The Water Code and Hawai'i's public trust water law provide a model for natural resource management for a simple reason: Water integrates everything that it touches. Water touches everything.

### The Evolving Nature of Hawai'i Water Law

#### *Chronology 1848–2022 (selected)*

| | |
|---|---|
| 1848 | Māhele |
| 1867 | *Peck v. Bailey*, 8 Haw. 658 |
| 1900 | Territory (Organic Act) |
| 1917 | *Carter v. Territory*, 24 Haw. 47 |
| 1921 | Hawaiian Homes Commission Act (HHCA) §§ 220 and 221, 42 Stat. 108 |
| 1929 | *City Mill v. Honolulu Sewer and Water Commission*, 30 Haw. 912 (groundwater) |
| 1930 | *Territory v. Gay*, 31 Haw. 376 |
| 1959 | Statehood |
| 1961 | Groundwater Use Act (Haw. Rev. Stat. chap. 177) |
| 1973 | *McBryde v. Robinson*, 54 Haw. 173 |
| 1978 | Constitutional Amendment (Haw. Const., art. XI, § 7) |
| 1979 | Pearl Harbor groundwater designated, Groundwater Use Act (Haw. Rev. Stat. chap. 177, repealed 1987) (SY: 225 mgd); |
| | Report to governor by "State Water Commission" |
| 1981 | Honolulu and Wailua groundwater designated under Ground Water Use Act (Haw. Rev. Stat. chap. 177, repealed 1987) |
| 1982 | Hawaii Instream Use Protection Act (windward O'ahu) (Haw. Rev. Stat. chap. 176D, repealed 1987); |
| | *Reppun v. Board of Water Supply*, 65 Haw. 531; |
| | *Robinson v. Ariyoshi*, 65 Haw. 641 (six certified questions) |
| 1985 | Legislature's Advisory Study Commission Report |
| 1986 | Hawai'i Senate (but not House) passed a water code bill for the first time; |
| | Water Code Roundtable |
| 1987 | Water Code passed, Haw. Rev. Stat. chap. 174C; |
| | Haw. Rev. Stat. chap. 176, 176D, & 177 repealed |
| 1988 | Administrative Rules adopted (Haw. Admin. R. chap. 13–167 to 13–171); |
| | Interim Instream Flow Standards ("IIFS") adopted statewide except windward O'ahu |

| | |
|---|---|
| 1989 | *Robinson v. Ariyoshi*, 887 F.2d 815 (9th Cir.) dismissed; |
| | Pearl Harbor uses documented. (Oahu Sugar Co. step down in sugar use); |
| | Sustainable yield revised: 225 mgd to 195 mgd |
| 1990 | County Water Use and Development Plans reviewed |
| 1991 | Water Code and HHCA amended to explicitly address DHHL's future water needs, including in plans and permits |
| 1992 | Windward O'ahu IIFS; |
| | Windward O'ahu groundwater designated; |
| | Molokai groundwater designated; |
| | Declarations of Water Use |
| 1994 | Waiāhole Ditch contested case initiated; |
| | Water Code Review Commission: Report to Legislature; |
| | Ewa Marina contested case initiated |
| 1995 | *Waiāhole* contested case hearings |
| 1996 | *Ko'olau Ag v. CWRM*, Haw. Sup. Ct. No. 18675 (Nov. 27, 1996); |
| | Windward O'ahu groundwater designation upheld; |
| | Waiāhole and Lā'ie contested case hearings |
| 1997 | Ewa groundwater management (May 14, 1997) |
| 1997 | *Waiāhole* contested case decision (CWRM) |
| 1998 | Hanalei River designated as American Heritage River |
| 1999 | Lā'ie (O'ahu) contested case decision (CWRM); |
| | Ewa Marina contested case decision (CWRM); |
| | Moloka'i Ranch contested case decision (CWRM) |
| 2000 | Waiāhole Ditch Combined Contested Case Hearing—*Water Use Permit Applications, Petitions for Interim Instream Flow Standard Amendments, and Petitions for Water Reservations, Appeal from the Commission on Water Resource Management* (Case No. CCH-OA95–1), 94 Haw. 97, 9 P.3d 409 (August 22, 2000) (Nakayama, J.) ("*Waiāhole I*") |
| 2001–2003 | Restoration of Hi'ilawe Stream (Hawai'i County); |
| | Repair of Hamakua Ditch (Hawai'i County); |
| | Designation of 'Īao aquifer (Maui) |

2004 — *In Re Wai'ola O Moloka'i, Inc.* and *Molokai Ranch,* Contested Case Hearing on Water Use, Well Construction, and Pump Installation Permit Applications (Case No. CCH-MO96–1), 103 Haw. 401, 83 P3d 664 (Jan. 29, 2004) (Levinson, J.). Appeal from CWRM Contested Case Hearing (Hawaiian Homes): First Hawai'i Supreme Court decision to interpret Hawaiian Homes Commission Act § 221 water rights;

Waiāhole Ditch Combined Contested Case Hearing—*Water Use Permit Applications, Petitions for Interim Instream Flow Standard Amendments, and Petitions for Water Reservations, Appeal from the Commission on Water Resource Management* (Case No. CCH-OA95–1), 105 Haw. 1, 93 P.3d 643 (June 21, 2004) (Nakayama, J.) ("*Waiāhole II*")

2006 — *Kelly v. 1250 Oceanside Partners,* 111 Haw. 205, 140 P.3d 985 (2006); DOH and Hawai'i County breached public trust responsibilities to protect coastal waters in South Kona, Hawai'i

2007 — *In Re Kukui, (Molokai),* 116 Haw. 481, 174 P.3d 320 (2007), Appeal from CWRM Contested Case Hearing (Hawaiian Homes): Hawai'i Supreme Court vacated and remanded Commission on Water Resource Management's decision involving Hawaiian homeland claims on Moloka'i;

*Waihe'e, Maui* (CWRM designates groundwater management areas due to rising chlorides)

2008 — Appraisal of Stormwater Reclamation and Reuse

2009 — Ko'olau Loa Watershed Management Plan;

Wai'anae Watershed Management Plan

2010 — Waiāhole Ditch Combined Contested Case—*Water Use Permit Applications, Petitions for Interim Instream Flow Standard Amendments, and Petitions for Water Reservations* (Case No. CCH-OA95–1) (October 13, 2010) (Memorandum Opinion). Supreme Court reversed and remanded Commission on Water Resource Management's third decision on the Waiāhole water case ("*Waiāhole III*");

| | |
|---|---|
| | Hawaii County Water Use and Development Plan |
| 2011 | Hawaii Rainfall Atlas (updated) |
| 2012 | *In Re Iao [Maui] Ground Water Management Water Use Permit Applications and Petition to Amend Interim Instream Flow Standards* (No. CCH-MA06–01). 128 Haw. 228, 287 P3d 129 (2012) (Nakayama, J.) ("*Nā Wai 'Ehā*"). Supreme Court reversed and remanded CWRM decision setting Instream Flow Standards for Nā Wai 'Ehā for failure to properly analyze public trust duties and traditional and customary rights; |
| | *In Re Petition to Amend Interim Instream Flow Stan dards for East Maui* (No. CAAP-10–0000161) (Nov. 30, 2012) (Memorandum Opinion) Intermediate Court of Appeals reversed and remanded CWRM decision denying contested case to set Instream Flow Standards in East Maui despite clear constitutional due process rights; |
| | Ko'olaupoko (O'ahu) Watershed Management Plan |
| 2013 | Pō'ai Wai Ola/West Kaua'i Watershed Alliance: Combined Petition to Amend IIFS for Waimea (Kaua'i) River and headwaters and tributaries, and Complaint/Dispute Resolution and Declaratory Order Against Waste (July 24, 2013); |
| | Kaloko-Honokohau National Park Service files petition to designate Keauhou Aquifer (North Kona, Hawai'i) as a groundwater management area (Sept. 13, 2013) |
| 2014 | *Kauai Springs v. Kaua'i Planning Commission,* 143 Haw. 141, 324 P.3d 951 (2014). Supreme Court vacated and remanded Kaua'i Planning Commission decision for failure to assess alternative water sources and carry out Public Trust duties in approving county zoning permits. The state was not a party, but zoning applications before county agencies now require greater scrutiny of water sources; |
| | *Nā Wai 'Ēha (West Maui)*—Contested Case Hearing (3 Phases):<br>1) Amended IIFS. Stipulated Agreement and Order<br>2) Appurtenant Rights |

|  |  |
|---|---|
|  | a. Determination of kuleana claims (Order 12/31/14)<br>b. Qualification phase<br>3) Water Use Permits;<br>Central Oahu Non-Potable Water Master Plan—Phase 2;<br>Red Hill contamination leaks |
| 2015 | *East Maui Petition to Amend Instream Flow Standards* (“IFS”);<br>*Contested Case Hearing;*<br>*Nā Wai ʻEhā (West Maui)*—Contested Case Hearing:<br>1) Appurtenant Rights—Qualification phase<br>2) Water Use Permits<br>3) Integration of 3 Phases and Commission Determination (2016);<br>*Keauhou Aquifer Designation petition* |
| 2016 | North Shore (Oʻahu) Watershed Management Plan;<br>Water Energy Nexus Report; Water Audit legislation |
| 2017 | *Waimea River, West Kauai Water Settlement Agreement;*<br>*A & B Revocable Permits (BLNR) for East Maui Streams:* 26 streams; 13 streams;<br>CWRM denies National Park Service Petition to Designate Keauhou Aquifer as a Water Management Area |
| 2018 | Waiʻoli, Kauaʻi Interim Instream Flow Standard (amended);<br>West Maui streams’ Interim Instream Flow Standard (amended) |
| 2019 | Water Resource Protection Plan (update);<br>Study of climate change impacts on recharge statewide;<br>Agricultural Water Use and Development Plan (update) |
| 2020 | Pearl Harbor Water Shortage Plan (update);<br>USGS statewide monitoring needs |
| 2021 | CWRM Environmental Assessment (EA) Exemption List |
| 2022 | Designation of Lahaina/West Maui Ground and Surface Water;<br>Restoration of five Molokaʻi streams;<br>CWRM takes position on Navy Red Hill Contamination; |

*Carmichael v. BLNR, A&B, Maui County, No. SCWC-16–0000071* (Haw. March 3, 2022). Renewing BLNR Revocable Permits for water requires Environmental Review;

A&B Water Lease application (continued)

## The Evolving Nature of Hawai'i Water Law

### *Conceptual Evolution: The Integrated Nature of Water*

1. Failure of the private ownership model
   Recognition of shared uses
   Riparian doctrine and correlative rights
   Reaffirmation of appurtenant (taro) rights (*McBryde*)
2. Hawai'i State Constitution (art. XI, §7)
3. Integration of ground and surface water (*Reppun*)
4. Public Trust (Haw. Sup. Ct., Robinson's sixth certified question)
   Sustainable yields: Overall supply, demand, and sustainability
5. Hawai'i Water Code (Haw. Rev. Stat. chap. 174C)
6. Hawai'i Commission on Water Resource Management—http://dlnr.hawaii.gov/cwrm/
7. Designating Water Management Areas (Haw. Rev. Stat. chap. 174C, part IV)
8. Stream protection:
   Interim Instream Flow Standards (Haw. Rev. Stat. §174C71)
   1990 Hawai'i Stream Assessment (Stream Protection and Management Plan)
   Wild and Scenic Rivers
9. Traditional and customary rights (Haw. Rev. Stat. §174C101(c))
10. Planning: Integrating land and water use (Haw. Rev. Stat. §174C31)
11. Hawaiian Home Lands (Haw. Rev. Stat. §174C101(a))
12. Alternative water sources: duty to consider
13. Conservation (Haw. Const., art. XI, §1; §7)
14. Wastewater Management (Honolulu, Ewa) (2010 Consent Decree)
15. Bulk heading tunnels (Kahana, O'ahu)
16. Watershed management:
    Public–Private Watershed Partnerships (on each island)
    EPA/DOH (CWA §319 studies)

17. GIS mapping
18. Water Quality Plans (Haw. Rev. Stat. §174C31)
    Clean Water Act of 1972 (33 U.S. C. §1251 et seq.)
    Stormwater reclamation (Wheeler)
    Total Maximum Daily Load ("TMDL") (CWA §303(d))
    Nonpoint Source Pollution Program (runoff)
    EPA delegation to State (July 2000 plan)
    Best Management Practices ("BMP")
    Ala Wai flood and canal plans
19. Integrated Resource Planning ("IRP") (State Water Code, Haw. Rev. Stat. §174C-31)
    CWRM/County
    Agriculture Plans (DOA)
    County Water Use and Development Plans—Revisited
    Stream Protection—Revisited
    State Projects Plan
    DOH—Water Quality
20. Wastewater Reuse: Integrated regional planning for Central O'ahu (http://dlnr.hawaii.gov/cwrm/planning/alternative/)
21. Reasonable-Beneficial Use (Haw. Rev. Stat §174C-3)
22. Economic uses of water; cost incentives
23. Public Trust analysis (*Waiāhole, 'Īao, Kauai Springs*)
24. Water-Energy Nexus
25. Full Life Cycle Accounting and Triple Bottom Line analysis
26. Climate change

## Notes

The author would like to thank Harris Manchester College, University of Oxford, and Sir Ralph Waller for their generosity in support of this work.

1. See L. Stephen Lau and John F. Mink, *Hydrology of the Hawaiian Islands* (Honolulu: University of Hawai'i Press, 2006).
2. The Navy's Red Hill fuel storage facility (O'ahu) is a dramatic example of expensive, aging 1940s infrastructure that poses a major threat to the Pearl Harbor aquifer drinking water.
3. See D. Kapuaala Sproat, "From Wai to Kanawai: Water Law in Hawai'i," in *Native Hawaiian Law: A Treatise,* edited by Melody K. MacKenzie, with Susan K. Serrano and D. Kapua'ala Sproat (Honolulu: Kamehameha Publishing, 2015), 522–610.

4. The chronology includes case citations to the decisions discussed in this chapter.
5. For more historical context, see Kamanamaikalani Beamer, *No Mākou ka Mana: Liberating the Nation* (Honolulu: Kamehameha Publishing, 2014); and Donovan C. Preza, "The Empirical Writes Back: Re-examining Hawaiian Dispossession Resulting from the Māhele of 1848" (Master's thesis, University of Hawai'i at Mānoa, 2010).
6. Principles adopted by the Board of Commissioners to Quiet Land Titles in Their Adjudication of Claims Presented to Them (August 20, 1846); adopted by the Hawai'i Legislature on October 26, 1846. L.1847 (emphasis added).
7. Hawaiian Homes Commission Act of 1920, 42 Stat. 108 (1921) ("HHCA").
8. Ironically, the HHCA section 221 granted the Hawaiian Homes Commission priority rights to water on public lands in a manner similar to (but different from) water rights of Native American tribes recognized by the U.S. Supreme Court in *Winters v. U.S.*, 207 U.S. 574 (1908). HHCA rights were established by federal statute, not by judicial interpretation.
9. *Territory v. Gay*, 31 Haw. 376 (1930).
10. *City Mill v. Honolulu Board of Water Supply*, 50 Haw. 912 (1929).
11. For an account of how the United States annexed Hawai'i's public lands, see Tom Coffman, *Nation Within: The Story of America's Annexation of the Nation of Hawaii* (Durham, NC: Duke University Press, 2003).
12. 1900 Organic Act, 31 Stat. 141, art. 66.
13. Haw. Const. art. V.
14. 1900 Organic Act, art. 66.
15. 1900 Organic Act, art. 82. Haw. Const. art. VI.
16. Principles adopted by the Board of Commissioners to Quiet Land Titles in Their Adjudication of Claims Presented to Them.
17. Joseph Sax, "Property Rights and the Economy of Nature: Understanding Lucas v. South Carolina Coastal Council," *Stanford Law Review* 45 (1993): 1433–1455; 1840 Hawaii Constitution, Laws of 1842 in *Fundamental Laws of Hawai'i* (1904), 29.
18. Principles adopted by the Board of Commissioners to Quiet Land Titles in Their Adjudication of Claims Presented to Them.
19. Jan Stevens, "The Public Trust: A Sovereign's Ancient Prerogative Becomes the People's Environmental Right," *University of California Davis Law Review* 14 (1984): 195–232.
20. *McBryde v. Robinson*, 54 Haw. 173 (1973); *Robinson v. Ariyoshi*, 65 Haw. 641 (1982) (six certified questions).

21. *Reppun v. Honolulu Board of Water Supply*, 65 Haw. 531 (1982).
22. An individual's contribution in labor to maintain the ʻauwai irrigation system was factored into the consideration.
23. *McBryde*, 54 Haw. at 186–187.
24. H. Bracton, *On the Laws and Customs of England*, translated by S. Thorne (1968), 39–40, cited in Stevens, "The Public Trust," 197.
25. *Illinois Central Railroad v. Illinois*, 146 U.S. 387 (1892) (deciding submerged land on Chicago's waterfront in Lake Michigan could not be alienated . . . even by the legislature).
26. *King v. Oahu Rail & Land Co.*, 11 Haw. 717 (1899). In this case, a company chartered by the government to build a railroad attempted to use the condemnation authority (granted to it by the government to obtain rights of way) to condemn submerged lands in Honolulu Harbor. Like the railroad in the *Illinois Central* case, using the government's own power to condemn public land showed just a little too much chutzpah for even the Republic of Hawaiʻi's Court in 1899.
27. Haw. Rev. Stat., sec.1–1 (1892).
28. *McBryde v. Robinson*, 54 Haw. 173, 189n15 (1973) ("It does seem a bit quaint in this age to be determining water rights on the basis of what land happened to be in taro cultivation in 1848. Surely any other system must be more sensible. Nevertheless, this is the law in Hawaiʻi, and we are bound to follow it. We invite the legislature to conduct a thorough re-examination of the area.").
29. Haw. Const. art. XVII.
30. Charles Reppun, Charlene Hoe, and Ron Albu played critical roles in formulating the constitutional amendment and seeing it through to adoption.
31. Haw. Rev. Stat., chap. 176 (repealed 1987).
32. *Reppun v. Honolulu Board of Water Supply*, 65 Haw. 531 (1982).
33. Administrative contested case hearings allow affected individuals and community organizations (who often lack the ability or resources to present their cases in court) to participate in evidentiary due process hearings within the department with the jurisdiction and responsibility for the resource. Haw. Rev. Stat., chap. 91. Integrated ongoing management is essential in the water context where future conduct is complex, prospective, and interactive. This "reimagination of administrative law" aims to make agency competence visible and break the binary thinking of the past that assumed (falsely) that the choice was either to limit public administration or not. See Elizabeth Fisher and Sidney Shapiro, *Administrative Competence: Reimagining Administrative Law* (Cambridge: Cambridge University Press, 2020).

34. *Robinson v. Ariyoshi,* 65 Haw. 641, 667–677 (1982) (six certified questions). The Hawai'i Court's response to the sixth question was prescient in framing the background of the public trust.
35. *Robinson v. Ariyoshi,* 65 Haw. 641, 677 (1982).
36. *Ariyoshi v. Robinson,* 477 U.S. 902 (1986). Deputy Attorneys General Steve Michaels and James Dannenberg were essential to the state's success in the *Robinson* litigation.
37. Chatham House rules applied. To increase open discussion, participants are free to use the information received, but neither the identity nor the affiliation of the speaker, nor that of any other participant, may be disclosed.
38. Ironically, it would be a midnight speech by an old-time estate trustee and sugar plantation manager Fred Trotter to a weary legislative conference committee that would remind legislators that while people in Hawai'i might disagree and fight each other in court, at least we all lived here and could sit in the same room and talk with each together. If the Code was not adopted, Trotter argued, then outside corporations with headquarters around the world would make decisions about Hawai'i water and natural resources. Hawai'i people would have no say. It worked.

    Underscoring the point, Bill Paty, the former Waialua sugar plantation manager, WWII war hero, president of the 1978 State Constitutional Convention, and chair of the State Board of Land and Natural Resources marshalled the bill through the legislature. And in the final denouement worthy of a Shakespearean drama, the Speaker of the House Richard Kawakami (Kaua'i), a longtime supporter of the Code, passed away while in office, thus guaranteeing the votes to honor his legacy.

    It would take one more twist of fate to secure the bill's passage. The conference committee report was short one signature just before midnight when the bill had to be decked in the clerk's basement office. The messenger raced down the steps with the final signature in hand as the bell struck midnight. After ten years of work, would the bill fail because the messenger could not get down one more flight of steps in time? Fate intervened. The legislature had not decked the state budget by midnight either. In time-honored fashion, the clerk unplugged and thus stopped the clock long enough for the elected officials to vote themselves a forty-eight-hour extension. Thus saved *from* the bell (and extinction), the Water Code passed.
39. See n. 38.
40. Haw. Rev. Stat., sec. 174C-31. See Commission on Water Resource Management, "Hawaii Water Plan," https://dlnr.hawaii.gov/cwrm/planning/hiwaterplan/.

41. See, e.g., "Hawaii County Water Use and Development Plan Update," August 2010, http://files.hawaii.gov/dlnr/cwrm/planning/wudpha2012.pdf.
42. Susan Miller and Carol Wilcox were pioneers in protecting Hawai'i's streams and instrumental in seeing protective measures included in the Water Code.
43. Haw. Rev. Stat., sec. 174C-101.
44. See Sproat, "From Wai to Kanawai," n. 4.
45. Marjorie Ziegler, Charles Reppun, Paul Achitoff, and Denise Antolini played critical roles in the *Waiāhole* proceedings.
46. Denise Antolini, "Water Rights and Responsibilities in the Twenty-First Century: A Forward to the Proceedings of the 2001 Symposium on Managing Hawai'i's Public Trust Doctrine," *University of Hawai'i Law Review* 24, no. 1 (2001): 1–39.
47. *In Re Waiāhole Contested Case Hearing,* 94 Haw. 97, 9 P.3d 409 (2000) ("*Waiāhole I*"). For a more detailed overview, see Sproat, "From Wai to Kanawai," 553–560. Justice Paula Nakayama wrote the decision for the Court. She addressed the historical, philosophical, legal, and practical questions directly and with rare understanding and nuance. She reminds us that Hawai'i's experience is rich and deep and that old wisdom is new again.
48. *Waiāhole I,* 94 Haw. at 154.
49. Water rights proceedings for 'Īao Valley, the Wailuku aquifer, and the East Maui watersheds are as long, as complicated, and as consequential as Waiāhole, but beyond the scope of this chapter.
50. Haw. Rev. Stat., sec. 174C-31 (1987).
51. See Commission on Water Resource Management, "Hawaii Water Plan."
52. These rights were reaffirmed in the Hawai'i Water Code, Haw. Rev. Stat., sec. 174C-31(n), -49(e), -101, and in Hawai'i Supreme Court decisions. *In Re Wai Ola O Molokai,* 279 P.3d 69 (2012); *In Re Molokai Ranch,* 83 P.3d 664 (2004).
53. Haw. Rev. Stat., sec. 174C-71 (1987).
54. A small membrane bio-reactor (MBR) unit could treat wastewater using photovoltaic (PV) systems to cut energy costs.
55. See Aqua Engineers, http://www.aquaengineers.com/.

## Bibliography

Antolini, Denise. "Water Rights and Responsibilities in the Twenty-First Century: A Forward to the Proceedings of the 2001 Symposium on Managing Hawai'i's Public Trust Doctrine." *University of Hawai'i Law Review* 24. no. 1 (2001): 1–39.

Beamer, Kamanamaikalani. *No Mākou ka Mana: Liberating the Nation*. Honolulu: Kamehameha Publishing, 2014.

Bracton, Henry de. *On the Laws and Customs of England*. Translated by S. Thorne. Cambridge, MA: Harvard University Press, 1968.

Coffman, Tom. *Nation Within: The Story of America's Annexation of the Nation of Hawaii*. Durham, NC: Duke University Press, 2003.

Fisher, Elizabeth, and Sidney Shapiro. *Administrative Competence: Reimagining Administrative Law*. Cambridge: Cambridge University Press, 2020.

Hawaiian Homes Commission Act of 1920, 42 Stat. 108.

Lau, L. Stephen, and John F. Mink. *Hydrology of the Hawaiian Islands*. Honolulu: University of Hawai'i Press, 2006.

Preza, Donovan C. "The Empirical Writes Back: Re-examining Hawaiian Dispossession Resulting from the Māhele of 1848." Master's thesis, University of Hawai'i at Mānoa, 2010.

Principles adopted by the Board of Commissioners to Quiet Land Titles in Their Adjudication of Claims Presented to Them (August 20, 1846). Adopted by the Hawai'i Legislature on October 26, 1846. L.1847.

Sax, Joseph. "Property Rights and the Economy of Nature: Understanding Lucas v. South Carolina Coastal Council." *Stanford Law Review* 45 (1993): 1433–1455.

Sproat, D. Kapua'ala. "From Wai to Kanawai: Water Law in Hawai'i." In *Native Hawaiian Law: A Treatise*, edited by Melody K. MacKenzie, with Susan K. Serrano and D. Kapua'ala Sproat, 522–610. Honolulu: Kamehameha Publishing, 2015.

Stevens, Jan. "The Public Trust: A Sovereign's Ancient Prerogative Becomes the People's Environmental Right." *U.C. Davis Law Review* 14 (1984): 195–232.

# PART II

# THREATS TO HAWAI‘I’S WATER FUTURES

Building upon the foundational relationships, belief systems, and governance of wai examined in part 1, part 2 of *Waiwai: Water and the Future of Hawai'i* focuses on critical threats to Hawai'i's water future. We have gathered thoughtful contributions from some of Hawai'i's foremost experts and leaders in confronting the issues that continue to affect our wai. We dive deeper into imagining futures where wai is not only protected but also cared for in ways that align with an 'Ōiwi worldview. Whether it be shifting community design and development plans to adapt to the various impacts of permanent sea level rise, or better understanding the changing of rainfall patterns on Hawai'i's water supply, it is now more necessary than ever that we begin to ask ourselves, what can we do? How do we prepare ourselves to solve these issues and to ensure that the future of water in Hawai'i is one that is balanced and just?

CHAPTER 5

# Climate Change and Water in Hawai'i

*Thomas W. Giambelluca*

Fresh water, found in our streams, lakes, groundwater aquifers, soil, and plants, is perhaps our most precious natural resource. In Hawai'i, isolated from all but local water sources and with climate changes that threaten water supply already occurring, we cannot take abundant fresh water availability for granted. Increasing temperatures will reduce water supplies while simultaneously increasing water needs. Rainfall, in decline for decades, will continue to change, leaving areas where current water use is high lacking unless we plan now to adapt to changing conditions.

Fresh water is not static—it moves constantly through a sequence of water flows called the hydrologic cycle. At large scales, this cycle is driven by the sun's energy, which evaporates water and moves it upward and horizontally in the air in the form of water vapor. Whenever air moves up, it cools. Air rises when winds are forced up windward mountain slopes or lifted up along the edges of an air mass (along cold fronts), when winds near the ground converge, or when a parcel of air is heated more than the air around it. In all these cases, air is caused to rise and cool. When air is cooled sufficiently, the water vapor in the air changes to liquid water or ice, forming clouds. Clouds produce precipitation, which brings the water back to the surface. On land, water flows through our environment are driven by precipitation, modified (reduced) by evaporation and water use by plants (transpiration), and influenced by storage and transmission of water by vegetation and soil, and routing on the surface and underground by soil and rock.

Precipitation is the single most important process controlling water availability, water shortages (droughts), and floods. In Hawai'i, precipitation is mainly in the form of rainfall, but direct interception of fog by vegetation can add substantial amounts of water to ecosystems and watersheds in the mountains. Snow is a very minor contributor, limited to the summits of the highest mountains, Mauna Kea and Mauna Loa. Average rainfall (figure 5.1) is spectacularly variable from place to place in Hawai'i, ranging from about eight inches (similar to large parts of the dry continental western U.S.) to around four hundred inches per year (the highest in the U.S.).[1]

Fog interception[2] in Hawai'i can add as much as forty inches per year to the water input in some mountain areas (figure 5.2).[3] This process is influenced by the amount of fog in the air, the wind speed, and the type and size of vegetation. Without vegetation, almost all the fog will simply move laterally and will not add any water to the land. Water reaching the land as rainfall or fog interception continues to move through various pathways back to the air by evaporating, into the soil, where some

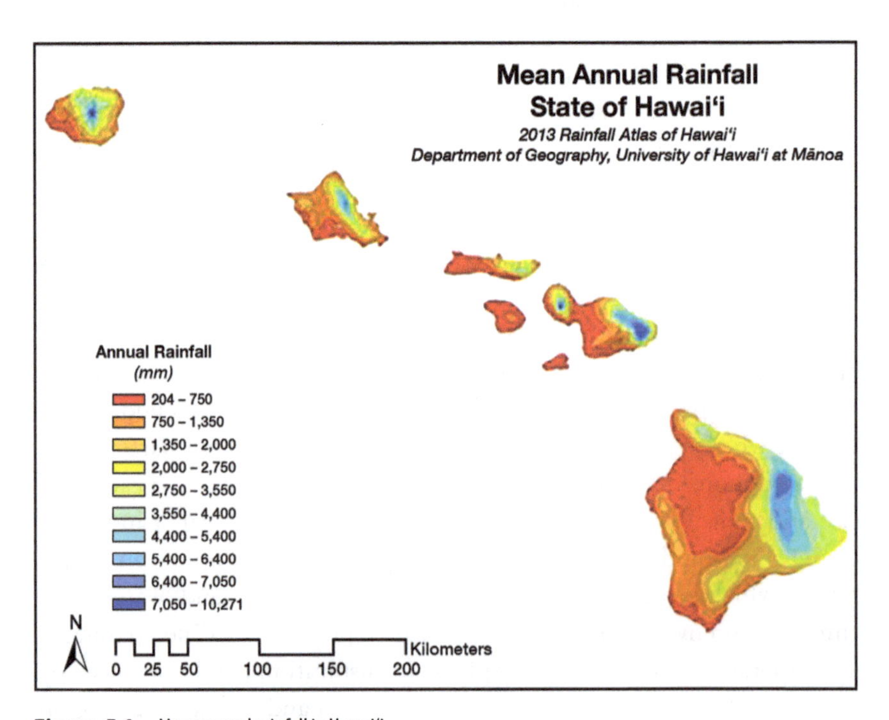


**Figure 5.1** Mean annual rainfall in Hawai'i.
*Source:* Giambelluca et al., "Online Rainfall Atlas of Hawai'i."

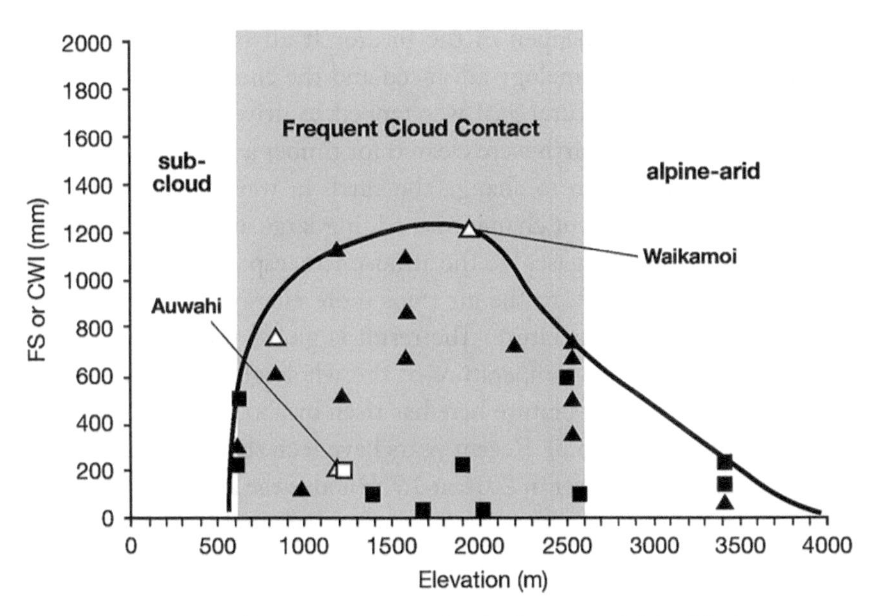


**Figure 5.2** Estimates of fog interception in Hawai'i based on fog screen catch (FS) and cloud water interception (CWI) from canopy water balance measurements.

*Source:* DeLay and Giambelluca, "History of Cloud Water Interception Research in Hawai'i."

can be taken up by plant roots and returned to the air through plant leaves, or over the land surface toward a nearby stream.

Some of the water entering the soil can move downward past the deepest plant roots, eventually adding itself to (recharging) the groundwater below or emerging in a stream channel. Groundwater and streams are especially important parts of the hydrological cycle because they are the sources of water for household use, agriculture, and, in the case of streams, support of freshwater plants and animals. In Hawai'i, windward mountain slopes with nearly constant rainfall and fog interception are the most important source areas for water, recharging groundwater aquifers and generating stream flow. Water received in these remote, extremely wet areas provides most of the domestic, agricultural, and commercial water supply.

## Climate Change and Water

Weather and climate exert dominant influences on the hydrological cycle, and climatic changes can alter Hawai'i's water flows. What changes are already occurring that might affect our water, and what changes are likely to happen in the future? It all starts with changes in temperature. As technology advanced and the energy stored in fossil fuels (oil, coal, and natural gas) was tapped to drive industrialization, and large areas of the earth were cleared for timber and to open land for farming, humans began to change the earth in ways never previously seen. The most important change was adding large amounts of carbon-containing greenhouse gases[4] to the atmosphere, especially carbon dioxide ($CO_2$). Adding $CO_2$ to the air traps more energy within the earth system, raising its temperature. The result is global warming, a steady increase in the average temperature of the whole globe. Hawai'i is no exception, and the temperature here has risen by about 0.9°F in the past hundred years[5] (figure 5.3). Recent years have seen sharp increases, with a new statewide record set in 2016 at 1.7°F above the average of the past hundred years.[6]

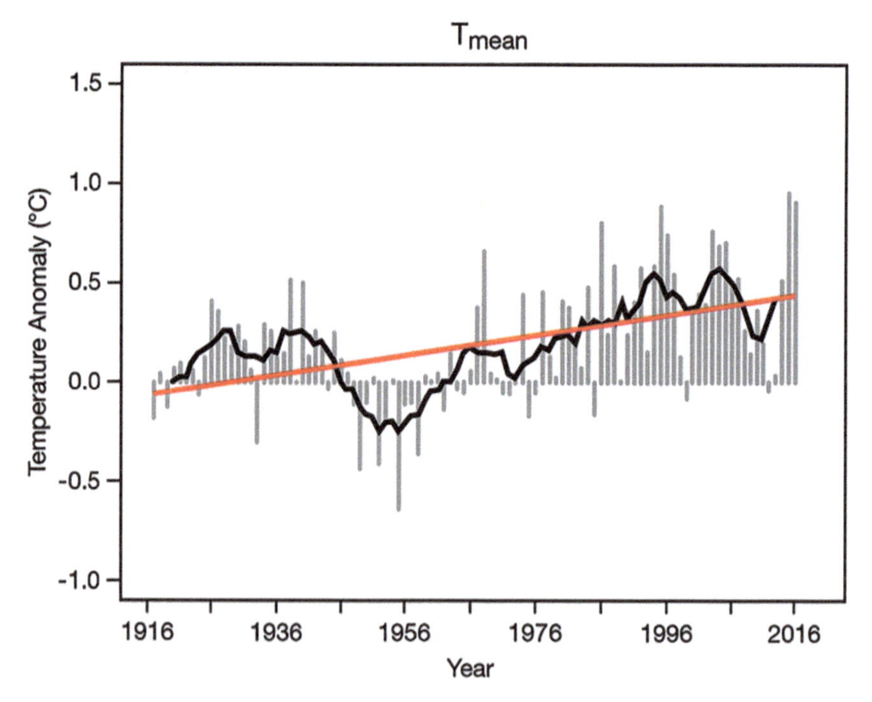


**Figure 5.3** The Hawai'i temperature index (HTI) showing the average temperature anomaly (departure from the 1944–1980 average) for selected long-term measuring stations in Hawai'i for the 100-year period 1916–2016. The red line shows the average temperature trend over that time, an increase of 0.05°C (0.09°F) per decade or 0.5°C (0.9°F) over 100 years.

*Source:* McKenzie, Giambelluca, and Diaz, "Regional Temperature Trends in Hawai'i: A Century of Change, 1917–2016."

Increasing temperature directly affects our local water cycle by increasing the rate of evaporation. In a hotter environment, more water will be needed to meet the needs of natural ecosystems, crops, and ornamental plants. In Hawai'i as elsewhere, the changing temperature is having enormous effects on other aspects of the climate, including patterns of cloud formation and precipitation, and could affect fog interception as well. Water on the islands could be affected by changes in cloud cover, not only because it relates to possible precipitation changes, but because clouds affect the amount of the sun's energy reaching the land, and that affects evaporation and transpiration. During the past twenty years or so, for example, cloud cover declined in high mountain areas of Hawai'i, allowing solar radiation to increase.[7] The drying effect of the increasingly sunny weather contributed to the decline of one of Hawai'i's most iconic native plants, the Haleakalā Silversword.[8,9]

One of the biggest questions raised by those concerned about climate change impacts in Hawai'i is how will our rainfall be affected. We have seen big ups and downs in rainfall in the past hundred years or more. Were they caused by global warming? What will the future bring—more floods such as we saw in 2018 on Kaua'i, or will we experience more extreme drought as in 2010? How will changing rainfall patterns affect groundwater recharge and stream flows? These questions have spurred researchers to study past changes and projections of future changes in Hawai'i's rainfall.

Using evidence from tree rings in the western U.S., a team of researchers developed a reconstruction of wintertime rainfall in Hawai'i stretching back five hundred years[10] (figure 5.4). This time series shows that our winter rainfall has gone through large variations in the past, even before the advent of the current human-driven changes in global climate. These natural variations undoubtedly had effects on the lives of the native Hawaiians before and after contact with Europeans. Notable, for example, is the prolonged drought that ended just prior to Cook's arrival in the islands (1750s–1770s; figure 5.5). It is also clear from this study that winter rainfall in Hawai'i has been in a long slow decline since the middle of the nineteenth century.

With rain gauges becoming more widely deployed in the islands by the early twentieth century, we can get a more accurate and detailed picture of how rainfall has changed over approximately the past ninety

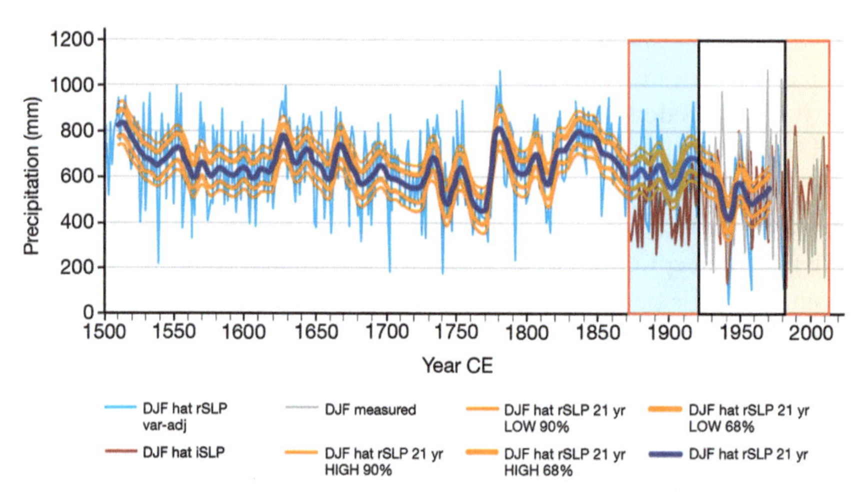


**Figure 5.4** A reconstruction of Hawai'i winter rainfall from the year 1500 to near present, based on remote proxy data from North America. Note the apparent long-term decrease beginning around 1840.
*Source:* Diaz et al., "A Five-Century Reconstruction of Hawaiian Islands Rainfall."

years. Abby Frazier analyzed data since 1920 to see how rainfall has changed in different areas of the islands.[11] She showed that it has generally gotten drier over time (figure 5.5), but not in a steady decline. Instead, most areas experienced alternating wet and dry periods lasting decades. One area, the western part of Hawai'i Island, has seen the biggest decrease in rainfall since 1920. In recent decades, most areas have been drier than normal. Is this an effect of climate change or simply part of the natural variability? According to a recent study, it's too soon to tell whether or not global climate change has already affected Hawai'i rainfall.[12]

To understand how temperature and rainfall might change in the future, we turn to climate models. Dozens of global climate models have been developed and are improved and refined by climate research groups around the world. These models are the main tools for looking at future changes in climate. Using scenarios of future emissions of greenhouse gases, based on different possible choices human society might make, global climate models provide projections of how climate might change. Making use of these global models for Hawai'i requires another analytical step called downscaling. The global model results are too coarse spatially to properly simulate the effects on the islands of climate processes or to represent the spatial details of our climate.

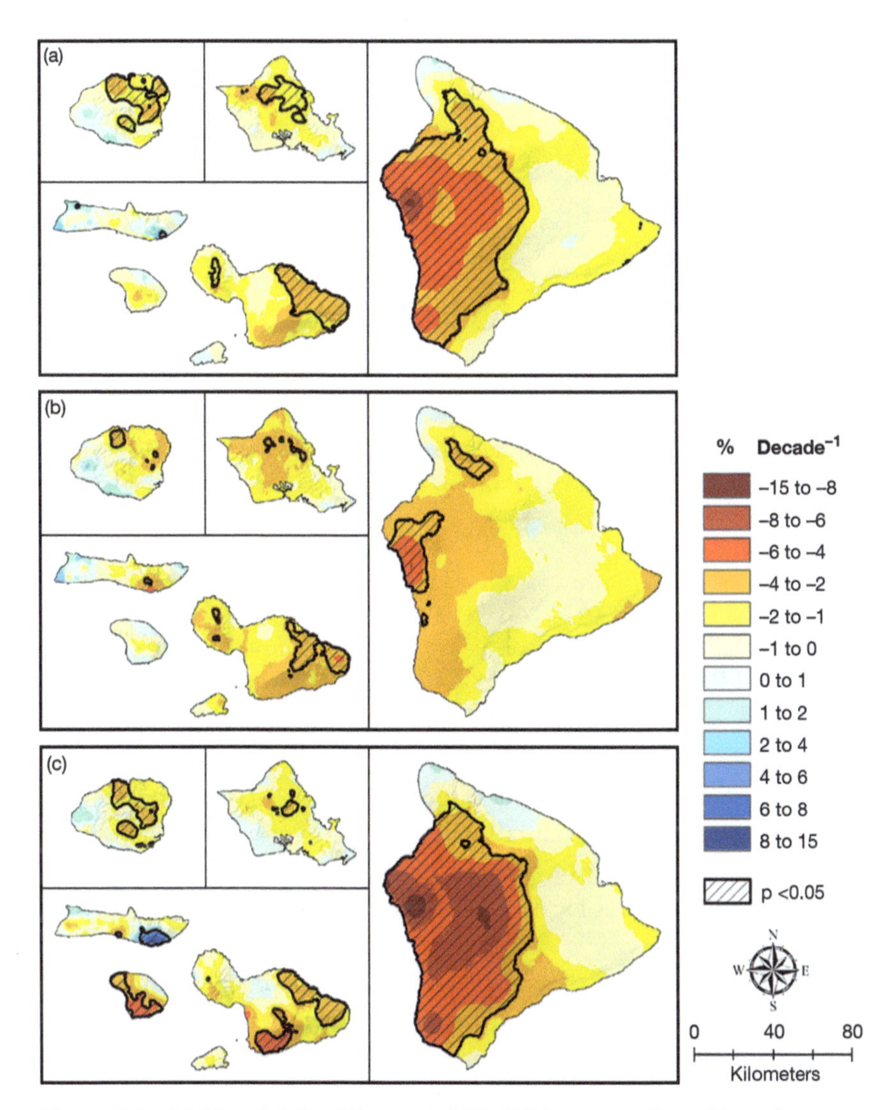


**Figure 5.5** Rainfall trends in Hawai'i for the period 1920–2012 in percent per decade: (a) annual; (b) November–April; (c) May–October.

*Source:* Frazier and Giambelluca, "Spatial Trend Analysis of Hawaiian Rainfall from 1920 to 2012."

Downscaling is an attempt to use the results of global model projections while bringing in finer-scale information about the islands' topography and land cover. Downscaled temperature projections indicate continued warming throughout the islands, with faster warming at higher elevations[13] (figure 5.6).

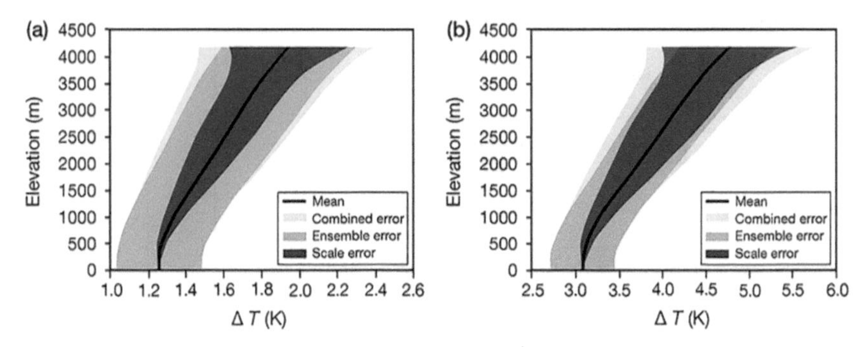


**Figure 5.6** Downscaled projected temperature change for Hawai'i as a function of elevation for (a) a moderate emissions scenario for the midcentury period (2040–2069); and (b) a business-as-usual emissions scenario (RCP 8.5) for the end-of-century period (2070–2099).

*Source:* Elison Timm, "Future Warming Rates over the Hawaiian Islands Based on Elevation-Dependent Scaling Factors."

So far, two groups of scientists have made downscaled projections of future rainfall for Hawai'i, each using a different downscaling method. The results differ significantly but generally point to small-to-moderate increases in rainfall in windward areas and small-to-severe decreases in leeward areas[14,15] (figure 5.7). This suggests that while rainfall in the most important groundwater recharge areas will remain high and perhaps increase somewhat, rainfall will decrease in the areas where most people live, where much of our agriculture is done, and where most of the tourism industry is located. This points to a big increase in water demand in the future.

Changing patterns of average rainfall will have significant effects on water supply and water use. But, other changes in rainfall characteristics might be even more important. For example, here and in other parts of the world, future rainfall is likely to come in the form of fewer, but more intense, storms. Even with little or no change in average rainfall, longer dry spells and more extreme heavy rainfall could have major impacts. If longer, more severe droughts are more common in the future, wildfires will become a bigger problem, natural ecosystems will be impacted—perhaps leading to more native plants and animals being lost, stream ecosystems being disrupted, and crops being damaged. With more intense rainfall during storms, floods will reach extreme levels not seen before.

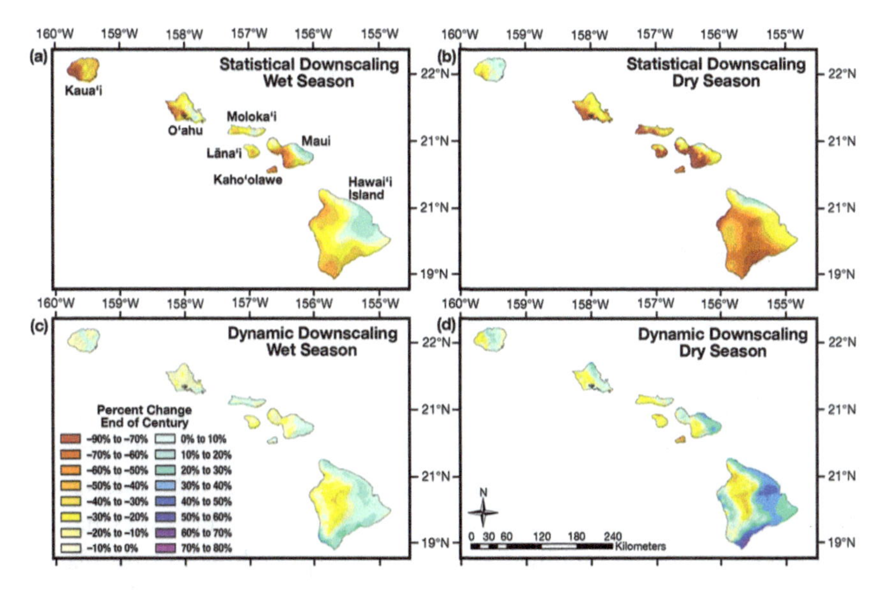


**Figure 5.7** Projected changes in rainfall in Hawai'i by the end of the century: (a) wet-season and (b) dry-season projected change based on statistical downscaling for 2017–2100 relative to 1978–2007; (c) wet-season and (d) dry-season projected change based on dynamical downscaling for 2080–2099 relative to 1990–2009. *Source:* Frazier et al., "The Influence of ENSO, PDO and PNA on Secular Rainfall Variations in Hawai'i."

While the details of future changes in Hawai'i's climate remain uncertain, it is clear that our climate will be different from what we have experienced in the past. Until recently, we could use the historical record of climate to infer the likelihood of damaging or dangerous weather events in the future. For example, structures such as bridges, culverts, and drainage ways have been designed to withstand the effects of floods, based on the probability of extreme high rainfall derived from historical observations. This approach was based on the assumption that future climate will lie within the same envelope of variability as seen in the past—the so-called stationarity assumption. With a changing climate, however, we can no longer make assumptions about future weather extremes based solely on what happened in the past. In 2018 alone, two record-setting rain events caused flooding on Kaua'i and O'ahu (April 13–15, associated with an upper-level low-pressure system) and on Hawai'i Island (August 22–25, associated with Hurricane Lane) that massively overwhelmed water control and conveyance structures designed to withstand extreme rain events. As stated in a landmark scientific paper,[16] because of climate change, "stationarity is dead." As we plan for the future, prudence demands that we account for the effects of climate changes that might occur, even if the details of those future changes are uncertain.

## Changes in Vegetation

Changes in temperature, cloudiness (sunniness), rainfall, and fog interception will have, and might already be having, effects on our water supply and water needs. Other changes in our islands are also suspected of having important effects on water flows. In particular, throughout the islands, we have seen dramatic shifts in plant species as the result of human activities (sandalwood harvesting, ranching, urban development, etc.), and the introduction of nonnative plants. Hawaiʻi's native ecosystems are being rapidly disrupted and displaced by invasive plants introduced intentionally, in some cases, and inadvertently, in others. The competitive advantages that make some plants invasive in our islands often include traits that allow them to grow faster than the native plants they are displacing. Fast growth rates tend to be accompanied by fast water use (transpiration) rates. Does invasion of our forests and other ecosystems by fast-growing nonnative plants cause water loss to increase and, therefore, groundwater recharge and stream flow to decrease? This remains an unanswered question in Hawaiʻi, but some evidence suggests that plant invasion is increasing plant water use and reducing our available water resources.[17] Some invasive plants might be less efficient at capturing fog, thus reducing water input.[18] Numerous government and nonprofit organizations are hard at work with goals of protecting native ecosystems and restoring areas degraded by direct human disturbance and invasive plants and animals. The extent to which these important conservation efforts improve water resources is not known, but several research efforts are underway to address this issue.[19]

Nonnative plants, especially albizia (*Falcataria moluccana*), a fast-growing tree now widespread in many areas of Hawaiʻi, can have significant negative effects on water flows in the watersheds where they proliferate. Albizia grows very quickly into very large trees. Its wood is fragile, causing trees to uproot, break, or drop large branches during storms. These resulting albizia logs create dams in streams during floods, increasing the inundated area. And when these dams can fail, waves of debris-laden water are triggered, with catastrophic effects downstream.[20]

## Planning to Adapt to Changes

### *Freshwater Council*

Given the challenges of climate change and species invasion, and recognizing the water needs of a growing population, led the Hawai'i Community Foundation to initiate a process to identify actions that we can take now to ensure future availability of clean fresh water. The Freshwater Council, launched in 2013, produced a "Blueprint for Action,"[21] with recommendations for concrete actions to ensure Hawai'i's water security out to the year 2030. Council members concluded that the combined effects of increased drought incidence, decreasing rainfall, higher evapotranspiration, and changes in land use translated to an uncertainty in water supply of 21 million gallons per day (mgd) of fresh water by the year 2030. In addition, they estimated that population growth would increase demand by 79 mgd by 2030, creating a potential shortfall of 100 mgd. The blueprint is designed to provide an additional 100 mgd of clean fresh water by the year 2030, by improving water conservation (+40 mgd), increasing groundwater recharge (+30 mgd), and reusing wastewater (+30 mgd). Recommended conservation actions include providing incentives for residential and commercial property owners to install rooftop catchment systems for landscape watering, invoking new requirements for leak detection in potable water systems, and mandating improvements in agricultural water use efficiency. To enhance recharge, the blueprint calls for watershed protection to reduce the negative impacts of nonnative species, and changes to urban stormwater control to increase opportunities for storage and infiltration, rather than drainage. To meet the goal of increasing wastewater reuse, the council recommends revising Department of Health water reuse guidelines, changing plumbing codes, and requiring large landscaped areas such as parks, schools, and golf courses to use treated wastewater for landscape irrigation.

### *Monitoring*

As the effects of climate change, land cover change, population increase, and other changes unfold with effects on water supply and demand, it is important that we carefully keep track of changes as they occur and improve predictions of what lies ahead. Maintaining and improving monitoring of climate, stream flow, groundwater, and land cover is a critically important element of any adaptation plan. Currently, we have fewer than half the number of rain gauge stations in Hawai'i we had fifty years ago. Many long-term rainfall and stream-flow gauges have been discontinued due to the demise of plantation agriculture and changing priorities for government spending. Measurements of other important aspects of climate, such as wind, humidity, and solar radiation are very sparse. Without an adequate network to measure these variables, it will be difficult to take timely action to ensure secure water supply as the climate changes.

In response to the critical need for better climate monitoring in Hawai'i, the Hawai'i Mesonet[22] was launched with funding from the National Science Foundation, the Honolulu Board of Water Supply, the Hawai'i Commission on Water Resource Management, and the Water Resources Research Center at the University of Hawai'i at Mānoa. When complete, the Hawai'i Mesonet will consist of approximately one hundred weather and climate monitoring stations, each measuring a full suite of weather elements. Station data are transmitted every fifteen minutes and ingested by the Hawai'i Climate Data Portal,[23] a data management, dissemination, and visualization tool, and made available to the National Oceanic and Atmospheric Administration data system, MADIS.[24]

Hawai'i faces daunting challenges to maintain access to adequate freshwater supplies and prepare for more extreme storms. These challenges call for bold action to expand climate and water monitoring, restore and protect watersheds, strengthen infrastructure, improve conservation and water use efficiency, and expand wastewater reuse. Meeting these challenges will require the efforts of all of us to build on current initiatives, such as those of the Freshwater Council. While these adaptation measures are urgently needed to reduce the impacts of climate change on water supply and flood hazards, it is essential that we also mitigate the causes of global climate change by pressing for government action to bring about rapid reductions in greenhouse gas emissions. This can be achieved by taxing carbon emissions, by incentivizing the development of alternative energy sources, and through individual efforts, such as recycling and the adoption of a plant-based diet, for example. We have an opportunity to avert the worst effects of climate change and prepare to deal with inevitable changes. But the time to act is running out.

## Notes

1. T. W. Giambelluca et al., "Online Rainfall Atlas of Hawai'i," *Bulletin of the American Meteorological Society* 94, no. 3 (2013): 313–316.
2. Fog or cloud water interception is a form of precipitation that does not show up in rain gauges. When wind blows fog through vegetation, tiny water droplets impact against the leaves and stems, and some of this water drips down to the ground, adding to the amount of water received from rain.
3. John K. DeLay and T. W. Giambelluca, "History of Cloud Water Interception Research in Hawai'i," in *Tropical Montane Cloud Forests: Science for Conservation and Management,* edited by L. A. Bruijnzeel, F. N. Scatena, and L. S. Hamilton, 332–341 (Cambridge: Cambridge University Press, 2010).
4. The term "greenhouse gases" refers to gases in the atmosphere that absorb long-wave radiation from the earth. These gases include naturally occurring compounds such as water vapor, carbon dioxide, and methane. Unlike the much more abundant gases, nitrogen and oxygen, these trace gases are powerful absorbers of radiation. In that way, they act like the glass windows in a greenhouse, allowing sunlight in, but absorbing and trapping outgoing long-wave radiation. Human activities have greatly increased the concentrations of some naturally occurring greenhouse gases and added others that do not occur naturally. These chemical changes in the atmosphere are the primary cause of contemporary global warming.
5. M. M. McKenzie et al., "Regional Temperature Trends in Hawai'i: A Century of Change, 1917–2016," *International Journal of Climatology* 39, no. 10 (2019): 3987–4001, 10.1002/joc.6053.
6. Ibid.
7. R. J. Longman et al., "Temporal Solar Radiation Change at High Elevations in Hawai'i," *Journal of Geophysical Research—Atmospheres* 119, no. 10 (May 2014): 6022–6033.
8. P. D. Krushelnycky, L. L. Loope, et al., "Climate-Associated Population Declines Threaten Future of an Iconic Plant," *Global Change Biology* 19, no. 3 (2013): 911–922.

9. P. D. Krushelnycky, F. Starr, et al., "Change in Trade Wind Inversion Frequency Implicated in the Decline of an Alpine Plant," *Climate Change Responses* 3, no. 1 (2016).
10. H. F. Diaz et al., "A Five-Century Reconstruction of Hawaiian Islands Rainfall," *Journal of Climate* 29, no. 15 (2016): 5661–5674.
11. A. G. Frazier and T. W. Giambelluca, "Spatial Trend Analysis of Hawaiian Rainfall from 1920 to 2012," *International Journal of Climatology* 37, no. 5 (April 2017): 2522–2531.
12. A. G. Frazier et al., "The Influence of ENSO, PDO and PNA on Secular Rainfall Variations in Hawai'i," *Climate Dynamics* 34, no. 12 (2017): 1928.
13. O. Elison Timm, "Future Warming Rates over the Hawaiian Islands Based on Elevation-Dependent Scaling Factors," *International Journal of Climatology* 37, no. S1 (August 2017): 1093–1104.
14. O. Elison Timm, T. W. Giambelluca, and H. F. Diaz, "Statistical Downscaling of Tainfall Changes in Hawai'i Based on the CMIP5 Global Model Projections," *Journal of Geophysical Research-Atmospheres* 120, no. 1 (January 2015): 92–112.
15. C. Zhang, Y. Wang, K. Hamilton, and A. Lauer. "Dynamical Downscaling of the Climate for the Hawaiian Islands. Part II: Projection for the Late Twenty-First Century," *Journal of Climate* 29, no. 23 (December 2016): 8333–8354.
16. P. C. D. Milly et al., "Stationarity Is Dead: Whither Water Management?" *Science* 319, no. 5863 (February 2008): 573–574.
17. A. Strauch et al., "Modeled Effects of Climate Change and Plant Invasion on Watershed Function Across a Steep Tropical Rainfall Gradient," *Ecosystems* 20 (2017): 583–600.
18. M. Takahashi et al., "Rainfall Partitioning and Cloud Water Interception in Native Forest and Invaded Forest in Hawai'i Volcanoes National Park," *Hydrological Processes* 25, no. 3 (January 2011): 448–464.
19. Several field studies by University of Hawai'i at Mānoa and Pacific Islands Water Science Center (USGS) researchers, either currently underway or in the planning stages, will address questions about the possible effects of replacement of native vegetation with nonnative plants on water loss by transpiration and the rate of water infiltration into the soil.
20. J. C. Watson, "Strategic Plan for the Control and Management of Albizia in Hawai'i," Hawai'i Invasive Species Council, 2018, https://dlnr.hawaii.gov/hisc/files/2018/01/Strategic-Plan-for-the-Control-and-Management-of-Albizia-In-Hawaii-1.pdf.

21. Hawai'i Fresh Water Initiative, "A Blueprint for Action: Water Security for an Uncertain Future," https://www.hawaiicommunityfoundation.org/file/cat/Fresh_Water_Blueprint_FINAL_062215_small.pdf (accessed June 1, 2024).
22. T. W. Giambelluca et al., "The Hawai'i Mesonet: Monitoring Weather and Climate across Steep Gradients in Mountainous Terrain" (conference paper, AGU Fall Meeting, Chicago, December 2022).
23. J. McLean et al., "Building a Portal for Climate Data—Mapping Automation, Visualization, and Dissemination," *Concurrency and Computation: Practice and Experience* 35, no. 18 (2021): e6727.
24. Meteorological Assimilation Data Ingest System, https://madis.ncep.noaa.gov/.

## Bibliography


DeLay, John K., and Thomas W. Giambelluca. "History of Cloud Water Interception Research in Hawai'i." In *Tropical Montane Cloud Forests: Science for Conservation and Management,* edited by L. A. Bruijnzeel, F. N. Scatena, and L. S. Hamilton, 332–341. Cambridge: Cambridge University Press, 2010.

Diaz, Henry F., Eugene R. Wahl, Eduardo Zorita, Thomas W. Giambelluca, and Jon K. Eischeid. "A Five-Century Reconstruction of Hawaiian Islands Rainfall." *Journal of Climate* 29, no. 15 (2016): 5661–5674. https://doi.org/10.1175/JCLI-D-15–0815.1.

Elison Timm, Oliver. "Future Warming Rates over the Hawaiian Islands Based on Elevation-Dependent Scaling Factors." *International Journal of Climatology* 37, no. S1 (August 2017): 1093–1104.

Elison Timm, Oliver, Thomas Giambelluca, and Henry Diaz. "Statistical Downscaling of Rainfall Changes in Hawai'i Based on the CMIP5 Global Model Projections." *Journal of Geophysical Research-Atmospheres* 120, no. 1 (January 2015): 92–112. https://doi.org/10.1002/2014JD022059.

Frazier, Abby G., and Thomas Giambelluca. "Spatial Trend Analysis of Hawaiian Rainfall from 1920 to 2012." *International Journal of Climatology* 37, no. 5 (April 2017): 2522–2531. https://doi.org/10.1002/joc.4862.

Frazier, Abby G., O. Elison Timm, T. W. Giambelluca, and H. F. Diaz. "The Influence of ENSO, PDO and PNA on Secular Rainfall Variations in Hawai'i." *Climate Dynamics* 34, no. 12 (2017): 1928. https://doi.org/10.1007/s00382-017-4003-4.

Giambelluca, Thomas W., Qi Chen, Abby G. Frazier, Jonathan P. Price, Yi-Leng Chen, Pao-Shin Chu, Jon K. Eischeid, and Donna M. Delparte. "Online Rainfall Atlas of Hawai'i." *Bulletin of the American Meteorological Society* 94, no. 3 (2013): 313–316.

Giambelluca, T. W., A. D. Nugent, Y. Tsang, A. G. Frazier, D. W., Beilman, H. Tseng, C. K. Shuler, C. Yap, and D. Giardina. "The Hawai'i Mesonet: Monitoring Weather and Climate across Steep Gradients in Mountainous Terrain." Conference paper, AGU Fall Meeting, Chicago, December 2022.

Hawai'i Fresh Water Initiative. "A Blueprint for Action: Water Security for an Uncertain Future." https://www.hawaiicommunityfoundation.org/file/cat/Fresh_Water_Blueprint_FINAL_062215_small.pdf. Accessed June 1, 2024.

Krushelnycky, P. D., L. L. Loope, T. W. Giambelluca, F. Starr, K. Starr, D. Drake, A. Taylor, and R. H. Robichaux. "Climate-Associated Population Declines Threaten Future of an Iconic Plant." *Global Change Biology* 19, no. 3 (2013): 911–922. https://doi.org/10.1111/gcb.12111.

Krushelnycky, P. D., F. Starr, K. Starr, R. J. Longman, A. G. Frazier, L. L. Loope, and T. W. Giambelluca. "Change in Trade Wind Inversion Frequency Implicated in the Decline of an Alpine Plant." *Climate Change Responses* 3, no. 1 (2016). https://doi.org/10.1186/s40665-016-0015-2.

Longman, R. J., T. W. Giambelluca, R. J. Alliss, and M. Barnes. "Temporal Solar Radiation Change at High Elevations in Hawai'i." *Journal of Geophysical Research-Atmospheres* 119, no. 10 (May 2014): 6022–6033. https://doi.org/10.1002/2013JD021322.

McKenzie, M. M., T. W. Giambelluca, and H. F. Diaz. "Regional Temperature Trends in Hawai'i: A Century of Change, 1917–2016." *International Journal of Climatology* 39, no. 10 (2019): 3987–4001. https://doi.org/10.1002/joc.6053.

McLean, J., S. B. Cleveland, M. Dodge, M. P. Lucas, R. J. Longman, T. W. Giambelluca, and G. A. Jacobs. "Building a Portal for Climate Data—Mapping Automation, Visualization, and Dissemination." *Concurrency and Computation: Practice and Experience* 35, no. 18 (2021): e6727. https://doi.org/10.1002/cpe.6727.

Milly, P. C. D., J. Betancourt, M. Falkenmark, R. M. Hirsch, Z. W. Kundzewicz, D. P. Lettenmeier, and R. J. Stouffer. "Stationarity Is Dead: Whither Water Management?" *Science* 319, no. 5863 (February 2008): 573–574. https://doi.org/10.1126/science.1151915.

Strauch, A., C. Giardina, R. Mackenzie, C. Heider, T. Giambelluca, E. Salminen, and G. Bruland. "Modeled Effects of Climate Change and Plant Invasion on Watershed Function across a Steep Tropical Rainfall

Gradient." *Ecosystems* 20 (2017): 583–600. https://doi.org/10.1007/s10021-016-0038-3.

Takahashi, M., T. W. Giambelluca, R. G. Mudd, J. K. DeLay, M. A. Nullet, and G. P. Asner. "Rainfall Partitioning and Cloud Water Interception in Native Forest and Invaded Forest in Hawai'i Volcanoes National Park." *Hydrological Processes* 25, no. 3 (January 2011): 448–464. https://doi.org/10.1002/hyp.7797.

Watson, John-Carl. "Strategic Plan for the Control and Management of Albizia in Hawai'i." Hawai'i Invasive Species Council, 2018. https://dlnr.hawaii.gov/hisc/files/2018/01/Strategic-Plan-for-the-Control-and-Management-of-Albizia-In-Hawaii-1.pdf.

Zhang, C., Y. Wang, K. Hamilton, and A. Lauer. "Dynamical Downscaling of the Climate for the Hawaiian Islands. Part II: Projection for the Late Twenty-First Century." *Journal of Climate* 29, no. 23 (December 2016): 8333–8354.

CHAPTER 6

# Sea Level Rise in Hawai'i

*Charles H. Fletcher III*

This chapter is dedicated to discussion of sea level rise (SLR) in Hawai'i, a particularly challenging phenomenon as the majority of our communities are ocean oriented, and located on the low-lying, relatively flat coastal plain near the shoreline. Their low elevation and proximity to the ocean make these locations highly vulnerable to the flooding and land loss associated with SLR, along with risk of damage related to severe weather coming from the ocean, land, and sky.

In this chapter I will address questions related to the magnitude, timing, risks, and potential societal responses associated with the phenomenon of SLR.

## Sea Level Rise Is a Permanent, Accelerating Feature of Our Physical Environment

The Intergovernmental Panel on Climate Change (IPCC) publishes reports every few years that provide a review of climate change science, impacts, and mitigation and adaptation options. The most recent report, Assessment Report 6 (AR6) published in 2021,[1] is the consensus of hundreds of scientists as well as the governments of every nation on Earth. They make the following important statement about the phenomenon of SLR and they evaluate their confidence in the truth of this statement as "high": "Sea level is committed to rise for centuries to millennia due to continuing deep-ocean warming and ice-sheet melt and will remain elevated for thousands of years (high confidence)."[2] In other words, glaciers will continue to melt and oceans will continue to thermally expand for hundreds to thousands of years just because of the global warming (1.5°C; 2.7°F)[3] we have already caused by burning fossil fuels and cutting down forests.

Applying this statement locally means that SLR is a permanent condition on the Hawaiian shoreline that will get worse with every year. SLR is not going away—the impacts are cumulative and accelerating.

Not only do we need to adapt to this reality with new engineering, architectural, and community design plans, we need to evaluate and potentially revise all of the laws and policies that guide how we live near the ocean. If you live on O'ahu, that means everything makai of the H1 between Diamond Head and Pearl Harbor will need an SLR adaptation plan. This includes Hickam Air Force Base, Honolulu International Airport, Honolulu Harbor, Iwilei, Kaka'ako, Waikīkī, Mō'il'ili, the entire Primary Urban Center of Honolulu, and elsewhere on similarly sited lands around every island.

In short, SLR is a physical hazard that poses the greatest economic, social, and environmental challenge faced in the history of the state. Figure 6.1, showing the flooding that would be caused by 6 ft of SLR (possible this century), illustrates what I mean.[4]

**Figure 6.1** Flooding of the Honolulu Primary Urban Center under 6 ft of sea level rise, considered likely with 1.5°C of global warming. Blue = hydrologically connected to the ocean; darker blue = greater depth; green = hydrologically "unconnected" areas that may also flood.

*Source:* NOAA, https://coast.noaa.gov/slr/.

### Redesigning Communities to Manage SLR Is a Rare, Generational Opportunity

A key aspect of managing this problem is to recognize that SLR adaptation opens opportunities for other types of activities that make communities more resilient. This is a rare moment that may come along only every few generations.

For instance, global warming increases the probability that Hawaiʻi will experience more intense tropical cyclones and rain events, as well as growing drought and summer heat. Certain types of building and energy design can help mitigate the impact of these challenges as well as reduce vulnerability to SLR. These include using enhanced design flood levels and freeboard in building architecture, rain catchment systems and pumped drainage, constructing shade and water features into open-area design on raised land or large pavilions, adding resilience centers in disadvantaged neighborhoods, and others. In short, adapting to SLR is an opportunity to improve our communities in many desirable ways.

With SLR comes growing vulnerability to flooding associated with tropical cyclones and tsunamis. It's critical to prepare for this with pre-disaster planning, updated building codes, resilience centers, hardened energy sources, and other safety steps. These actions offer opportunities to not only increase resiliency and adaptation, but also to reduce and sequester atmospheric carbon dioxide and methane, the cause of global warming. For instance, with soil amendments, carbon-negative cement, and building materials that use recycled plastics.

Holistically redesigning our communities with these challenges in mind also presents opportunities to mitigate other harmful impacts. Restoring coastal ecology and clean nearshore waters begins with reducing polluted runoff, and expanding enhanced protections. Neighborhood resilience centers, and nature-based solutions offer opportunities to rejuvenate cultural practices and address social justice concerns by providing affordable housing and resources for those who do not want housing.

Sea level rise, when viewed in this light, brings with it a number of opportunities that we should be vigilant to identify, and commit to act on.

### How High Will the Ocean Rise and When?

Sea level rise is one of the main consequences of global warming.[5] Knowing how fast SLR can develop, given a scenario of future greenhouse gas emissions, is crucial to inform both mitigation and adaptation choices.

Global warming has now reached 1.5°C (2.7°F) above the natural background. Annual greenhouse gas emissions continue to set new records, and the level of atmospheric carbon dioxide climbs higher every year.[6] Warming is accelerating,[7] and is on track to reach 3°C (5.4°F) before the end of the century.[8,9,10]

How much will sea level rise as a result of today's level of global warming? As a reference for you, the last time it was this warm was during the Eemian Interglacial (ca. 125,000 years ago),[11] when global mean sea level (GMSL) was more than 6 m (20 ft) higher than present (figure 6.2).[12] Research indicates that on multiple occasions over the past three million years, when global temperatures increased 1 to 2°C (1.8 to 3.6°F), melting polar ice sheets and warming oceans caused global sea levels to rise at least 6 m (20 ft) above present levels.[13] However, keep in mind that these levels of flooding took centuries to reach; 6 m of SLR is not considered a likely prospect for the twenty-first century.

The global rate of SLR is accelerating already.[14] According to IPCC AR6,[15] GMSL rose faster in the twentieth century than in any prior century over the last three millennia, with a 0.15 to 0.25 m (0.5 to 0.8 ft)

**Figure 6.2** Fossil coral head, Ka'ena Point, O'ahu. Approximately 125,000 years ago, global sea level was over 6 m (20 ft) higher than today.

*Source:* Charles H. Fletcher III.

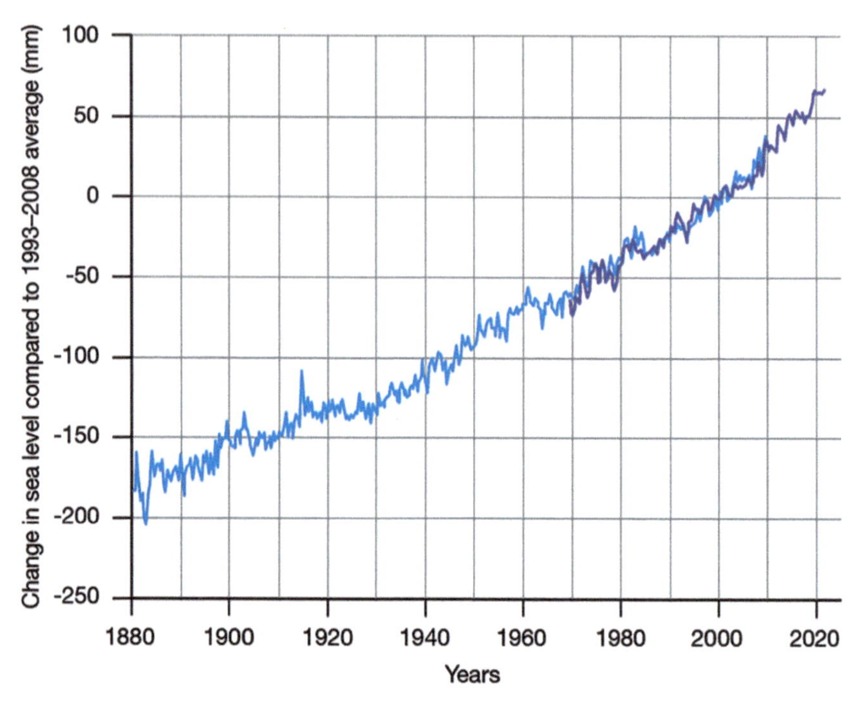


**Figure 6.3** Seasonal (3-month) global mean sea level estimates (light blue line) and University of Hawai'i Fast Delivery sea level data (dark blue). The values are shown as change in GMSL in millimeters compared to the 1993–2008 average.

*Source:* NOAA Climate.gov image based on analysis and data from Philip Thompson, University of Hawai'i Sea Level Center, https://www.climate.gov/news-features/understanding-climate/climate-change-global-sea-level.

rise over the period 1901–2018. The rate of SLR (figure 6.3) has accelerated since the late 1960s, with an average rate of 1.6 to 3.1 mm (0.06 to 0.12 in) per year over the period 1971–2018, increasing to 3.2 to 4.2 mm (0.13 to 0.17 in) per year over the period 2006–2018.

What is causing SLR? Ocean thermal expansion (38%) and mass loss from glaciers (41%) dominate the total sea level change from 1901 to 2018. The contribution of Greenland and Antarctica to GMSL rise was four times larger during 2010–2019 than during 1992–1999. Because of the increased ice-sheet mass loss, the total loss of land ice was the largest contributor to GMSL rise over the period 2006–2018.[16]

By 2050, relative to the period 1995–2014, GMSL is projected to rise between 0.15 to 0.29 m (0.5 to 1.0 ft), and by 2100 between 0.28 to 1.01 m (1.0 to 3.3 ft).[17] This SLR is primarily caused by thermal expansion and mass loss from glaciers and ice sheets. These projections do not include ice-sheet-related processes that are characterized by deep uncertainty; higher amounts of rise could be caused by disintegration of marine ice shelves around Antarctica, and faster-than-projected ice discharge from Greenland.

The IPCC has introduced a new high-end risk scenario, stating that a global rise "approaching 2 m by 2100 and 5 m by 2150" cannot be ruled out due to deep uncertainty in ice sheet processes.[18] Other studies suggest that melting of the Antarctic ice sheet, which has tripled in the past five years,[19] could raise global sea level by up to 3 m (10 ft) by the year 2300 and continue for thousands of years thereafter.[20]

Disparate estimates of future SLR are the result of unsettled science. Every new research paper pushes forward our understanding of how the world's ice and oceans will respond to future warming. But development decisions in the coastal zone are being made every day. Planners in expanding communities cannot wait for the science to clear up.

With this in mind, researchers engaged a set of experts in a poll.[21] Questions were designed to trigger assessments of how high GMSL might rise by the end of this century. In the scenario where greenhouse gas emissions continue unchecked, the study concluded there is a 10% chance of sea level exceeding 2 m (6.6 ft) by 2100.

Upon release of the study, one of the authors commented: "If you knew there was a 10% chance a plane would crash, you wouldn't get on it. It's the same with sea level rise." This was an allusion to building in the coastal zone and how it would be wise to design long-lived projects such that they avoided flooding associated with 2 m (6.6 ft) of future SLR.

There is additional guidance for community planners discussed later in this chapter. But first, let's understand the different types of impacts that are caused by SLR.

### Sea Level Rise Impacts

Of the threats posed by SLR, there are two fundamental categories of negative impacts:

1. Low-frequency, relatively rare, potentially catastrophic events that will become more damaging because of SLR and which may be life-threatening and may constitute disasters (tsunamis and hurricane storm surge).

2. High-frequency, relatively common events that typically are not life-threatening and do not constitute disasters. These include the following: high tide flooding consisting of groundwater inundation, hydrostatic flooding, storm drain backflow, seasonal high wave flooding, compound flooding, and coastal erosion.

It is important to plan for both categories. These impacts, all of which have already been observed on Hawai'i's shoreline, will affect coastal communities and ecosystems. Careful planning based on the best available science that identifies the location, magnitude, and timing of these impacts will aid effective adaptation, and help avoid hasty, expensive, and environmentally damaging emergency interventions.

Modeling that provides spatially accurate visualizations of impacts is especially important for developing implementable policy approaches and for initiating long-term planning. Given the uncertainty that still exists around the timing and severity of impacts, scenario-based modeling can be a useful tool to allow planners and policy makers to evaluate a range of risk scenarios and select from them as appropriate to the nature and scale of the project at hand.

For instance, it will be important for major infrastructure projects with long life spans to plan for both low- and high-frequency impacts of SLR. Depending on the specifics of the project, it might be acceptable for short-term, minor development projects to only plan for high-frequency impacts.

## Low-Frequency Events

### *Tsunami*

Residents of Hawai'i are no strangers to tsunamis. In fact, the island archipelago lies like a bull's-eye in the center of the "Ring of Fire," a geologic region that embodies the earthquake-, volcanic-, landslide-, and tsunami-prone area of the Pacific Ocean. All sides of the Pacific are lined with coasts where Earth's lithospheric plates meet to form violent faults that are capable of generating tsunamis that travel to our shores with the potential to cause devastation.

The most destructive tsunami to strike Hawai'i occurred on April 1, 1946, after an earthquake measuring 7.4 on the Richter scale struck the ocean floor off the Aleutian Islands of Alaska.[22] Waves traveled at 500 mph. In Hilo, the death toll was high: 173 were killed, 163 injured,

488 buildings were demolished, and 936 more were damaged. Damage at the time was estimated to be $25 million ($375 million today). Witnesses told of waves inundating streets, homes, and storefronts. Many victims were swept out to sea by receding water.

Tsunamis are caused by sudden movement of the seafloor that generates a wave, or actually a series of waves that travel across the ocean until they reach a coast. Seafloor movements may include shaking by an earthquake, faulting, land sliding, or submarine volcanic eruptions. Tsunamis reaching a coastline generate a rapidly rising water level that floods the coast like a river flowing over the shoreline. Breaking waves may ride atop the water, generating additional damage to the high-velocity flow of the inundation.

The geography of the shoreline plays an important role in the form of the tsunami. Tsunami waves may be very large in embayments, actually experiencing amplification in long funnel-shaped bays. Fringing and barrier reefs appear to have a mitigating influence on tsunamis by dispersing the wave energy. Within Kāne'ohe Bay, O'ahu, protected by a barrier reef, the 1946 tsunami reached only 60 cm (2 ft) in height, while at neighboring Mōkapu Head the wave crest exceeded 6 m (20 ft). Despite complex differences in the geography and orientation of Hawai'i's many coastlines, several locations have historically been subject to severe tsunami impacts, including Hilo Bay (Hawai'i Island), Kahului Bay (Maui Island), and Kaiaka Bay (O'ahu Island).

At sea level on the coast, there is no safe place during a tsunami. On low-lying shorelines and coastal plains that characterize so much of Hawai'i, a tsunami may occur as a rapidly growing high tide that rises over several minutes and inundates low coastal lands. Water flows rapidly landward, carrying anything that is lightweight, and in severe cases destroying buildings and sweeping up people as it progresses. The return of these flood waters to the sea also causes much damage.

As sea level rises over the course of the twenty-first century, tsunamis that previously might have been considered minor will increase their depth and the distance they penetrate inland and thus the damage they may potentially cause.

### *Storm Surge*

When a hurricane strikes, storm surge poses an enormous threat to life and property. For instance, Hurricane 'Iniki (1992) caused 6 fatalities and $3.1 billion in damage. Much of this damage occurred directly, or indirectly, as a result of storm surge.

Storm surge is a rapid rise of water that is pushed onto the coast by a hurricane. Four mechanisms contribute to storm surge:

1. Wind blows onto the coast and piles up water (potentially more than 85% of the surge);
2. Waves push water inland faster than it can drain off (called wave setup, potentially 5%–10% of the surge);
3. Low atmospheric pressure of a hurricane raises the ocean level near the eye (potentially 5%–10% of the surge);
4. The stage of tide at the time the surge hits plays an important role because at high tide the water depth can increase an additional 0.3 to 1.0 m (1 to 3.3 ft); this can spell the difference between moderate damage and catastrophic damage.

The level of flooding associated with storm surge depends on several factors: the size and intensity of a hurricane; the angle that it approaches the coast; the slope of the seabed at the coastline; the slope of the land at the coast; and how fast the hurricane is moving.

Winds in a hurricane near Hawai'i rotate counterclockwise. Hence, the forward right quarter of a hurricane approaching a coastline from the sea is the most dangerous area. The wind is blowing onshore, the speed of the storm adds to the wind speed, and the storm surge is greatest.

Storm surge moves with the forward speed of the hurricane—typically 16 to 24 kph (10 to 15 mph). This wind-driven water has tremendous power because 0.76 $m^3$ (1 $yd^3$) of sea water weighs 784 kilograms (1,728 pounds)—almost a ton. Hence, a 30 cm (1 ft) deep storm surge can sweep a car off the road.

Compounding the destructive power of the rushing water is the large amount of floating debris that typically accompanies the surge. Trees, pieces of buildings, and other debris float on top of the storm surge and act as battering rams that can batter any buildings unfortunate enough to stand in the way.

Since the nineteenth century, global sea level has risen an average of 0.20 m (8 in). Sea level rise is expected to accelerate through this century as the climate continues to warm. With rising seas, storm surge is amplified. That means a storm today could create more extensive flooding than an identical storm a century ago, with sometimes catastrophic damage to our homes and critical infrastructure. Even if there were no immediate increase in the strength of hurricanes from climate change, continued SLR means storm surge will be more destructive.

How are hurricanes near Hawai'i changing? This question has not been definitively answered, but we do know that around the world the zone where hurricanes form is shifting poleward—away from the equator—which means storms that used to pass south of the Big Island are now more likely to hit our shores.[23] Major hurricanes have become 15% more likely over the past forty years,[24] and climate models project an increase in hurricanes near Hawai'i[25] (figure 6.4).[26]

We know that as sea level rises, it will cause higher coastal inundation levels for hurricane storm surge. The proportion of hurricanes reaching Category 4 and 5 levels will likely increase. Hurricanes have already become bigger and more destructive in the U.S.[27] Overall, we can expect hurricanes to be bigger, rainier, and intensifying more rapidly, to have higher wind speeds, and for storm surge to cause more flooding because of SLR.

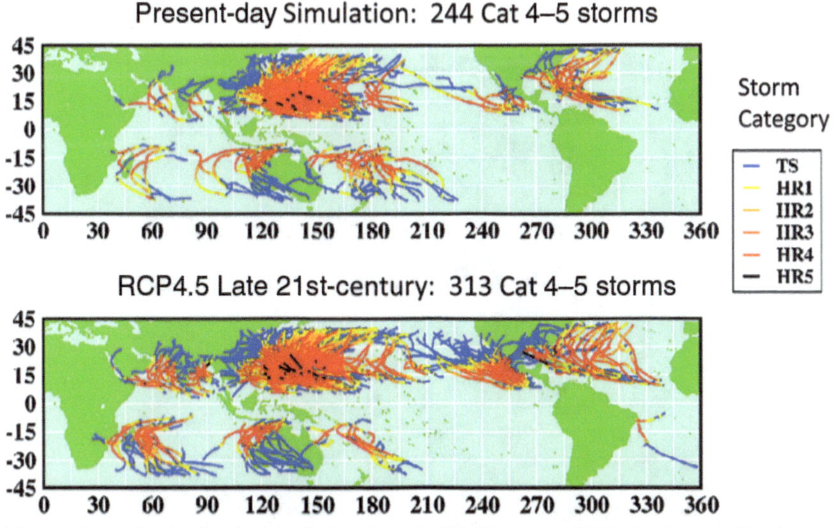


**Figure 6.4** Simulated present-day (top) and projected late 21st-century (bottom) Category 4–5 tropical cyclone tracks. Colors indicate storm category (legend). This model result clearly indicates an increase in Category 2, 3, and 4 hurricanes in Hawaiian waters.

*Source:* NOAA, Geophysical Fluid Dynamics Laboratory, "Global Warming and Hurricanes," https://www.gfdl.noaa.gov/global-warming-and-hurricanes/.

## High-Frequency Events: High Tide Flooding

The first evidence that a coastal community is experiencing the impacts of SLR is high tide flooding, also known as a *king tide*. The scientific term for a king tide is a perigean spring tide. King tides in the Hawaiian Islands tend to occur during the summer (July and August) and winter months (December and January) in conjunction with new moons and full moons. We see king tides when the moon is at its closest point to the Earth in its monthly orbit, so the gravitational pull is stronger, and, the sun, the moon, and Earth are in alignment. Which means that the sun's and moon's individual gravitational pulls work together, producing the highest high tides of the year, or the king tides.

During king tides, waves run up the shoreline much further than usual, wetlands are flooded by standing water, storm drains fill with seawater, and buildings that have basements often show groundwater flooding (figure 6.5). The first time this is noticed, it is usually ignored. But take note—this is the "tip of the spear" of SLR.

Sea level rise scenarios show most of the action taking place in the second half of this century because they consist of exponentially rising curves that gain steam and really start accelerating after 2050. These

**Figure 6.5** Testing the salinity of groundwater inundation in a Waikīkī building. The water was brackish.

*Source:* Habel et al., "Development of a Model to Simulate Groundwater Inundation."

curves represent permanent flood scenarios. However, arriving decades ahead of this permanent inundation is the phenomenon of high tide flooding. High tide flooding takes several simultaneous forms: storm drain backflow, groundwater inundation, hydrostatic flooding,[28] and high wave run-up. Compound flooding of this type, if it occurs on roads, sidewalks, major intersections, and especially if it is raining, can interrupt urban transportation in serious and costly ways (figure 6.6).

One of the impacts of sea level rise that is not widely appreciated is that flooding will not only cross the shoreline from the sea, it will come out of the ground.[29] This phenomenon is known as *groundwater inundation*.[30] In the Waikīkī area, and probably other coastal settings in Hawai'i, the water table lies at an elevation somewhat above high tide. The water table rises and falls with the tides and with weather patterns; thus, it is intimately and immediately connected to the surface of the sea.

**Figure 6.6** The El Niño of 2015 produced high sea surface temperatures around Hawai'i which raised sea level due to thermal expansion. During the perigean spring tide (July and August), Honolulu experienced storm drain backflow and groundwater inundation. The warm ocean also fueled intense rains, which led to flooding if it rained during high tide. Rapidly developing urban flooding brought traffic to a standstill in the late afternoon, at the height of commuting hour. Photo credit: Dennis Oda.

*Source: Honolulu Star Advertiser*, October 8, 2015, https://www.staradvertiser.com/2015/10/08/hawaii-news/kokua-line/record-rainfall-responsible-for-hawaiian-telcom-outages-3/.

This connectivity decreases with distance from the shoreline, but within several blocks of the ocean, the position of the water table can be strongly correlated to marine processes (i.e., tides, waves, etc.). Thus, a rise in sea level related to global warming means that the coastal water table will likely rise to the point of breaking through the ground surface and creating new wetlands—not a desirable feature in an urban setting at the foundation of a building or road, or in an ecosystem that serves as a refuge for endangered species that are unaccustomed to saturated soil and free-standing water bodies (figure 6.7).[31]

Analysis[32] of the impacts of groundwater inundation in urban Honolulu found that 1 m (3.3 ft) of SLR generates groundwater inundation across 23% of the region, threatening $5 billion of taxable real estate and 48 km (30 mi) of roadway. Analysis of current conditions reveals that 86% of 259 active cesspool sites in the study area are likely inundated suggesting that cesspool effluent is currently entering coastal waters and will flood the ground surface as sea level rises.

Researchers found that groundwater flooding will occur regardless of seawall construction, impacting pedestrians, commercial and recreation activities, tourism, transportation and infrastructure. It will require innovative planning and intensive engineering efforts to accommodate standing water in the streets.

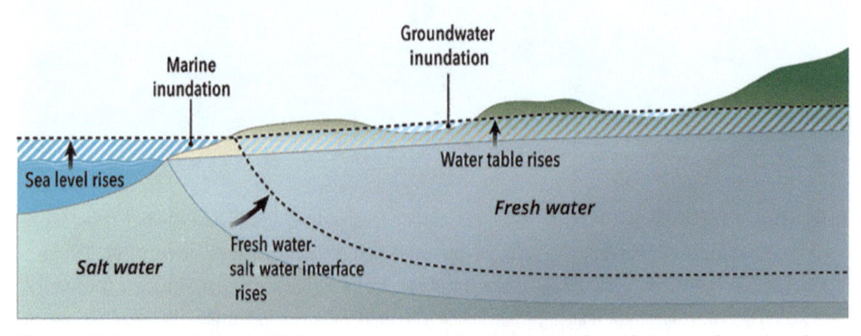


**Figure 6.7** Groundwater inundation. Low-lying coastal areas may be vulnerable to groundwater inundation, which is localized coastal-plain flooding due to a rise of the groundwater table caused by sea level rise.
*Source:* Charles H. Fletcher III.

Modeling of this problem shows that most of the area in the Honolulu urban region that is flooded by SLR is not connected to the ocean. Inundation is a result of water flowing out of storm drainage pipes along roadsides, the water table breaking through the land surface to create wetlands, and the collection of runoff into pools because it cannot drain into the ground or the ocean.

Unfortunately, at least some of the groundwater in the Honolulu urban corridor is polluted (figure 6.8). Poor land use and waste management practices including contaminant spills, leaking underground storage tanks, and cesspool use in areas lacking adequate unsaturated space, have contributed to a high level of contamination in groundwater and nearshore coastal waters. Groundwater inundation brings

**Figure 6.8** A film of oil floats on the water table under Nimitz Highway in Honolulu.
*Source:* Honolulu Board of Water Supply.

these contaminants to the surface. Focused efforts will be required to limit public contact with, and/or heavily remediate contaminated groundwater.

In summary, we have identified multiple forms of high-frequency SLR flooding, and each will have its own adaptation response. These flood types are:

1. *Hydrostatic Flooding,* also called direct marine inundation and "bathtub flooding."
    a. Adaptation includes seawalls, and flood proofing, raising the land surface with fill, or moving developed assets to higher ground.
2. *Storm Drain Backflow,* where the urban drainage system is flooded by seawater that discharges directly onto streets and sidewalks.
    a. Adaptation includes converting gravity-fed drainage to pumped drainage, installing backflow preventers, raising street elevation, and directing runoff to retention areas.
3. *Groundwater Inundation,* where the water table rises with SLR and breaks the ground surface creating a wetland, or expanding existing wetlands.
    a. Adaptation includes localized pumping of the water table (requires specific hydrologic conditions), raising the ground surface and raising developed assets, moving assets to higher ground.
4. *Compound Flooding,* which occurs when it rains and the normal means of drainage (into the ground and into storm drains) are blocked by high tide.
    a. Adaptation includes pumped drainage, raising the ground surface and raising developed assets, moving assets to higher ground, directing flow to retention features using landscaped elevation.

## Seasonal High Wave Flooding

A major impact related to SLR includes flooding by predictable high waves that enter Hawaiian waters every year. These seasonal high waves, or "swells," are responsible for the renowned surfing culture originated in our Indigenous past and for which Hawaiʻi is world-famous today.

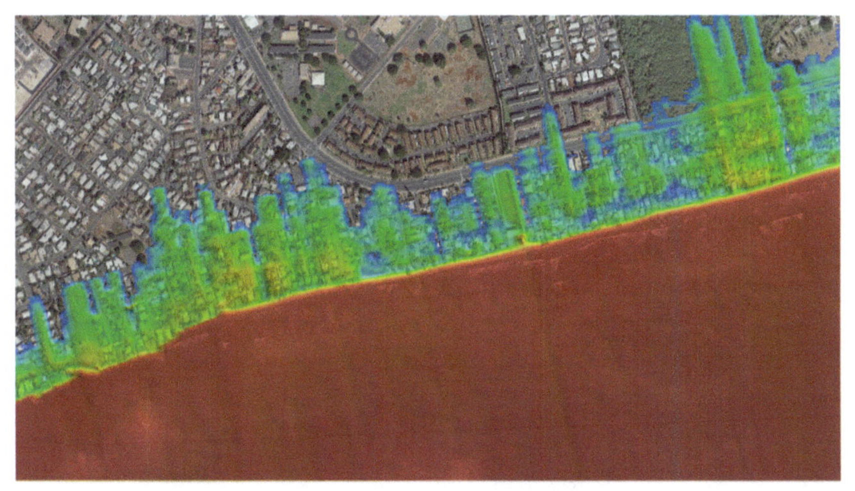

**Figure 6.9** Annual wave inundation, Ewa Beach, under 0.98 m (3.2 ft) of SLR.
*Source:* Charles H. Fletcher III.

On shores facing to the south, high waves produced by storminess in the southern hemisphere arrive every summer, and on shores facing to the north, high waves produced by storminess in the North Pacific arrive every winter. At present, these seasonal high waves are responsible for occasional flooding of the first row of beachfront homes and certain sections of coastal roads.

However, SLR is changing this. Modeling the characteristics of these annual waves under higher sea levels reveals that between 91 to 122 cm (3 to 4 ft) of SLR, wave flooding moves from being damaging to being catastrophic (figure 6.9). This threshold has been called a critical point.[33] It is the sea level height beyond which flooding rapidly accelerates and threatens an entire beachfront community.[34] A critical point can be thought of as a "tipping point," a limit at which some aspect of the climate system irretrievably shifts to a new state.

## Beaches—A Special Management Consideration

Communities located in most coastal settings engage in some form of tourism economy, and from a tourist point of view, beaches are the most desirable natural asset these communities have. From a resident's point of view, beaches are endemic to the lifestyle in Hawai'i, a place of cultural practice, food harvesting, recreation, family bonding and memory making, and a basic necessity for access to the ocean and its resources.

However, as sea level rises, coastal erosion increasingly threatens beachfront land.[35] During SLR, beaches, which are by definition environments located at the edge of the ocean, are forced to retreat landward. In doing so, beachfront land is eroded—this is called *coastal erosion.* If this land has not been altered by landscaping and other forms of development, and if it is sand rich (such as with a natural dune field), erosion of this land will supply the migrating beach with its lifeblood, sand. Any attempt to stop coastal erosion, such as with a wall, will damage, narrow, and eventually destroy the beach (figure 6.10).[36]

A widely held view is that erosion can be managed. This is true, but only if you are willing to damage the beach. Any attempt to slow erosion, divert it, or stop it altogether is simply an effort to stop beach migration, condemning the coastal zone to wave scour and drowning. If beach preservation, public access, and open space are the goals of any management program, there is only one way to manage coastal erosion—get out of the way and let nature take its unimpeded course.

If left alone to migrate landward with the rising ocean levels, beaches will survive SLR just fine, as they have for 20,000 years since global sea level was 130 m (400 ft) lower at the culmination of the last

**Figure 6.10** Rocky Point, Sunset Beach, O'ahu. Efforts to protect homes with shoreline hardening has caused beach narrowing and loss.
*Source:* Habel et al., "Development of a Model to Simulate Groundwater Inundation."

ice age. As coastal managers, we have to figure out how to get out of the way. However, seawalls are often used to protect homes and roads from the effects of erosion. This causes beaches to disappear.[37] This practice is known as armoring or hardening and while it provides a temporary reprieve from coastal erosion, armoring on retreating beaches eventually leads to beach narrowing and beach loss.[38]

Beach loss represents the destruction of a critical natural ecosystem (think monk seals and sea turtles), the loss of a valuable tourism asset, and a blow against the lifestyle that Hawai'i residents seek in their lives. But most importantly, beach loss is theft from Hawai'i's children.

Hawai'i beaches belong to our children. Every weekend, families from every ahupua'a descend to the nearest beach. Here is where Hawai'i's children receive their first lessons in observing winds, rain, clouds, and the sun. It is on beaches that children learn how to be safe in the ocean, how to swim, and how to read the waves. Beaches belong to children. With SLR, continuation of Hawai'i's shameful history of seawall construction is nothing less than brazen theft of a critical resource from a future generation. A generation that is fully justified in viewing our management of coastal resources as the act of criminals.

Because coastal erosion can lead to negative outcomes, it is useful to model future erosion to identify where it is likely to strike, which beaches are most threatened, and to provide data for strategic planning.[39] Simply responding to coastal erosion by building seawalls is reactive and does not constitute good planning. For example, sections of the world-famous Sunset Beach on O'ahu are characterized by chronic erosion.[40] In 2015, there were eighteen beachfront homes (10% of the beachfront community) directly threatened by erosion. Using a model of erosion associated with SLR, it is possible to estimate the developing erosion threat to beachfront homes at Sunset.[41]

With 30 cm (1 ft) of sea level rise, the number of buildings threatened by erosion at Sunset Beach swells to 109, comprising 60% of the beachfront community. If even a portion of these buildings were to build seawalls, it would spell the swift end of Sunset Beach, an iconic stretch of shore that is world-famous and a critical part of the tourism allure of the Hawaiian Islands. Sunset Beach is also home to endangered monk seals that haul onto the sand to rest and to give birth to their pups. Green sea turtles, another endangered species, also use the beach for resting.

## Sunset Beach Is a Cassandra Metaphor

How we manage the erosion that threatens over a dozen homes today is likely to foretell the future of all beaches in Hawai'i over the next decade or so. Do we value our surfing culture, customary fishing and gathering practices, cultural protocols and wa'a traditions? Are we willing to put the beach, and public trust, first?

Projections of future erosion at Sunset Beach send a clear message that unless authorities prepare soon with comprehensive policies designed to protect the beach, near-term SLR is going to trigger a cascade of seawall construction. This will cause serious environmental and economic problems. It is incumbent upon authorities that are responsible for protecting natural coastal environments to consider this future and develop planning strategies to avoid the worst effects.

Doesn't Hawai'i have laws that protect beaches? Yes, at multiple levels. The state Coastal Zone Management (CZM) Program in the Office of Planning is periodically recertified, and annually receives millions of dollars from the National Oceanic and Atmospheric Administration (NOAA) National Coastal Zone Management Program.[42] The goals of the Hawai'i CZM program mirror those of the national office from which they receive funding. Hawai'i CZM objectives and policies are embodied in a law that was passed by the Hawai'i legislature in 1977 called "Chapter 205A, Hawai'i Revised Statutes."[43] Multiple objectives are listed: provide recreation, preserve open space, protect coastal ecosystems, reduce exposure to coastal hazards, and, notably—*protect beaches.*

Despite the intent of this law, research[44] reveals that shoreline hardening has caused over 21.5 km (13 mi) of beach loss statewide, posing threats to the economy, coastal ecology, federally protected species (monk seals and green sea turtles), local culture and lifestyle, and the general ocean safety training of Hawai'i's children.

Coastal erosion is a major problem in Hawai'i. Data[45] reveal that 70% of beaches on O'ahu, Maui, and Kaua'i experience an erosional trend. On O'ahu, shoreline hardening has caused over 17.3 ± 1.5 km (10 mi) of beach narrowing and over 10.4 ± 0.9 km (6 mi) of total beach loss.

As global SLR continues to accelerate, it is essential that we establish exit strategies for buildings, roads, and other types of coastal development. The undefined status of these assets stands in the way of fulfilling our destiny as caretakers and stewards of the treasured places that belong to our children.

## Managing SLR

Because no single physical model accurately represents all major processes contributing to SLR, five scenarios (figure 6.11) have been developed under a multiagency federal task force that provide both global mean and local relative SLR scenarios to 2150.[46] While these scenarios are not model projections, they are physical assessments based on peer-reviewed research and they do take into account new understanding of the behavior of processes that contribute to sea level rise that have not been considered in previous studies.

These scenarios have been developed for community planning to frame risk tolerance for use by decision-makers. The choice of which scenario to use for planning should be determined through an assessment of risk tolerance associated with any individual project or asset. Each of the five scenarios is defined by a target value of global mean SLR by 2100 as follows: Low (0.3 m, 1 ft), Intermediate Low (0.5 m, 1.6 ft), Intermediate (1 m, 3.3 ft), Intermediate High (1.5 m, 5 ft), and High (2 m, 6.6 ft). These are then regionalized to provide the scenarios at individual tide gauges.

Projections from the Low and Intermediate Low scenarios are already lower than the observed acceleration of global mean SLR, which

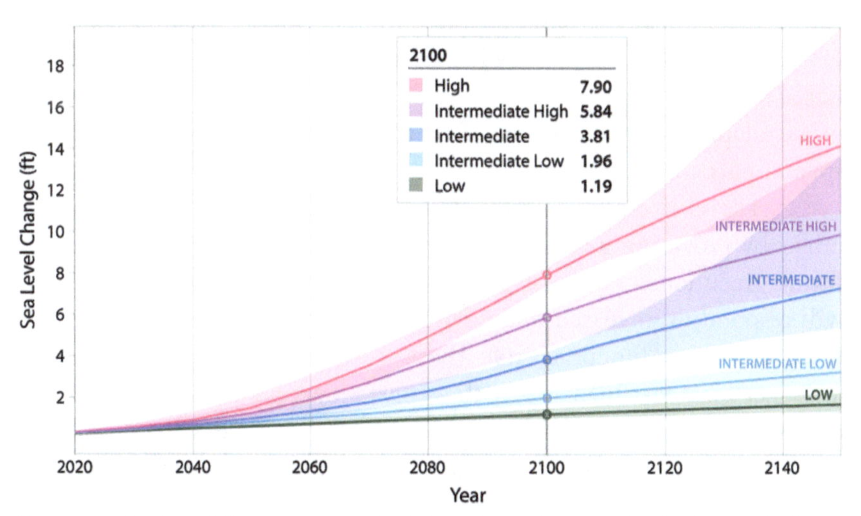


**Figure 6.11** Future SLR planning scenarios for the Honolulu Tide Gauge, developed by Sweet et al., "Global and Regional Sea Level Rise Scenarios for the United States." Median values are provided for each scenario, along with likely ranges represented by shaded regions showing the 17th–83rd percentile ranges. See https://sealevel.nasa.gov/task-force-scenario-tool for additional explanation.

**Table 6.1** Sea level rise (m, ft) scenarios 2030–2150 for Honolulu

Median sea level projection values (m, ft) for 2030–2150 for five sea level scenarios, relative to year 2000

| Decade | Low | Intermediate Low | Intermediate | Intermediate High | High |
|---|---|---|---|---|---|
| 2030 | 0.12, 0.38 | 0.13, 0.43 | 0.14, 0.47 | 0.16, 0.52 | 0.17, 0.56 |
| 2040 | 0.15, 0.50 | 0.18, 0.60 | 0.21, 0.68 | 0.24, 0.79 | 0.27, 0.90 |
| 2050 | 0.18, 0.61 | 0.24, 0.80 | 0.29, 0.97 | 0.37, 1.23 | 0.45, 1.49 |
| 2060 | 0.22, 0.72 | 0.30, 1.00 | 0.40, 1.32 | 0.57, 1.86 | 0.73, 2.39 |
| 2070 | 0.26, 0.85 | 0.37, 1.21 | 0.53, 1.75 | 0.83, 2.71 | 1.07, 3.52 |
| 2080 | 0.29, 0.96 | 0.44, 1.45 | 0.69, 2.28 | 1.13, 3.69 | 1.49, 4.90 |
| 2090 | 0.33, 1.08 | 0.52, 1.70 | 0.91, 2.97 | 1.44, 4.74 | 1.94, 6.36 |
| 2100 | 0.36, 1.19 | 0.60, 1.96 | 1.16, 3.81 | 1.78, 5.84 | 2.41, 7.90 |
| 2150 | 0.52, 1.70 | 0.99, 3.24 | 2.21, 7.26 | 3.00, 9.86 | 4.30, 14.12 |

*Source:* Sweet et al., "Global and Regional Sea Level Rise Scenarios."

is on trajectory to reach 0.7 m (2.3 ft) by the year 2100.[47] Hence, the Intermediate, Intermediate High, and High interagency scenarios represent a more realistic basis for planning and modeling impacts. Each scenario provides planners with SLR targets at decadal frequency (table 6.1). These values offer planners a time frame for assessing appropriate adaptation strategies based on the risk tolerance and planned life span of particular projects.

Sea level rise scenarios assist with planning in the face of uncertainty by providing a range of possible futures that represent potential future greenhouse gas emissions, and how Earth's biophysical processes will respond to increased temperatures (e.g., thawing permafrost, rainforest ecosystem function, etc.). These scenarios assume that the rate of ice-sheet mass loss increases with a constant acceleration; however, this might not be the case, and so it is possible, for example, to be on the Intermediate scenario early in the century but the High scenario or beyond late in the century.

Many communities across the U.S. have adopted the federal SLR scenarios for their planning strategies. Given that community planning is risk averse, and that investments in public infrastructure projects are expensive and typically designed to last a long time, planning on the basis of Intermediate or higher scenarios is appropriate. Using the Intermediate or higher SLR scenario as a benchmark, adding 0.3 m (1 ft) of freeboard to accommodate a king tide, and adding an additional 0.3 m (1 ft) of freeboard to accommodate heavy rainfall at high tide when there is no drainage capacity in the coastal zone would provide an even greater buffer against flood damage.

### Meeting the Challenge of SLR Will Take a Unified Response

After years of discussion, awareness building among government staff and the public, and developing confidence in the scientific messaging about climate change, Hawai'i is now taking vulnerability to SLR seriously. Modeling by Anderson et al.[48] was used in the "Hawai'i Sea Level Rise Vulnerability and Adaptation Report" (SLR Report)[49] to determine that developed assets and land value exposed to erosion, groundwater inundation, and/or high seasonal waves sums to over $4 billion, including 22 km (13.7 mi) of roadway. Approximately 2000 residents will be displaced. At about 1 m (3.2 ft) of SLR, a total of $19 billion of developed assets and land value are exposed, with over 19,800 residents displaced and 61 km (38 mi) of roadway at risk.

Publication of the SLR Report led Mayor Alan Arakawa of Maui County to sign a proclamation acknowledging climate change and its impacts, and calling on the planning department to propose rule changes to the Maui, Moloka'i, and Lana'i planning commissions to include sea level rise in their shoreline setback calculations. The proclamation directed "county departments to use the Report in their plans, programs and capital improvement decisions, to mitigate impacts to infrastructure and critical facilities triggered by sea level rise." Since that first step, other counties and state agencies have updated their plans to reflect the reality and urgency of SLR planning.

For each individual asset threatened by SLR, a decision-making process must be engaged to decide on the specific adaptation strategy to be employed (even if it is to abandon the asset), a budget scoped and funded, and a construction project detailed and completed. Adaptation is, in short, a massive operation. We are well along the pathway to success. But the hardest part of the journey still lies ahead.

## A Sustainable and Resilient Future

Preparing for the challenges that accompany SLR and other forms of climate change is an opportunity to design and build ideal communities that reflect our island heritage. Resilient, sustainable communities that are focused on a reciprocal relationship with nature, clean water, social equity and justice, Indigenous ways of knowing and doing, and economic affordability will provide our children and their children with a future in which they can thrive.

## Notes

1. IPCC, "Summary for Policymakers," 3-32.
2. Ibid., 21.
3. Gillett et al., "Constraining Human Contributions," 207–212.
4. IPCC, "Ocean, Cryosphere and Sea Level Change," 1211–1362.
5. Park et al., "Future Sea-Level Projections," 636.
6. Nunez, "Carbon Dioxide Levels."
7. Reports indicate that warming could reach 1.5°C by 2030, ten years earlier than projected by IPCC, and 2°C by 2045 (Xu, Ramanathan, and Victor, "Global Warming Will Happen Faster than We Think").
8. Raftery et al., "Less than 2°C Warming by 2100 Unlikely," 637–641.
9. Wang et al., "Climate Change of 4°C Warming above Pre-industrial Levels," 757–770.
10. United Nations Environment Programme, "Emissions Gap Report 2022."
11. Hoffman et al., "Regional and Global Sea Surface Temperatures," 276–279.
12. Kopp et al., "Probabilistic Assessment of Sea Level," 863–867.
13. Dutton et al., "Sea-Level Rise due to Polar Ice-Sheet Mass Loss," 6244.
14. Nerem, Frederikse, and Hamlington, "Extrapolating Empirical Models."
15. IPCC, "Ocean, Cryosphere and Sea Level Change."
16. Ibid.
17. Ibid.
18. IPCC, "Summary for Policymakers," 3-32.
19. IMBIE Team, "Mass Balance of the Antarctic Ice Sheet," 219–222.
20. Golledge et al., "The Multi-millennial Antarctic Commitment," 421–425.
21. Bamber et al., "Ice Sheet Contributions to Future Sea-Level Rise," 11195–11200.

22. Fletcher et al., *Atlas of Natural Hazards.*
23. Sharmila and Walsh, "Recent Poleward Shift of Tropical Cyclone Formation," 730–736.
24. Kossin et al., "Global Increase in Major Tropical Cyclone Exceedance," 11975–11980.
25. Murakami et al., "Projected Increase in Tropical Cyclones near Hawaii," 749–754.
26. Knutson et al., "Global Projections of Intense Tropical Cyclone Activity," 7203–7224.
27. Grinsted, Ditlevsen, and Christensen, "Normalized US Hurricane Damage," 23942–23946.
28. Hydrostatic flooding is passive water flow, as if the ocean had no waves at high tide, across a shoreline into lower areas.
29. Bjerklie et al., "Preliminary Investigation of the Effects of Sea-Level Rise."
30. Rotzoll and Fletcher, "Assessment of Groundwater Inundation," 477–481.
31. Habel et al., "Development of a Model to Simulate Groundwater Inundation," 122–134.
32. Ibid.
33. Kane et al., "Critical Elevation Levels for Flooding," 1679–1687.
34. Anderson et al., "Modeling Multiple Sea Level Rise Stresses," 1–14.
35. Romine et al., "Beach Erosion under Rising Sea-Level," 1321–1332.
36. Summers et al., "Failure to Protect Beaches," 427–443.
37. Ibid.
38. Fletcher et al., "Beach Loss along Armored Shorelines," 209–215.
39. Tavares et al., "Risk of Shoreline Hardening," 13633.
40. Fletcher et al., "National Assessment of Shoreline Change."
41. Anderson et al., "Doubling of Coastal Erosion," 75–103.
42. See "About the National Coastal Zone Management Program," https://coast.noaa.gov/czm/about/.
43. See "§205A-2 Coastal Zone Management Program; Objectives and Policies," https://www.capitol.hawaii.gov/hrscurrent/Vol04_Ch0201-0257/HRS0205A/HRS_0205A-0002.htm.
44. Fletcher et al., "National Assessment of Shoreline Change."
45. Fletcher et al., "Beach Loss along Armored Shorelines," 209–215.
46. Sweet et al., "Global and Regional Sea Level Rise Scenarios."
47. Nerem, Frederikse, and Hamlington, "Extrapolating Empirical Models."
48. Anderson et al., "Modeling Multiple Sea Level Rise Stresses."
49. Hawai'i Climate Change Mitigation and Adaptation Commission, "Hawai'i Sea Level Rise Vulnerability and Adaptation Report."

## Bibliography

Anderson, Tiffany R., Charles H. Fletcher, Matthew M. Barbee, L. Neil Frazer, and Bradley M. Romine. “Doubling of Coastal Erosion under Rising Sea Level by Mid-century in Hawaii.” *Natural Hazards* 78 (2015): 75–103.

Anderson, Tiffany R., Charles H. Fletcher, Matthew M. Barbee, Bradley M. Romine, Sam Lemmo, and Jade Delevaux. “Modeling Multiple Sea Level Rise Stresses Reveals up to Twice the Land at Risk Compared to Strictly Passive Flooding Methods.” *Scientific Reports* 8, no. 1 (2018): 1–14.

Bamber, Jonathan L., Michael Oppenheimer, Robert E. Kopp, Willy P. Aspinall, and Roger M. Cooke. “Ice Sheet Contributions to Future Sea-Level Rise from Structured Expert Judgment.” *Proceedings of the National Academy of Sciences* 116, no. 23 (2019): 11195–11200.

Bjerklie, David M., John R. Mullaney, Janet Radway Stone, Brian J. Skinner, and Matthew A. Ramlow. “Preliminary Investigation of the Effects of Sea-Level Rise on Groundwater Levels in New Haven, Connecticut.” U.S. Geological Survey Open-File Report 2012–1025, 2012. http://pubs.usgs.gov/of/2012/1025/.

Dutton, Andrea, Anders E. Carlson, Anthony J. Long, Glenn A. Milne, Peter U. Clark, R. DeConto, Ben P. Horton, S. Rahmstorf, and Maureen E. Raymo. “Sea-Level Rise due to Polar Ice-Sheet Mass Loss during Past Warm Periods.” *Science* 349, no. 6244 (2015): aaa4019.

Fletcher, Charles H., Eric E. Grossman, Bruce M. Richmond, and Ann E. Gibbs. *Atlas of Natural Hazards in the Hawaiian Coastal Zone.* Denver, CO: U.S. Geologic Survey, 2002. https://pubs.usgs.gov/imap/i2761/i2761.pdf.

Fletcher, Charles H., Robert A. Mullane, and Bruce M. Richmond. “Beach Loss along Armored Shorelines on Oahu, Hawaiian Islands.” *Journal of Coastal Research* (1997): 209–215.

Fletcher, Charles H., Bradley M. Romine, Ayesha S. Genz, Matthew M. Barbee, Matthew Dyer, Tiffany R. Anderson, S. Chyn Lim, Sean Vitousek, Christopher Bochicchio, and Bruce M. Richmond. “National Assessment of Shoreline Change: Historical Shoreline Change in the Hawaiian Islands.” U.S. Geological Survey Open-File Report 2011–1051, 2012. https://pubs.usgs.gov/of/2011/1051.

Gillett, N. P., M. Kirchmeier-Young, A. Ribes, H. Shiogama, G. C. Hegerl, and R. Knutti. “Constraining Human Contributions to Observed Warming since the Pre-industrial Period.” *Nature Climate Change* 11, no. 3 (2021): 207–212.

Golledge, Nicholas R., Douglas E. Kowalewski, Timothy R. Naish, Richard H. Levy, Christopher J. Fogwill, and Edward G. W. Gasson.

"The Multi-millennial Antarctic Commitment to Future Sea-Level Rise." *Nature* 526, no. 7573 (2015): 421–425.

Grinsted, Aslak, Peter Ditlevsen, and Jens Hesselbjerg Christensen. "Normalized US Hurricane Damage Estimates Using Area of Total Destruction, 1900-2018." *Proceedings of the National Academy of Sciences* 116, no. 48 (2019): 23942–23946.

Habel, Shellie, Charles H. Fletcher, Kolja Rotzoll, and Aly I. El-Kadi. "Development of a Model to Simulate Groundwater Inundation Induced by Sea-Level Rise and High Tides in Honolulu, Hawaii." *Water Research* 114 (2017): 122–134.

Hawai'i Climate Change Mitigation and Adaptation Commission. "Hawai'i Sea Level Rise Vulnerability and Adaptation Report." Prepared by Tetra Tech, Inc. and the State of Hawai'i Department of Land and Natural Resources, Office of Conservation and Coastal Lands, under the State of Hawai'i Department of Land and Natural Resources Contract No: 64064, 2017.

Hoffman, Jeremy S., Peter U. Clark, Andrew C. Parnell, and Feng He. "Regional and Global Sea-Surface Temperatures during the Last Interglaciation." *Science* 355, no. 6322 (2017): 276–279.

IMBIE Team, A. Payne, and B. D. Vishwakarma. "Mass Balance of the Antarctic Ice Sheet from 1992 to 2017." *Nature* 558, no. 7709 (2018): 219–222. https://doi.org/10.1038/s41586–018–0179-y.

Intergovernmental Panel on Climate Change (IPCC). "Ocean, Cryosphere and Sea Level Change." In *Climate Change 2021: The Physical Science Basis. Working Group I Contribution to the Sixth Assessment Report of the Intergovernmental Panel on Climate Change,* 1211–1362. Cambridge: Cambridge University Press, 2023. https://doi.org/10.1017/9781009157896.011.

Intergovernmental Panel on Climate Change (IPCC). "Summary for Policymakers." In *Climate Change 2021: The Physical Science Basis. Working Group I Contribution to the Sixth Assessment Report of the Intergovernmental Panel on Climate Change,* 3–32. Cambridge: Cambridge University Press, 2023. https://doi.org/10.1017/9781009157896.001.

Kane, Haunani H., Charles H. Fletcher, L. Neil Frazer, and Matthew M. Barbee. "Critical Elevation Levels for Flooding due to Sea-Level Rise in Hawai'i." *Regional Environmental Change* 15 (2015): 1679–1687.

Knutson, Thomas R., Joseph J. Sirutis, Ming Zhao, Robert E. Tuleya, Morris Bender, Gabriel A. Vecchi, Gabriele Villarini, and Daniel Chavas. "Global Projections of Intense Tropical Cyclone Activity for the Late Twenty-First Century from Dynamical Downscaling of CMIP5/RCP4. 5 Scenarios." *Journal of Climate* 28, no. 18 (2015): 7203–7224.

Kopp, Robert E., Frederik J. Simons, Jerry X. Mitrovica, Adam C. Maloof, and Michael Oppenheimer. "Probabilistic Assessment of Sea Level during the Last Interglacial Stage." *Nature* 462, no. 7275 (2009): 863–867.

Kossin, James P., Kenneth R. Knapp, Timothy L. Olander, and Christopher S. Velden. "Global Increase in Major Tropical Cyclone Exceedance Probability over the Past Four Decades." *Proceedings of the National Academy of Sciences* 117, no. 22 (2020): 11975–11980.

Murakami, Hiroyuki, Bin Wang, Tim Li, and Akio Kitoh. "Projected Increase in Tropical Cyclones Near Hawaii." *Nature Climate Change* 3, no. 8 (2013): 749–754.

Nerem, R. S., T. Frederikse, and B. D. Hamlington. "Extrapolating Empirical Models of Satellite-Observed Global Mean Sea Level to Estimate Future Sea Level Change." *Earth's Future* 10, no. 4 (2022): e2021EF002290.

Nunez, Christina. "Carbon Dioxide Levels Are at a Record High. Here's What You Need to Know." *National Geographic* 13 (2019).

Park, Jun-Young, Fabian Schloesser, Axel Timmermann, Dipayan Choudhury, June-Yi Lee, and Arjun Babu Nellikkattil. "Future Sea-Level Projections with a Coupled Atmosphere-Ocean-Ice-Sheet Model." *Nature Communications* 14, no. 1 (2023): 636.

Raftery, Adrian E., Alec Zimmer, Dargan M. W. Frierson, Richard Startz, and Peiran Liu. "Less than 2°C Warming by 2100 Unlikely." *Nature Climate Change* 7, no. 9 (2017): 637–641.

Romine, Bradley M., Charles H. Fletcher, L. Neil Frazer, and Tiffany R. Anderson. "Beach Erosion under Rising Sea-Level Modulated by Coastal Geomorphology and Sediment Availability on Carbonate Reef-Fringed Island Coasts." *Sedimentology* 63, no. 5 (2016): 1321–1332.

Rotzoll, Kolja, and Charles H. Fletcher. "Assessment of Groundwater Inundation as a Consequence of Sea-Level Rise." *Nature Climate Change* 3, no. 5 (2013): 477–481.

Sharmila, S., and K. J. E. Walsh. "Recent Poleward Shift of Tropical Cyclone Formation Linked to Hadley Cell Expansion." *Nature Climate Change* 8, no. 8 (2018): 730–736.

Summers, Alisha, Charles H. Fletcher, Daniele Spirandelli, Kristian McDonald, Jin-Si Over, Tiffany Anderson, Matthew Barbee, and Bradley M. Romine. "Failure to Protect Beaches under Slowly Rising Sea Level." *Climatic Change* 151 (2018): 427–443. https://doi.org/10.1007/s10584-018-2327-7.

Sweet, W. V., B. D. Hamlington, R. E. Kopp, C. P. Weaver, P. L. Barnard, D. Bekaert, W. Brooks, M. Craghan, G. Dusek, T. Frederikse, G.

Garner, A. S. Genz, J. P. Krasting, E. Larour, D. Marcy, J. J. Marra, J. Obeysekera, M. Osler, M. Pendleton, D. Roman, L. Schmied, W. Veatch, K. D. White, and C. Zuzak. "Global and Regional Sea Level Rise Scenarios for the United States: Updated Mean Projections and Extreme Water Level Probabilities Along U.S. Coastlines." NOAA Technical Report NOS 01. Silver Spring, MD: National Oceanic and Atmospheric Administration, National Ocean Service, 2022. https://oceanservice.noaa.gov/hazards/sealevelrise/noaa-nostechrpt01-global-regional-SLR-scenarios-US.pdf.

Tavares, Kammie-Dominique, Charles H. Fletcher, and Tiffany R. Anderson. "Risk of Shoreline Hardening and Associated Beach Loss Peaks before Mid-century: O'ahu, Hawai'i." *Scientific Reports* 10, no. 1 (2020): 13633.

United Nations Environment Programme. "Emissions Gap Report 2022: The Closing Window—Climate Crisis Calls for Rapid Transformation of Societies." Nairobi, 2022. https://www.unep.org/emissions-gap-report-2022.

Wang, Xiaoxin, Dabang Jiang, and Xianmei Lang. "Climate Change of 4°C Global Warming above Pre-industrial Levels." *Advances in Atmospheric Sciences* 35 (2018): 757–770.

Xu, Yangyang, Veerabhadran Ramanathan, and David G. Victor. "Global Warming Will Happen Faster than We Think." *Nature* 564, no. 7734 (2018): 30–32.

CHAPTER 7

# Climate Change

## Global Implications for Hawai'i

*Charles H. Fletcher III*

News from the front lines of global climate change is both good and bad.

The good news is that a true worldwide sustainable energy transition is finally taking place. Investment in clean energy (figure 7.1) now exceeds the amount of money that pours into new oil and gas projects. Not only will this provide humanity with clean and renewable forms of energy, it also means that the very worst global warming projections are increasingly unlikely to take place. On the other hand, global warming impacts are worse than anticipated and pose serious threats to human health and security. Clean energy investment must dramatically accelerate to recover a world where future communities can thrive.

As governments accelerate their investments in solar, wind, geothermal, and hydroelectric forms of energy, we should see enormous co-benefits in human health, social equity, and a growing respect for nature's rhythms that could redefine global economics. Amid all the doom and gloom of climate change news, there is definitely reason for optimism and hope.

Unfortunately, there is also bad news. The clean energy transition is not moving fast enough to prevent devastating climate change impacts to communities housing billions of people who will be forced to move to safer and more secure locations. Although $1.7 trillion was invested in renewable energy in 2023, on a net zero pathway designed to stop global warming, energy transition funding needs to average far more, about $4.55 trillion annually from 2023 through 2030, and more thereafter.

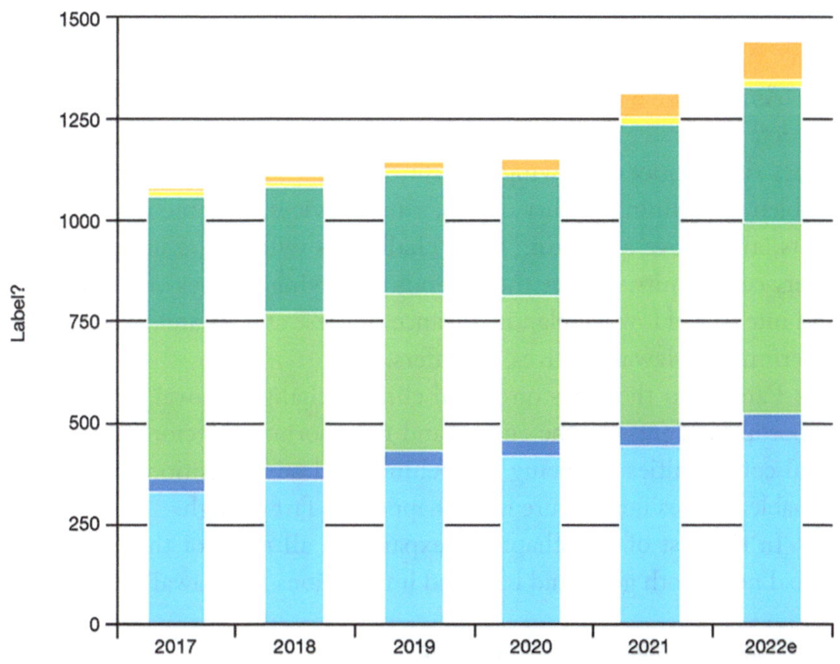


**Figure 7.1** Annual clean energy investment: light blue = renewable power; dark blue = nuclear; light green = energy efficiency and other end uses; dark green = grids and storage; yellow = low-carbon fuels and carbon capture and storage; orange = electric vehicles. Investment in carbon-free forms of energy totaled a record $1.7 trillion in 2023, a 31% jump from the previous year.

*Source:* IEA, "World Energy Investment 2022."

In addition, slightly over $1 trillion was invested in unabated fossil fuel supply and power, of which around 15% was to coal and the rest to oil and gas. In other words, we have entered a type of "climate purgatory" where record amounts of cash are being spent on clean energy, but records are also being set with investments in dirty fossil fuels. This rising demand for new forms of energy is driven by many sectors, chiefly new power-hungry digital data centers serving the growing AI industry, new EV charging stations, rising standards in underdeveloped nations, and more.

Critically, researchers are seeing early evidence of a potentially irreversible breakdown in foundational global natural systems. Rising heat, wildfire, and drought threaten terrestrial photosynthesis and its capacity to pull carbon out of the air. The massive circulation systems of air and ocean that govern weather, rain, and air temperature and humidity are changing in dangerous ways. Forests, marine ecosystems, and plant and animal communities everywhere are under attack by a combination of changing precipitation and temperature conditions, invasive species, pollution, disease, and biodiversity loss to insatiable human consumption.

As a remote and isolated archipelago, Hawai'i will experience the impacts of these global patterns in complex ways—supply line disruptions, rising prices, growing exposure to diseases and mental health problems, declining quality of goods and services, extreme weather impacts, and above all—heat. These challenges will not be unique to us. Every community will see these trends to varying degrees of immediacy and intensity. How we plan in advance of these shocks and stresses will determine if Hawai'i thrives, or suffers.

Parallel to the news on global climate change, Hawai'i is making important progress in the energy and transportation sectors. But some local communities are being left behind, and our transition into a sustainable and resilient future is not happening fast enough.

In the rest of this chapter, I expand on all three of these themes: global news both good and bad, and implications for Hawai'i.

## Fossil Fuel Expansion Is Coming to an End

After 200 years of fossil fuel expansion, we are at a turning point in the global energy system.

According to analysis released by the International Energy Agency (IEA),[1] after remaining flat for several years, global clean energy spending is finally ramping up. In 2023, investments of more than $1.7 trillion poured into clean energy, including renewable power, nuclear, grids, storage, low-emission fuels, efficiency improvements, and end-use renewables and electrification. That's more than ever before, and more than the $1.1 trillion spent on new oil and gas projects.

The positive momentum behind clean energy investment is not distributed evenly across countries or sectors, highlighting issues that policymakers will need to address to ensure a broad-based and secure transition to a global clean energy economy. More than 90% of the increase in clean energy investment since 2021 has taken place in advanced economies (the U.S., Japan, the European Union) and China. Comparing 2023 with the data for 2021, annual clean energy investment has risen much faster than investment in fossil fuels over this period (24% vs. 15%).

In 2023, for the first time, global investments[2] in low-carbon energy exceeded capital aimed at expanding fossil fuels. Solar generation across the EU rose by a record 24%, helping to avoid billions in rising gas costs driven by the Russian war on Ukraine. Some 20 EU nations sourced a record share of their power from solar, including the Netherlands, Spain, and Germany. For the first time ever, wind and solar supplied more of the EU's electricity than any other power source.[3] Continued growth in wind and solar, followed by expansion of hydroelectricity and nuclear generation, forced fossil fuel power generation to drop by an unprecedented 20% in 2023—double the previous record observed in 2020.

However, while clean energy investment jumped 24% between 2021 and 2023, it needs to be far higher to get the world on a path to net zero emissions by 2050, considered the target year for stopping global warming. To be consistent with a net zero pathway, global energy transition funding needs to average about $4.55 trillion per year from 2023 through 2030. And during the 2030s, annual investment would need to push even higher, to $6.88 trillion.

Clean energy momentum has been led by renewable power and EVs, with important contributions from other areas such as batteries, heat pumps, and nuclear power. In 2023, low-emissions power accounted for almost 90% of total investment in electricity generation. Solar is the star performer, and more than $1 billion per day went into solar investments ($380 billion for the year as a whole), pushing this spending above that in the oil industry for the first time.

Consumers are investing in more electrified end uses. Demand for electric cars is booming, with sales leaping by more than one-third in 2023 after a record-breaking 2022. As a result, investment in EVs has more than doubled since 2021, reaching $130 billion in 2023. Global sales of heat pumps have seen double-digit growth since 2021.

Renewable energy is getting cheaper. The world is set to add as much renewable power in the next five years as it did in the previous 20 years. Between 2010 and 2019, the costs for solar energy fell by 85%, and solar is ~33% cheaper than natural gas. Wind energy costs fell by 55%, and lithium-ion batteries by 85%.[4] In 99% of cases, it is more expensive to keep coal-fired power plants in the U.S. running than to build an entirely new solar or wind energy operation nearby.[5]

Investment is pouring into renewable energy at a record rate. Consider the following: 1) for the first time in history, the growth in global energy demand will now "almost entirely be met by renewables" according to the IEA; 2) jobs in clean energy are increasing and will more than double by 2030, from 6 million to 14 million; 3) transitioning to 100% renewables by 2050 yields $12 trillion in energy savings; and 4) the global market for clean energy manufacturing will be $650 billion by 2030, triple today's market for clean energy manufacturing.

Around the world, we also see that new countries are passing climate laws. Now, 91% of the world's gross domestic product is covered by countries with a net zero emissions target. In their 2023 Energy Outlook, long considered an industry standard report on the status of the global energy market, British Petroleum defines a "Central Scenario" that projects that global carbon emissions will peak in the 2020s and, by around 2050, descend to around 30% below 2019 levels.

Although the evidence is strong that a global transition to clean energy is emerging, as we discuss later in this chapter, the annual rate at which fossil fuels are being replaced by clean forms of energy needs to more than triple, and overall, global ambition to create a clean energy system needs to accelerate.

## A Vast and Dangerous Planetary Experiment

Global warming has now reached 1.5°C above the global preindustrial average air temperature.[6] By raising the average temperature of the atmosphere (figure 7.2), humans are engaged in a vast and dangerous

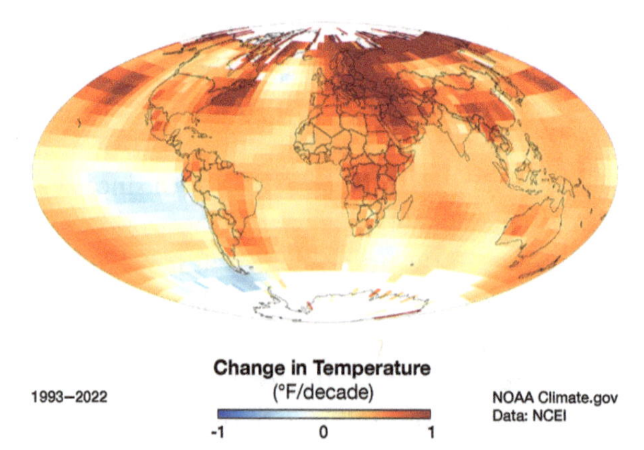


**Figure 7.2** Global temperature trends, 1993–2022. Earth's temperature has risen by an average of 0.14°F (0.08°C) per decade since 1880, or about 2–2.3°F (1.1–1.3°C) in total. The rate of warming since 1981 is more than twice as fast: 0.32°F (0.18°C) per decade.

*Source:* NOAA "Climate Change, Global Temperature," https://www.climate.gov/news-features/understanding-climate/climate-change-global-temperature.

planetary experiment that puts at risk the environments and ecosystems that sustain life on Earth. Burning coal, natural gas, and oil to make electricity and heat, to power manufacturing, to run cars, ships, and airplanes, and to produce food releases over forty billion tons of carbon dioxide ($CO_2$) and other greenhouse gases into the air each year.

$CO_2$ is a powerful greenhouse gas that traps heat that would otherwise escape to space. Year after year, adding $CO_2$ to the atmosphere is like adding extra blankets on your bed when you sleep; they trap heat, keeping you warm. In the case of climate change, too warm.

Having entered the air, $CO_2$ naturally cycles between the atmosphere, the oceans, soil and rock, and plants and animals. Its removal occurs across multiple time scales. About 50% of an increase in $CO_2$ will be removed from the atmosphere within 30 years, and a further 30% will be removed within a few centuries. The remaining 20% could linger in the atmosphere for many thousands of years causing essentially permanent climate change.[7]

The average concentration of $CO_2$ in the air has risen from a natural level of about 277 ppm (parts per million) to over 420 ppm, the highest in human history.[8] Sixty years ago, in the first decade of measuring atmospheric $CO_2$ at the Mauna Loa Observatory, the annual mean growth rate was less than 0.9 ppm per year. Over the most recent decade, the growth rate was about 2.4 ppm per year. The year 2023 marked the 12th consecutive year that $CO_2$ increased by more than 2 ppm, the fastest sustained rate of increase in 64 years of monitoring.[9]

Today's rate of $CO_2$ accumulation is about ten times faster than the most rapid event of any time since 66 million years ago, when an asteroid impact caused the extinction of the dinosaurs.[10] The last time $CO_2$ levels were this high was during the Pliocene Climatic Optimum, between 4.1 and 4.5 million years ago. At that time Earth's climate was radically different; the global average temperature was 2–3°C warmer, beech trees grew near the South Pole, there was no Greenland ice sheet, no West Antarctic ice sheet, and there is evidence that global sea level was as much as 25 m (82 ft) higher than today.[11]

Methane ($CH_4$), 25 to 80 times more powerful at trapping heat than $CO_2$ (but with a shorter residence time), is another greenhouse gas that is rapidly accumulating in Earth's atmosphere. $CH_4$ is responsible for about one-third of near-term global warming, and emissions have set record highs for four years in a row. The largest sources of $CH_4$ are wetlands, freshwater areas, agriculture, fossil-fuel extraction, landfills and waste, and fires.[12] In 2024, methane concentration in the atmosphere exceeded 1,932 ppb (parts per billion), almost triple the preindustrial level of 700 ppb, for the first time in human history. Global methane emissions slowed at the beginning of the twenty-first century, but began a rapid acceleration around 2007 (figure 7.3).

Carbon isotopic signatures since 2007 suggest that much of the recent increase in $CH_4$ is due to microbial decomposition of organic matter, rather than the combustion of fossil fuels.[13] Potential explanations include rising emissions from landfills and reservoirs, growing livestock herds, and increasing activity by microbes in warming and expanding wetlands. If, as one study suggests,[14] it is true that microbes are responsible for the majority (85%) of the growth in emissions since 2007, then it raises the possibility that warming itself is contributing to the increase, potentially through mechanisms such as increasing the productivity of tropical wetlands.

Is the increasing wetland production of $CH_4$ nature's response to over 1°C of global warming? If so, it may mean that climate change is taking on characteristics that are beyond our ability to control.

## Global Warming Causes Devastating Impacts

There are two problems with trapping heat in the atmosphere. 1) Warming is happening so fast that plants and animals cannot adapt to the resulting changes and many ecosystems are being damaged and species are going extinct. 2) Air that is warmer and with more $CO_2$ triggers several processes that are dangerous for people: sea level rise, bigger and harsher storms, declining nutritional value of plants we use as food, spreading diseases and parasites, expanding drought and flooding, more frequent and powerful heat waves, and other effects.

According to the Intergovernmental Panel on Climate Change (IPCC) 6th Assessment Report (AR6), it is unequivocal that human influence has warmed the atmosphere, ocean, and land.[15] The scale of recent changes across the climate system is unprecedented over hundreds to thousands of years. Weather and climate extremes are being affected by human-induced climate change in every area across the globe. Extreme weather events and their particular attribution to human influence

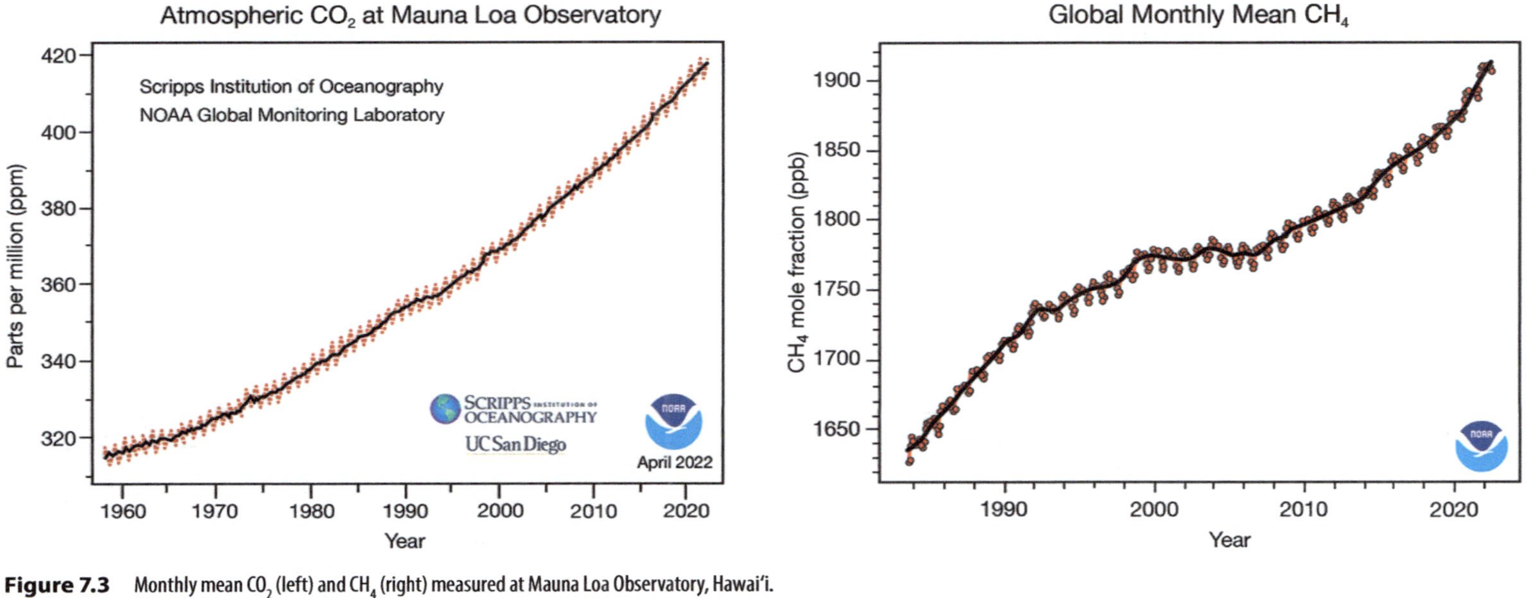


**Figure 7.3** Monthly mean $CO_2$ (left) and $CH_4$ (right) measured at Mauna Loa Observatory, Hawai'i.

*Source:* NOAA, Global Monitoring Laboratory, Carbon Cycle Greenhouse Gases, https://gml.noaa.gov/ccgg/.

have strengthened since the last IPCC report in 2013. AR6 warns that the climate crisis affects everyone. It projects:

1. Increasing deaths and physical and mental illness due to greater extreme weather, including drought, storms, heat, and wildfires;
2. Growing breadbasket failures and agricultural losses;
3. Increases in heat-related morbidity and mortality for people and wildlife;
4. Destruction for small island settlements;
5. Debilitating drought and declining freshwater resources on a regional scale;
6. Severe health effects due to increasing epidemics, particularly from diseases spread by insects and other animals;
7. Loss of nature, including marine and terrestrial biodiversity, ecological complexity, and faunal intactness.

Global warming risks food and water availability, with the global land area and human population in conditions of extreme to exceptional drought more than doubling by 2100 under a scenario of continued emissions.[16] Climate change threatens natural ecosystems[17] that provide life-sustaining resources, human security, livable conditions for communities, and the stability of one-third of the human population.

Even under current levels of warming, the daily lives of at least 3.3 billion people are "highly vulnerable" to climate change, and now, people are 15 times more likely to die from extreme weather than in years past.[18] In addition, at 2°C global warming, models project that up to 3 billion people may suffer "chronic" water scarcity.

Several studies show that the emergence of heat and humidity too severe for human tolerance has already begun. By the year 2070, expansion of the hottest and most inhospitable locations in the Sahara Desert will displace approximately 3 billion people, making the whole of North Africa, the Middle East, India, South-East Asia, Northern Australia, half of South and Central America inhospitable to human life (figure 7.4). Most of these people will leave their homelands because food and water will be scarce. Where will they go? How will they be received? Who will meet their needs?[19]

Today, one in three people are exposed to deadly heat stress. This number is projected to increase up to 75% by the end of the century.

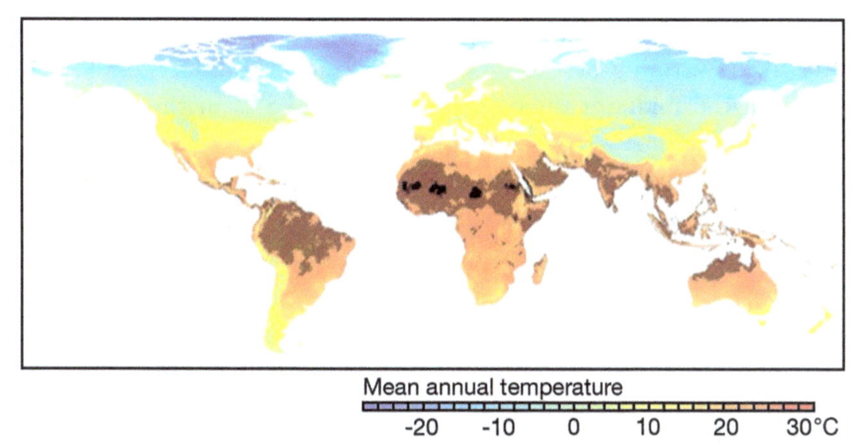


**Figure 7.4** In the current climate, unlivable conditions (mean annual temperatures >29°C) are restricted to the small dark areas in the Sahara region. In 2070, such conditions are projected to occur throughout the shaded area following a scenario of continued emissions. Absent migration, that area would be home to 3.5 billion people in 2070. Background colors represent the current mean annual temperatures.
*Source: Washington Post,* May 4, 2020, https://www.washingtonpost.com/weather/2020/05/04/human-climate-niche/.

Serious flooding threatens an additional 500,000 people, and up to 1 billion people living on coasts will be exposed to sea level rise–related flooding by 2050. The changing climate drives the spread of diseases in people, crops, livestock, and wildlife. Even if warming is held below 1.6°C by 2100, 8% of today's farmland will be unfit for food production. By 2050, declining food production and nutrient losses are projected to cause severe stunting affecting 1 million children in Africa alone, and 183 million additional people to go hungry.

## Progress on 1.5°C

In 2015, at the annual Conference of the Parties (COP) of the United Nations Framework Convention on Climate Change (UNFCCC), 196 nations and territories agreed to emissions reductions aimed at limiting global warming to less than 2°C, and to pursue efforts to end warming before 1.5°C.[20] In 2018, publication of a special report[21] on the impacts of 1.5°C of warming alerted the world to the fundamental risks associated with continued warming.

Greenhouse gas emission restrictions over the past decade mean that high-end emission scenarios leading to more than 4°C warming this century are no longer likely.[22] Shared Socioeconomic Pathway

SSP2–4.5[23] is an appropriate reflection of current policies and leads to very likely warming of 1.2–1.8°C in the near term (2021–2040), 1.6–2.5°C in the midterm (2041–2060), and 2.1–3.5°C in the long term (2081–2100).[24]

These projections are in line with a data-driven analysis that uses historical temperature observations to predict the time until UNFCCC thresholds are reached.[25] In the SSP2–4.5 scenario, the central estimate for the 1.5°C global warming threshold is between 2033 and 2035, including a $\pm 1\sigma$ range of 2028 to 2039. In addition, research suggests a substantial probability of exceeding the 2°C threshold even in the low (SSP1–2.6) climate forcing scenario. Given substantial evidence of accelerating risks to natural and human systems at 1.5°C and 2°C, these results are further evidence for high-impact climate change over the next three decades.

To have a 50% chance of avoiding more than 1.5°C of warming, global emissions will have to essentially end in three decades. This means emissions must peak at the latest before 2025 and then fall to 48% below 2019 levels in 2030, reaching net zero by the early 2050s; at the same time, methane emissions will also need to be reduced by about one-third.[26] Unfortunately, globally, governments still plan to produce more than double the amount of fossil fuels in 2030 than would be consistent with stopping warming at 2°C.

In total, 153 nations responsible for 75% of global greenhouse gas emissions have pledged to reach net zero emissions between 2050 and 2070. Made under the UNFCCC, these pledges, if implemented in full and on time, may limit total global warming to 1.4–2.8°C by the end of the century However, on-the-ground policies to limit greenhouse gas emissions are advancing at a snail's pace.[27]

The United Nations Environment Programme (UNEP) 2023 "Gap Report"[28] identified a widespread failure to improve emission-cutting pledges among signatories to the 2015 agreement. Emission reductions of 45% by 2030 are needed to keep 1.5°C in play. However, since COP26 in Glasgow (2021), nations have shaved just 1% off their projected greenhouse gas emissions for 2030, and in 2022 and 2023, COPs 27 and 28 ended with no increase in ambition.

Current UNFCCC pledges for action will lead to long-term temperature increases of 2.4–2.6°C. However, on-the-ground climate policies put the world on track for a median warming of 2.8°C (figure 7.5). According to the Gap Report, there is currently "no credible pathway"

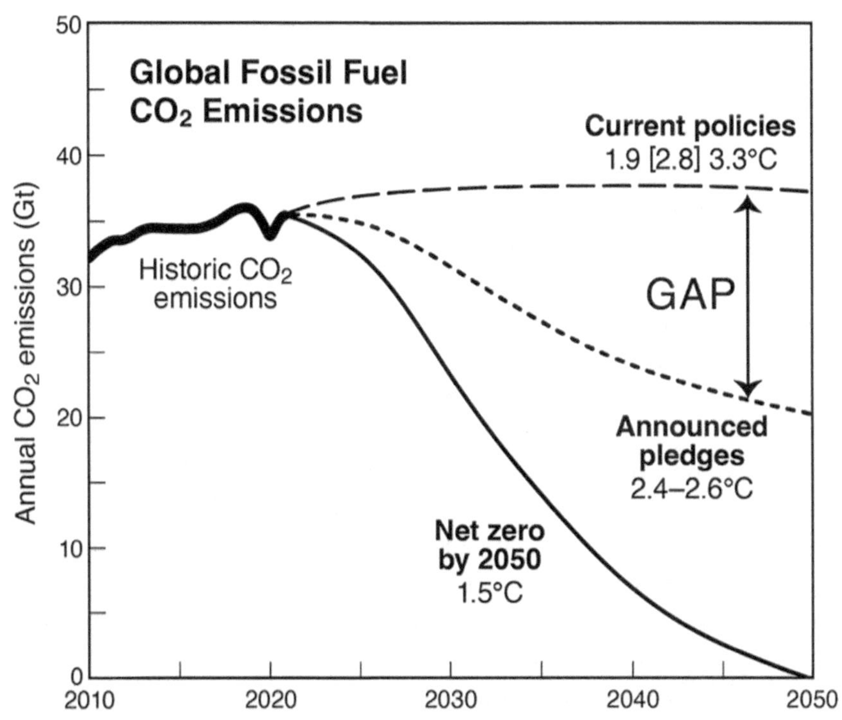


**Figure 7.5** (Solid) Limiting warming to 1.5°C requires cutting emissions to net zero by 2050. (Dotted) If implemented in full and on time, emission cuts pledged under the UNFCCC may limit total global warming to 2.4–2.6°C by the end of the century. (Dashed) Current policies and actions result in a median warming 2.8°C by 2100. *Source:* Charles H. Fletcher III.

to keep the rise in global temperatures below the key threshold of 1.5°C. Incremental change is "no longer an option." Avoiding dangerous levels of warming will require a "wide-ranging, large-scale, rapid and systemic transformation" involving much faster changes to electricity supplies, industry, transportation, and buildings. As well as protection of natural landscapes, changes to diet and farming, and cutting carbon from food supply chains. Additionally, the financial system must be reformed to provide investments of $4–$6 trillion U.S. dollars per year.

## Challenges to UNFCCC Goals

Planned cuts in global emissions are inadequate for protecting human security. On their own, currently implemented government policies will increase global surface temperature to a level of severe and dangerous global heating that can only be avoided with a massive rollout of carbon dioxide removal technology and large-scale reforestation that is nowhere in evidence today. However, underreporting of emissions,[29] growing methane concentrations,[30] and decreases in natural carbon storage[31] suggest that limiting global temperatures may be even more challenging than indicated by the UNEP analysis.

On average, global emissions are underreported 23%, with 70% underreporting of energy-related methane emissions alone. Methane is rising at a rate faster than any other time in recorded history, and carbon dioxide continues to increase at historically high rates. In addition, the terrestrial biome, historically responsible for sequestering about 30% of anthropogenic carbon dioxide emissions, has already neared, and temporarily crossed, a photosynthetic thermal maximum beyond which terrestrial carbon storage will grow increasingly unstable· Observations and theoretical analyses indicate that climate warming will lead to reduction of carbon assimilation capacity and eventually heat damage, as multiple studies suggest that a variety of ecosystems are operating at or near thermal thresholds.[32] Models project that drawdown of $CO_2$ by photosynthesis may decrease as much as 50% by 2040.

Studies[33] also report that global carbon loss from tropical forests has doubled in only 20 years, and that since 2016 the Brazilian portion of the Amazon Forest has been a net greenhouse gas source. Eighty-three percent of forest carbon loss results from agriculture, revealing that strategies to reduce forest loss are not successful, and that amplified carbon losses from deforestation are undercounted in Earth system models.

Combined, the impact of these barriers on slowing and ultimately reducing greenhouse gas concentrations in Earth's atmosphere is significant, and not accounted for in warming projections made by the UNFCCC on the basis of pledged nationally determined contributions. Jointly, increasing carbon loss from natural systems, crossing thermal limits to photosynthesis, underreporting of global emissions, and the ambition gap between pledged reductions and actual on-the-ground policies, constitute a significant, complex, and multisectoral threat to stabilizing Earth's climate at a livable level.

## Planetary Boundaries

In his book *Earth in Mind,*[34] David Orr writes that on a typical day we lose about 300 $km^2$ (116 $mi^2$) of rain forest to logging (one acre per second), 186 $km^2$ (72 $mi^2$) of land to encroaching deserts, and numerous species to extinction. Other sources tell us that in a day, the world's human population increases by more than 200,000,[35] we add over 110 million tons[36] of carbon dioxide to the atmosphere (over 40 billion tons per year), and we burn an average of 84.4 million barrels of oil (1,000 barrels *per second*[37]). By the end of the day, Earth's fresh water, soil, and ocean are more acidic,[38] its natural resources more depleted, its temperature a little hotter,[39] and the diversity of plant and animal life more damaged.

Unless you pay close attention to the scientific literature, you may not be aware of the stress that human communities have put on the natural systems of this planet, or of the rising potential for out-of-control climate change.[40] Alone, the expanding human population places an enormous burden on Earth resources and ecosystems. With the addition of climate change, critical systems that support life are being pushed to the very edge of annihilation.

The world has warmed only 1.5°C above the preindustrial background, yet we already see nearly one-third of the world population exposed to deadly heat waves,[41] a ninefold increase in large North American wildfires,[42] record-setting regional-scale megadrought,[43] animal and plant extinctions projected to increase twofold to fivefold in coming decades,[44] a pattern of genetic diversity loss,[45] and a weakened global ecosystem[46] described in one paper with over 15,000 co-authors, as "pushed to its breaking point."[47]

Massive systems of ice, ocean and air circulation, photosynthesis, and ecological relationships (collectively, biophysical systems)[48] are shifting toward instability, perhaps irreversibly.[49] These systems are the pillars of life on Earth allowing for hydrologic, geochemical, and temperature regimes that permit stable natural plant and animal communities, human food production, clean water, work, and safe human development.

Slow emerging changes such as deep ocean warming, stratification, and sea level rise are committed to continue even in the scenario where net zero emissions are reached. Because of this committed impact, climate actions must extend centuries, perhaps longer. At these time scales, preparation for high impact, low probability risks—such as an abrupt slowdown of Atlantic Meridional Overturning Circulation, ecosystem change such as savannahization of rain forest, or irreversible ice sheet loss—should be fully integrated into long-term solutions.[50]

AR6 assigns *high confidence* in the potential for abrupt, irreversible (extending decades or longer) change in permafrost carbon, West Antarctic ice sheets and shelves, Arctic winter sea ice, global sea level rise, and ocean acidification and deoxygenation. The Atlantic and Southern Ocean overturning circulation, and the global monsoon are assigned *medium confidence* in the potential for abrupt change. Abrupt, irreversible change in these systems may accelerate negative impacts, and potentially place aspects of climate change beyond our control.[51]

## Warming Impacts in Hawai'i

Hawai'i is facing unavoidable, costly, and dangerous impacts from climate change that threaten our future socioeconomic viability. As an isolated and remote group of islands without the capacity to easily exchange critical resources such as fresh water, food, or medical supplies with neighboring states, Hawai'i is especially vulnerable to the impacts of climate change.

Four major categories illustrate the breadth of risk to Hawai'i: air, precipitation, land and ocean ecosystems, and human communities.

### *Air Temperature*

From 1950 to 2020, observed changes in annual near-surface air temperature for 5 long-term reporting stations in Hawai'i (Līhu'e, Honolulu, Ho'olehua, Kahului, Hilo) show that temperatures across the islands have increased at rates ranging 0.11–0.22°C (0.2–0.4°F) per decade (figure 7.6).[52]

Since 1950, temperatures across the Hawaiian Islands have risen by about 1.11°C (2°F), with a sharp increase in warming over the last decade. Temperatures in Honolulu have increased by 1.4°C (2.6°F) over the period 1950–2020 and have consistently been above the 1951–1980 average since 1975. Statewide, the number of hot days and the number of very warm nights were well above average during the 2015–2020 period, with values more than double the respective long-term averages (figure 7.7). In Hawai'i, the rate of temperature increase is greatest at high elevations, far exceeding the global average rate of change. The annual number of days below freezing is decreasing over time, as is the daily temperature range, largely due to nighttime warming.

Historically, temperatures in Hawai'i have been tightly coupled with the decadal variability of the atmospheric circulation and sea surface temperature anomalies in the Pacific Basin (known as the Pacific

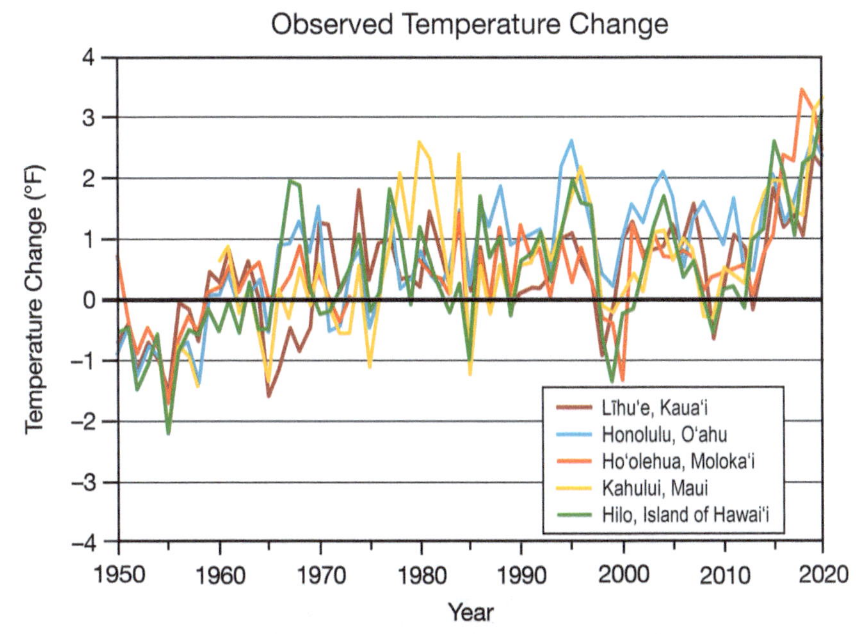


**Figure 7.6** Observed changes (compared to the 1951–1980 average: horizontal black line) in annual near-surface air temperature for 5 long-term reporting stations in Hawai'i from 1950 to 2020: Līhu'e, Kaua'i (red line); Honolulu, O'ahu (blue line); Ho'olehua, Molokai'i (orange line); Kahului, Maui (yellow line); and Hilo, Island of Hawai'i (green line).

*Source:* Stevens et al., "Hawai'i State Climate Summary 2022."

Decadal Oscillation); however, since the 1970s, increasing temperatures are more consistent with an increase in the frequency of the trade wind inversion (a layer marking the top surface of cumulus clouds where temperature increases with height) and a decrease in the frequency of trade winds (steady, persistent northeasterly winds).

Warming air temperatures have multiple impacts in Hawai'i. They lead to heat waves, expanded pathogen ranges and invasive species, thermal stress for native flora and fauna,[53] increased electricity demand, increased wildfire, potential threats to human health, and increased evaporation, which both reduces water supply and increases demand. Rapid warming at highest elevations impedes precipitation, the source of Hawai'i's fresh water.

### *Rainfall*

In Hawai'i, rainfall patterns are complex and vary according to season and location. A drier season typically develops from May through October, when warm, steady trade winds cause light to moderate showers.

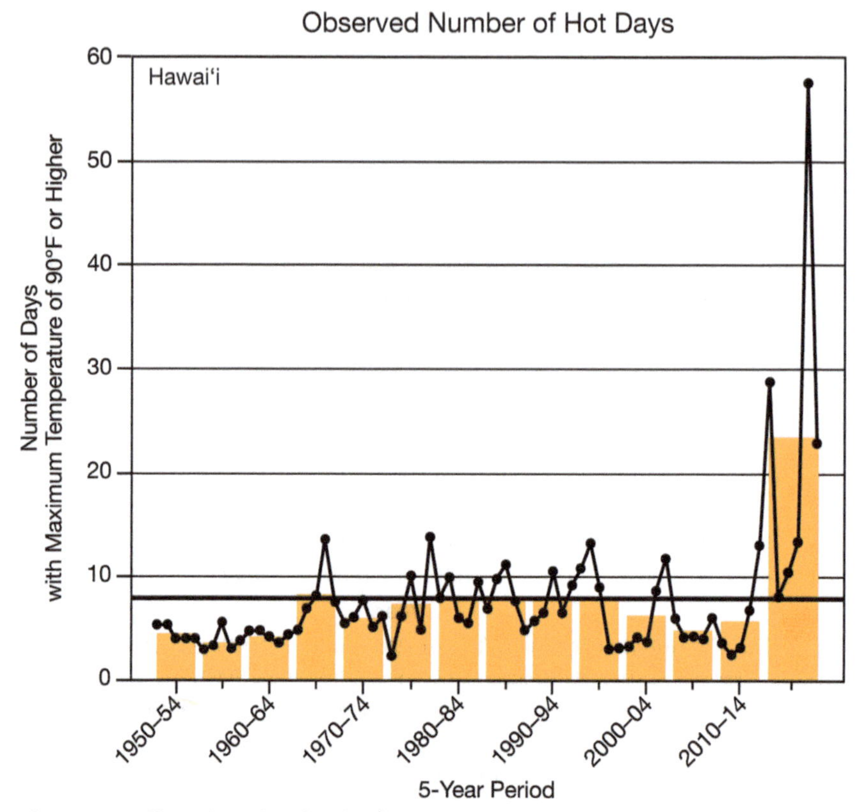


**Figure 7.7** Observed annual number of hot days (maximum temperature of 90°F or higher) for Hawai'i from 1950 to 2020. Dots show annual values. Bars show averages over 5-year periods (last bar is a 6-year average). The horizontal black line shows the long-term (entire period) average of 7.9 days. The number of hot days increased dramatically during the 2015–2020 period, with a multiyear average more than double the long-term average. *Source:* Stevens et al., "Hawai'i State Climate Summary 2022."

A wet season runs from November through April, with weaker and less frequent trade winds and a significant amount of rain from mid-latitude storms.

The interaction of mountainous terrain, persistent trade winds, heating and cooling of the land, and other factors results in dramatic differences in average rainfall over short distances. Total annual rainfall sometimes exceeds 762 cm (300 in) along the windward slopes of mountains, while it averages less than 50 cm (20 in) in leeward coastal areas and on the highest mountain slopes.[54]

Overall, Hawai'i has seen a decline in rainfall over the past century, with widely varying precipitation patterns on each island (figure 7.8).

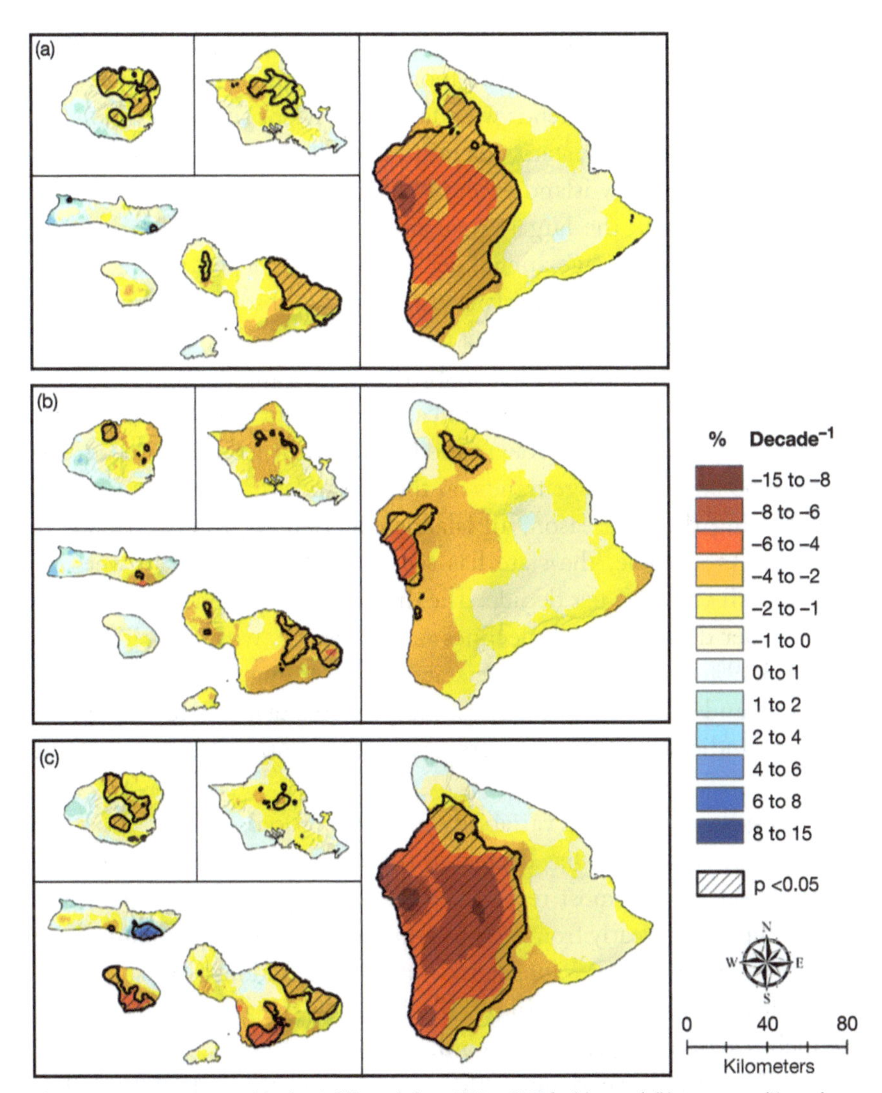


**Figure 7.8** Percent-per-decade rainfall trends from 1920 to 2012 for (a) annual, (b) wet season (November–April), and (c) dry season (May–October). Statistically significant trends are indicated with black hatching.
*Source:* Frazier and Giambelluca, "Spatial Trend Analysis of HI Rainfall."

The period since 2008 has been particularly dry.[55] Heavy rainfall events and droughts have become more common, increasing runoff, erosion, flooding, and water shortages.[56] Consecutive wet days and consecutive dry days are both increasing in Hawai'i.[57]

From 1920 to 2012, over 90% of the state experienced drying trends, with Hawai'i Island experiencing the largest significant long-term declines in annual rainfall, particularly in the Kona region. The highest rates of drying during dry season months were found in high-elevation areas, while wet season trends were fairly consistent across elevation.[58]

Averaged annually, statewide rainfall has declined 1.78% per decade since 1920, and every island except Moloka'i experienced a net decline in annual rainfall. The largest percent declines have been on Hawai'i Island, where the average annual decline was 2.22% per decade. On an island-wide basis, Maui experienced the largest absolute decline in annual rainfall since 1920. The Kona region of Hawai'i Island has the largest declines in the state, with statistically significant declines in annual rainfall of 2–15% per decade. Significant downward trends also occurred in mountainous regions of Kaua'i and O'ahu, while leeward areas mostly showed no annual average trends on these islands.[59]

During the wet season, all islands experienced negative trends in rainfall. As a whole, the state has seen a declining trend in wet season rainfall of 1.56% per decade. The dry season experienced the largest percent per decade rainfall declines of any season, with Hawai'i Island losing 3.19% per decade and the state overall losing 2.45% per decade. Every island had negative trends except Moloka'i, where the spatial patterns indicate significant increases in the southeastern part of the island.[60]

Despite great spatial variability in precipitation amounts across the islands, annual rainfall has decreased throughout the island chain, particularly during the most recent two decades (figure 7.9). Annual precipitation varies greatly from year to year; however, overall amounts have decreased since 1950 at 5 long-term monitoring stations (Līhu'e, Honolulu, Ho'olehua, Kahului, Hilo). Hilo has seen the greatest decrease of 14 inches across the period of record.[61]

In 10 of the 15 years since 2007, wet-season precipitation was below average, with 4 of the remaining 5 years being very near average. All of the 17 substantially above average wet years occurred prior to 2006 (figure 7.10).

Hawai'i has historically experienced drier than normal conditions during the El Niño wet season and greater than normal rainfall during the La Niña wet season. However, a drying trend during La Niña years is evident since 1956. Moreover, El Niño events have occurred more frequently over the last two decades. Larger total acres burned by wildfires are more likely to occur in the year following an El Niño event.[62]

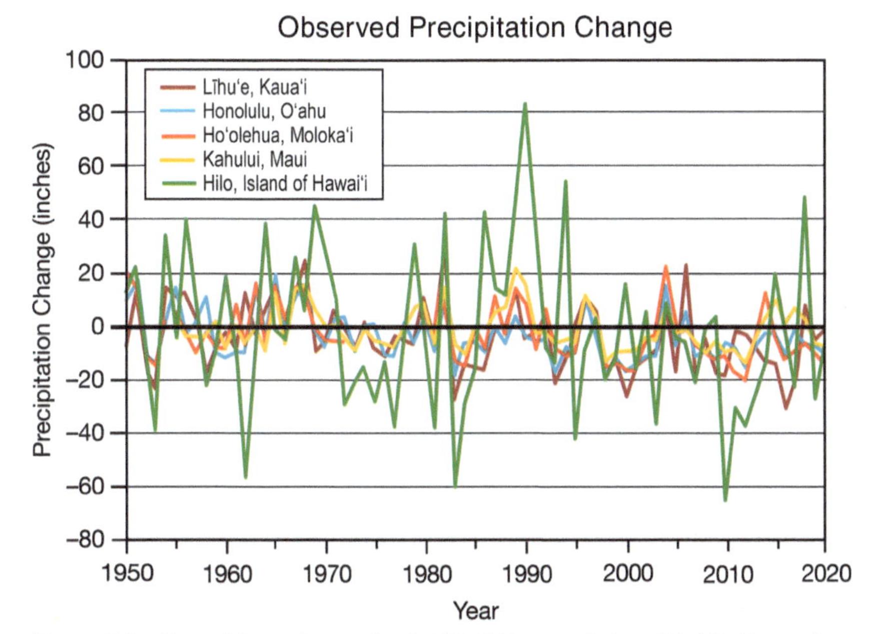


**Figure 7.9** Observed changes (compared to the 1951–1980 average: horizontal black line) in annual precipitation for 5 long-term reporting stations in Hawai'i from 1950 to 2020: Līhu'e, Kaua'i (red line); Honolulu, O'ahu (blue line); Ho'olehua, Moloka'i (orange line); Kahului, Maui (yellow line); and Hilo, Island of Hawai'i (green line).

*Source:* Stevens et al., "Hawai'i State Climate Summary 2022."

The changing relationship between La Niña and Hawai'i rainfall and the increasing El Niño frequency seem to have contributed to long-term drought since 1980. The Island of Hawai'i has experienced the largest significant long-term declines in annual and dry-season rainfall, with Hilo experiencing a decrease in annual precipitation of 14 inches since 1950. An increase in the frequency of the trade wind inversion is also linked to a decrease in precipitation at high elevations.

The number of consecutive dry days across the major Hawaiian Islands has increased since the 1950s. An increase in drought conditions has been seen in recent years, particularly at high elevations. In 2010, more than 40% of the Hawaiian Islands experienced severe, extreme, or exceptional drought conditions. Such conditions lead to a lack of usable water and an increased risk of fire. The number of 3-inch extreme precipitation events has been near or below average since 1990, with areas at the highest elevations experiencing the largest downward trend.

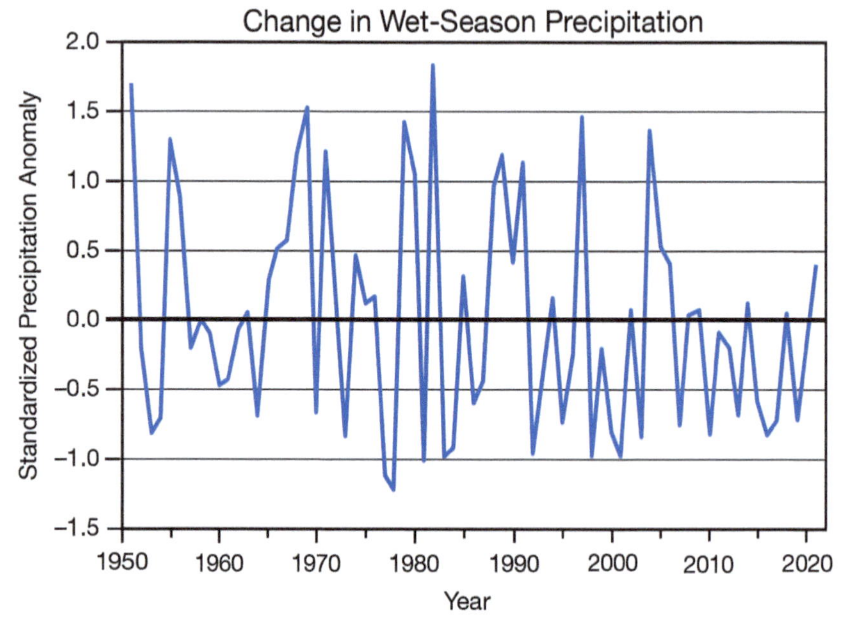


**Figure 7.10** Time series of wet-season Hawai'i precipitation for 1950–2021. These values represent precipitation variations over different climate regions of the Hawaiian Islands. A normalization technique is applied to each individual station, and a regional value is then computed as the arithmetic average of all station values. The time series represents the variations in the regional standardized precipitation anomalies. The year axis label indicates the ending year of the period (e.g., 1990 is November 1989 through March 1990). A persistent dry pattern has been in place since 2007.

*Source:* Stevens et al., "Hawai'i State Climate Summary 2022."

Regionally, extreme precipitation events have become less frequent for Oʻahu and Kaua'i but more frequent for the Island of Hawaiʻi.

### *Streamflow*

Streamflow in Hawaiʻi has declined over approximately the past century, consistent with observed decreases in rainfall.[63] This indicates declining groundwater levels in watersheds.

### *Trade Winds*

Average daily wind speeds are slowly declining in Honolulu and Hilo, while remaining steady across western and south Pacific sites.[64] Already, trade winds interact with ridgelines in ways that produce less cloud cover and less rainfall, resulting in reduced water supply and higher water demand.

### *Tropical Cyclones*

More frequent tropical cyclones are projected for the waters near Hawai'i. This is not necessarily because there will be more storms forming in the east Pacific; rather, it is projected that storms will follow new tracks that bring them into the region of Hawai'i more often.[65]

### *El Niño—Southern Oscillation*

As air temperature rises, the impact of El Niño events grows. During the strong El Niño of 2015, where Honolulu set 11 days of record heat, the local energy utility was compelled to issue emergency public service announcements asking residents and businesses to curtail escalating air-conditioning use that stressed the electrical grid. Climate models project that there will be increasing frequency and strength of El Niño and La Niña events as a result of continued warming in the twenty-first century.

Frequency of intense El Niño events is projected to double in the twenty-first century, with the likelihood of extreme events occurring roughly once every decade.[66] Strong El Niño events are associated with extreme rainfall and flooding, drought, high heat, extreme tides, active hurricane seasons, high sea surface temperatures and coral bleaching, extraordinarily high waves on north-facing shores, and compound events such as intense rain at high tide, which lead to urban flooding.

Models project a near doubling in the frequency of future extreme La Niña events, from one in every 23 years to one in every 13 years. Approximately 75% of the increase occurs in years following extreme El Niño events, thus projecting more frequent swings between opposite extremes from one year to the next.[67]

### *Terrestrial Ecosystems*

Natural ecosystems are the key to aquifer recharge. Indigenous vegetation that captures cloud water is responsible for nearly 40% of groundwater recharge. Yet, Hawai'i's abundant flora and fauna are particularly susceptible to the harmful effects of climate change. Hawai'i is home to 31% of the nation's plants and animals listed as threatened or endangered, and less than half of the landscape on the islands is still dominated by native plants.

Increases in the frequency of events such as wildfires, tropical cyclones, drought periods, hurricanes—all exacerbated by climate change—threaten the balance of Hawai'i's delicate ecosystems. Studies indicate that endemic and endangered birds and plants are highly vulnerable to climate change and are already showing shifting habitats.[68] Even under moderate warming, 10 of 21 existing native forest bird species are projected to lose over 50% of their range by 2100. Of those, 3 may lose their entire ranges and 3 others are projected to lose more than 90% of their ranges, making them of high concern for extinction.[69] Warming air temperatures also bring mosquito-borne diseases to previously safe upland forests, driving several native bird species toward extinction.[70]

### *Sea Surface Temperature*

Because of climate change, the oceans are becoming warmer and more acidic. Over 90% of the heat trapped by greenhouse gases since the 1970s has been absorbed by the oceans, and today the oceans absorb heat at twice the rate they did in the 1990s. Globally averaged, sea surface temperatures (SSTs) have already increased by 1.0°C (2.1°F) over the past 100 years, with half of this rise occurring during the 1990s.

A marine heatwave developed in the waters around Hawai'i in 2019 (figure 7.11). Heat records were set across Hawai'i, and by October 2019, Honolulu had recorded 45 days of record-high temperature, including 29 days from June to August, equal to more than two record highs per week. Beginning August 10, Honolulu hit 32.2°C (90°F) each of the next 37 days. Nighttime lows also set records. From 1950 to 2018, only 14 nights failed to drop below 26.6°C (80°F); 2019 featured 19 such nights, while 2019 saw the 1st, 2nd, and 3rd hottest calendar days on record in Honolulu. Honolulu hit 35°C (95°F) on the final day of August, a record for hottest August temperature in a century, and tied the record for hottest year-round temperature.

Of four long-running weather monitoring stations in Hawai'i, three saw their warmest summer on record in 2019. Only the station in Hilo did not. In Līhu'e, August 24 to September 12 set daily heat records. In July, August, and September, 48 days set record highs, 44 nights set record-high lows, and zero days or nights set record lows. Over 300 records were tied or broken in 2019. Only 5 of these were for record lows, revealing a strong warming shift in median temperature across Hawai'i. The cause of this extraordinary summer was a record-setting marine heat wave, the result of weak atmospheric circulation that produced very calm wind patterns.

North Central Pacific averaged SST trends follow the globally averaged trend. Over the period 2018–2023, almost the entire tropical

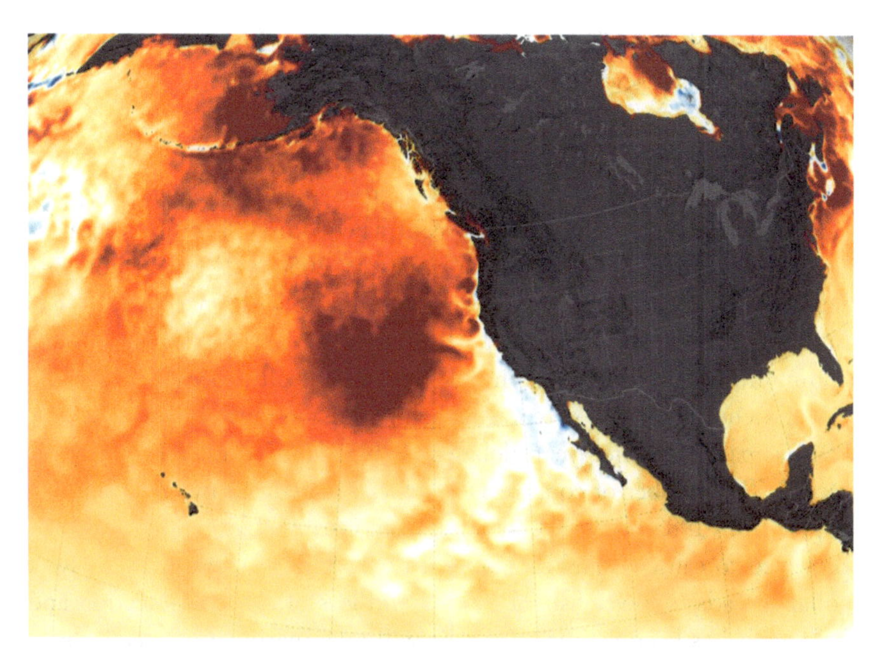

**Figure 7.11** Sea surface temperature anomalies across the northeastern Pacific in August 2019. Red colors depict areas that were hotter than average for the same month for the period 1985–2012; blues were colder than average.

*Source:* NASA Earth Observatory, "Marine Heat Wave Returns to the Northeast Pacific," https://earthobservatory.nasa.gov/images/145602/marine-heat-wave-returns-to-the-northeast-pacific.

Pacific, in particular areas along the equator, have seen temperatures warmer than the 30-year average.[71] Sea surface temperature increase has also intensified in areas of tropical cyclone genesis relevant to Hawai'i, suggesting a connection with strengthened storminess. Increased heat and evaporation contribute to a more extreme hydrological cycle and more extreme weather in particular hurricanes. More frequent tropical cyclones are also projected for waters near Hawai'i because of the new tracks that storms will likely follow as a result of climate change.

### *Ocean Acidification*

In addition to warming, the accumulation of $CO_2$ in the atmosphere causes the ocean to become more acidic. This is because of the chemical reaction that occurs when water bonds with $CO_2$. Ocean acidification interferes with the natural processes of marine organisms and ecosystems. It reduces the ability of marine organisms to build shells and other hard structures. Nearly 30 years of oceanic pH measurements, based on data collected from Station ALOHA, Hawai'i, show a roughly 8.7% increase in ocean acidity over this time.[72]

### *Coral Bleaching*

Ocean warming and acidification are projected to cause annual coral bleaching in some areas, like the central equatorial Pacific Ocean, as early as 2030 and affect almost all reefs by 2050.[73] This will not only devastate local coral reef ecosystems but will also have profound impacts on ocean ecosystems in general. Ultimately it will threaten the human communities and economies that depend on a healthy ocean.[74]

Extended periods of coral bleaching first occurred in Hawai'i in 2014 as part of a three-year global-scale bleaching event, the longest on record. The number of coral reefs in Hawai'i impacted by bleaching has tripled from 1985 to 2012. Globally, ocean warming and acidification are projected to cause annual bleaching on over 98% of reefs by 2050. Reef collapse leads to lower fisheries yields and loss of coastal protection and habitat. It has been estimated that it will take marine ecosystems thousands of years to recover from these climate-related upheavals. Scientists have concluded that when seas are hot enough for a sustained amount of time, nothing can protect coral reefs. The only hope for securing a future for coral reefs is urgent and rapid action to reduce global warming.

### *Sea Level Rise*

Climate change is contributing to a rising sea level to which Hawai'i is particularly vulnerable. The frequency of high tide flooding in Honolulu has increased from 6 to 11 days per year since the 1960s.[75] Under the NOAA Intermediate global mean sea level rise scenario (1 m by 2100), high tide flooding frequency is projected to reach 50 days per year by 2028, and to reach 100 days per year by 2038.[76]

Due to global gravitational effects, estimates of future sea level rise in Hawai'i and other Pacific islands are about 20% higher than the global mean.[77] With a projected 0.98 m (3.2 ft) of sea level rise, 25,800 acres will experience chronic flooding, erosion, and/or high wave impacts. One-third of this land is designated for urban use, and impacts include 61 km (38 mi) of major roads, and more than $19 billion in assets.[78]

The Hawai'i Department of Transportation estimates that under 1 m (3.3 ft) of sea level rise, 10–15% of the state's highway system would be affected, worth a total of $15 billion. This figure assumes the state will need $7.5 million for every mile of highway road that must either be raised, pushed back, or relocated entirely to escape erosion and flooding in the next 50 to 100 years—and $40 million for every mile of bridge.

### *Coastal Erosion*

Over 70% of beaches in Hawai'i are in a state of chronic erosion. This is likely related to long-term sea level rise as well as coastal hardening.[79] Coastal hardening of chronically eroding beaches caused the combined loss of 9% (13.4 mi, 21.5 km) of the length of sandy beaches on Kaua'i, O'ahu, and Maui.[80]

### *Cultural Practice*

Indigenous populations will be disproportionally impacted by climate change due to their strong ties to place and greater reliance on natural resources for sustenance.[81] About 550 Hawaiian cultural sites are exposed to chronic flooding with a sea level rise of 1 m.[82] Sea level rise impacts on traditional and customary practices (including fishpond maintenance, cultivation of salt, and gathering from the nearshore fisheries) have been observed.[83] Because of flooding and sea level rise, Indigenous practitioners have had limited access to the land where salt is traditionally cultivated and harvested since 2014. Detachment from traditional lands has a negative effect on the spiritual and mental health of the people.[84] In Hawai'i, climate change impacts, such as reduced streamflow, sea level rise, saltwater intrusion, episodes of intense rainfall, and long periods of drought, threaten the ongoing cultivation of taro and other traditional crops and interfere with the livelihood and security of Pacific communities.[85]

Studies have documented significant and harmful effects on the Native Hawaiian community and their traditional and customary rights and practices. Climate change impacts on Native Hawaiians can be categorized in three general ways: impacts on upland forests; impacts on traditional agriculture; and impacts on coastal and nearshore waters.

Hawaiians considered the uplands the realm of the gods or "wao akua" due to its importance in the ecosystem and for being a source of physical and spiritual nourishment. With the rapid decline of Hawai'i's native forests, Hawai'i's native species have also perished and this both limits the perpetuation of cultural knowledge across generations and severs the connection between Hawaiians and natural and cultural resources. Hawaiian practices in the upland that will be affected by climate change also include the collection of timber and medicinal plants, and collection related to traditional hula. Although cultural practitioners continue to collect these materials, it is now difficult to find the necessary resources.

In the "wao kanaka," or the realm of man, traditional Hawaiian agriculture practices are at high risk of becoming unsustainable in the future due to climate change. Changing rainfall, diminished streamflow, rising temperatures, and rising sea levels causing saltwater intrusion threaten Hawaiian farming practices and food security.

Climate change also poses significant chances of disrupting or preventing numerous traditional and customary Hawaiian practices at the coastline and in nearshore areas on Oʻahu and Maui. Climate change has the potential to impact the practice of burying and the already buried Hawaiian remains in soft sand dunes, cultivating and collecting sea salt, gathering marine life, using and maintaining fishponds, fishing, and paddling and sailing in the open ocean.

### *Food*

Hawaiʻi is exposed to significant food insecurity due to its geographically isolated nature.[86] The closest port is over 37,000 kilometers away in Oakland, California, which categorizes the Hawaiian population as one of the most food import dependent in the world. The state of Hawaiʻi imports approximately 90% of its food and over 73% of its energy.[87] Hawaiʻi's 1.4 million residents and 10 million annual visitors are uniquely vulnerable to food distribution interruptions due to natural disasters (which are exacerbated by climate change), and other economic and social disturbances.

Surpassing the 1.5°C (2.7°F) warming threshold established by the United Nations threatens global food security. Crop failure due to drought, flood, or damaging events in the course of a growing season increases disproportionately between 1.5°C and 2°C (2.7–3.6°F). Limiting global warming to 1.5°C (2.7°F) would reduce the risk of simultaneous crop failure for food resources such as maize, wheat, and soybean by 26%, 28%, and 19%, respectively.[88]

In sum, we know that the present damage and future risk to coastal communities such as Hawaiʻi posed by rising sea levels, declining rainfall, expanding drought, increased risk of landfalling tropical cyclones, flooding, environmental loss, threats to fundamental Indigenous identity, and food and water impacts that put in question the long-term viability of the Hawaiian Islands as a socioeconomic entity are caused by global warming and related impacts.

These local trends are all expected to accelerate as the air continues to warm. Future projections include more frequent tropical cyclones and extreme El Niño events, 50% of native forest bird species will lose over half their range, and as early as 2030, almost all reefs in Hawai'i are expected to experience annual bleaching.

With roughly 1 m (3.3 ft) of sea level rise projected this century, 25,800 acres in Hawai'i will experience chronic flooding, erosion, salinization, and high wave impacts. One-third of this land is designated for urban use. Impacts include 61 km (38 mi) of major roads, and more than $19 billion in assets.[89]

Heat waves, declining rainfall and increasing rain intensity, growing storminess, impacts to ocean ecology, separation from cultural practices, growing social disparity, drought, displaced global populations, and loss of marine and terrestrial ecosystems—these are challenges placed in our future by a convergence of global climate change, local land management informed by nonindigenous practices, and a failure to recognize that resource exploitation is unsustainable.

The future of the living systems on our planet and the future of Hawai'i are dependent on our collective efforts to change the status quo. We must radically shift course while reestablishing healthy relationships with the natural resources around us. We must redesign the extractive and waste-creating systems of our world with new kinds of design principles that are circular and regenerative.

The water cycle itself is one of the most powerful examples of a circular system that has continued since the origins of life on our planet. How can we create design principles that mimic the water cycle? These conceptual guidelines can inform decision-making leading to real-world community designs. Connected greenways, placemaking, and hazard reduction are examples. If our communities are going to continue supporting a high quality of life, they must be renovated on the basis of design principles that recognize that water, food, culture, social equity, weather, and ecosystems are mutually interdependent and rapidly changing in unique ways. Linear approaches to building communities in Hawai'i will no longer work to our advantage. In fact, quite the opposite: they will work to our disadvantage.

## Notes

1. IEA, "World Energy Investment 2022."
2. BloombergNEF, "Global Low-Carbon Energy Technology Investment."
3. Jones, "European Electricity Review."
4. "Data Point."
5. Budryk, "99 Percent of Coal Plants."
6. Lindsey and Dahlman, "Climate Change."
7. Denman et al., "Couplings between Changes in the Climate System and Biogeochemistry," 501.
8. NOAA, "Carbon Dioxide."
9. NOAA, "Increase in Atmospheric Methane."
10. Zeebe, "Anthropogenic Carbon Release Rate."
11. Dumitru et al., "Constraints on Global Mean Sea Level."
12. Allen, "Cause of the 2020 Surge."
13. Tollefson, "Scientists Raise Alarm."
14. Lan et al., "Improved Constraints on Global Methane Emissions."
15. IPCC, "Summary for Policymakers," in *Climate Change 2021.*
16. Pokhrel et al., "Global Terrestrial Water Storage."
17. Nolan et al., "Past and Future Global Transformation."
18. IPCC, "Summary for Policymakers," in *Climate Change 2022.*
19. Xu et al., "Future of the Human Climate Niche."
20. United Nations Framework Convention on Climate Change, https://unfccc.int.
21. IPCC, "Summary for Policymakers," in *Global Warming of 1.5°C.*
22. Hausfather and Peters, "Emissions."
23. Shared Socioeconomic Pathways (SSPs) are scenarios of projected socioeconomic global changes up to 2100. They are used to derive greenhouse gas emissions scenarios with different climate policies.
24. IPCC, "Summary for Policymakers," in *Climate Change 2022.*
25. Diffenbaugh and Barnes, "Data-Driven Predictions."
26. Dvorak et al., "Estimating the Timing of Geophysical Commitment."
27. Climate Action Tracker, "Glasgow's 2030 Credibility Gap."
28. United Nations Environment Programme, "Emissions Gap Report 2023."
29. Mooney et al., "Countries' Climate Pledges Built on Flawed Data."
30. NOAA, "Increase in Atmospheric Methane."
31. Duffy et al., "How Close Are We."
32. Still et al., "No Evidence."
33. Feng et al., "Doubling of Annual Forest Carbon Loss."
34. Orr, *Earth in Mind.*
35. Answers.com, "How Much Does World Population Increase Each Day?" http://wiki.answers.com/Q/How_much_does_world_population_increase_each_day.

36. See CO2.Earth: "What the World Needs to Watch," https://www.co2.earth/.
37. Tertzakian, *A Thousand Barrels a Second.*
38. The U.S. Geological Survey has found that mining and burning coal, mining and smelting metal ores, and use of nitrogen fertilizer are the major causes of chemical oxidation processes that generate acid in the Earth-surface environment. These widespread activities have increased carbon dioxide in the atmosphere and resulted in increased acidity of oceans; increased acidity of freshwater bodies and soils because of acid rain; increased acidity of freshwater streams and groundwater due to drainage from mines; and increased acidity of soils due to added nitrogen to crop lands. See Rice and Herman, "Acidification of Earth."
39. See USDA, "USDA Announces Streamlined Disaster Designation Process."
40. Ripple et al., "Many Risky Feedback Loops."
41. Mora et al., "Global Risk of Deadly Heat."
42. Abatzoglou and Williams, "Impact of Anthropogenic Climate Change."
43. Williams, Cook, and Smerdon, "Rapid Intensification of the Emerging Southwestern North American Megadrought."
44. Wiens, "Climate-Related Local Extinctions."
45. Exposito-Alonso et al., "Genetic Diversity Loss in the Anthropocene."
46. Díaz et al., "Pervasive Human-Driven Decline."
47. Ripple et al., "World Scientists' Warning."
48. Lenton et al., "Climate Tipping Points."
49. McKay et al., "Exceeding 1.5°C Global Warming."
50. World Meteorological Organization, "State of the Global Climate."
51. Hansen et al., "Ice Melt, Sea Level Rise and Superstorms."
52. Stevens et al., "Hawai'i State Climate Summary 2022."
53. Fortini, "A Landscape-Based Assessment."
54. Stevens et al., "Hawai'i State Climate Summary 2022."
55. Frazier and Giambelluca, "Spatial Trend Analysis of HI Rainfall."
56. Kruk et al., "On the State of the Knowledge of Rainfall Extremes."
57. Kruk et al., "On the State of the Knowledge of Rainfall Extremes."
58. Frazier and Giambelluca, "Spatial Trend Analysis of HI Rainfall."
59. Ibid.
60. Ibid.
61. Stevens et al., "Hawai'i State Climate Summary 2022."
62. Ibid.
63. Bassiouni and Oki, "Trends and Shifts in Stream Flow in Hawai'i."
64. Marra and Kruk, "State of Environmental Conditions in Hawai'i."
65. Murakami et al., "Projected Increase in Tropical Cyclones Near Hawai'i."
66. Cai et al., "Increased Frequency of Extreme El Niño Events."

67. Ibid.
68. Jacobi et al., "Baseline Land Cover."
69. Fortini et al., "Large-Scale Range Collapse of HI Forest Birds."
70. Paxton et al., "Collapsing Avian Community on a Hawaiian Island."
71. Marra and Kruk, "State of Environmental Conditions in Hawai'i."
72. Ibid.
73. Van Hooidonk et al., "Opposite Latitudinal Gradients."
74. Marra and Kruk, "State of Environmental Conditions in Hawai'i."
75. Ibid.
76. Thompson et al., "Statistical Model for Frequency of Coastal Flooding."
77. Marra and Kruk, "State of Environmental Conditions in Hawai'i."
78. Hawai'i Climate Change Mitigation and Adaptation Commission, "Hawai'i Sea Level Rise Vulnerability and Adaptation Report."
79. Summers et al., "Failure to Protect Beaches under Slowly Rising Sea Level."
80. Fletcher et al., "National Assessment of Shoreline Change."
81. Yeo, "5 Ways Climate Change Harms Indigenous People."
82. Hawai'i Climate Change Mitigation and Adaptation Commission, "Hawai'i Sea Level Rise Vulnerability and Adaptation Report."
83. Sproat, "An Indigenous People's Right to Environmental Self-Determination."
84. Akutagawa et al., "Health Impact Assessment."
85. Sproat, "An Indigenous People's Right to Environmental Self-Determination."
86. Miles, "If We Get Food Right."
87. Gaupp, "Increasing Risks."
88. Ibid.
89. Hawai'i Climate Change Mitigation and Adaptation Commission, "Hawai'i Sea Level Rise Vulnerability and Adaptation Report."

## Bibliography

Abatzoglou, J. T., and A. P. Williams. "Impact of Anthropogenic Climate Change on Wildfire across Western U.S. Forests." *Proceedings of the National Academy of Sciences* 113, no. 42 (2016): 11770–11775. https://doi.org/10.1073/pnas.16071711.

Akutagawa, M., Elizabeth Cole, Tressa P. Diaz, Tanaya Dutta Gupta, Clare Gupta, Shaelene Kamakaala, Maile Taualii, and Angela Fa'anunu. "Health Impact Assessment of the Proposed Mo'omomi Community-Based Subsistence Fishing Area." Kohala Center, 2016. http://hdl.handle.net/10125/46016.

Allen, George H. "Cause of the 2020 Surge in Atmospheric Methane Clarified." *Nature* 617, no. 7940 (2022): 413–414. https://doi.org/10.1038/d41586-022-04352-6.

Bassiouni, M., and D. S. Oki. "Trends and Shifts in Stream Flow in Hawai'i, 1913–2008." *Hydrological Processes* 27, no. 10 (May 2013): 1484–1500. http://dx.doi.org/10.1002/hyp.9298.

BloombergNEF. 2023. "Global Low-Carbon Energy Technology Investment Surges Past $1 Trillion for the First Time." BloombergNEF, January 26, 2023. https://about.bnef.com/blog/global-low-carbon-energy-technology-investment-surges-past-1-trillion-for-the-first-time/.

Budryk, Zack. "99 Percent of US Coal Plants Are More Expensive than New Renewables Would Be: Report." *The Hill,* January 30, 2023. https://thehill.com/policy/energy-environment/3836301-99-percent-of-u-s-coal-plants-are-more-expensive-than-new-renewables-would-be-report/.

Cai, Wenju, Simon Borlace, Matthieu Lengaigne, Peter van Rensch, Mat Collins, Gabriel Vecchi, Axel Timmermann, et al. "Increased Frequency of Extreme El Niño Events due to Greenhouse Warming." *Nature Climate Change* 4 (2014): 111–116. https://doi.org/10.1038/nclimate2100.

Cai, Wenju, Guojian Wang, Agus Santoso, Michael J. McPhaden, Lixin Wu, Fei-Fei Jin, Axel Timmermann, et al. "Increased Frequency of Extreme La Niña Events under Greenhouse Warming." *Nature Climate Change* 5 (2015): 132–137. https://doi.org/10.1038/nclimate2492.

Climate Action Tracker. "Glasgow's 2030 Credibility Gap: Net Zero's Lip Service to Climate Action." Warming Projections Global Update, November 2021. https://climateactiontracker.org/documents/997/CAT_2021-11-09_Briefing_Global-Update_Glasgow2030CredibilityGap.pdf.

"Data Point: Clean Energy Costs Are Falling." *Economist Impact,* May 30, 2022. https://impact.economist.com/sustainability/net-zero-and-energy/data-point-clean-energy-costs-are-falling.

Denman, Kenneth L., Guy Brasseur, Amnat Chidthaisong, Philippe Ciais, Peter M. Cox, Robert E. Dickinson, et al. "Couplings between Changes in the Climate System and Biogeochemistry." In *Climate Change 2007: The Physical Science Basis. Contribution of Working Group I to the Fourth Assessment Report of the Intergovernmental Panel on Climate Change,* 499–587. Cambridge: Cambridge University Press, 2007.

Díaz, S., J. Settele, E. S. Brondízio, H. T. Ngo, J. Agard, A. Arneth, P. Balvanera, et al. "Pervasive Human-Driven Decline of Life on Earth Points to the Need for Transformative Change." *Science* 366, no. 6471 (2019): eaax3100. https://doi.org/10.1126/science.aax3100.

Diffenbaugh, Noah S., and Elizabeth A. Barnes. "Data-Driven Predictions of the Time Remaining until Critical Global Warming Thresholds Are Reached." *Proceedings of the National Academy of Sciences* 120, no. 6 (2023): e2207183120. https://doi.org/10.1073/pnas.2207183120.

Duffy, Katharyn A., Christopher R. Schwalm, Vickery L. Arcus, George W. Koch, Liyin L. Liang, and Louis A. Schipper. "How Close Are We to the Temperature Tipping Point of the Terrestrial Biosphere?" *Science Advances* 7, no. 3 (2021): eaay1052. https://doi.org/10.1126/sciadv.aay1052.

Dumitru, Oana A., Jacqueline Austermann, Victor J. Polyak, Joan J. Fornós, Yemane Asmerom, Joaquín Ginés, Angel Ginés, and Bogdan P. Onac. "Constraints on Global Mean Sea Level during Pliocene Warmth." *Nature* 574, no. 7777 (2019): 233–236. https://doi.org/10.1038/s41586-019-1543-2.

Dvorak, M. T., K. C. Armour, D. M. W. Frierson, C. Proistosescu, M. B. Baker, and C. J. Smith. "Estimating the Timing of Geophysical Commitment to 1.5 and 2.0°C of Global Warming." *Nature Climate Change* 12, no. 6 (2022): 547–552. http://doi.org/10.1038/s41558-022-01372-y.

Exposito-Alonso, M., T. R. Booker, L. Czech, L. Gillespie, S. Hateley, C. C. Kyriazis, P. L. M. Lang, et al. "Genetic Diversity Loss in the Anthropocene." *Science* 377, no. 6613 (2022): 1431–1435. https://doi.org/10.1126/science.abn5642.

Feng, Yu, Zhenzhong Zeng, Timothy D. Searchinger, Alan D. Ziegler, Jie Wu, Dashan Wang, Xinyue He, et al. "Doubling of Annual Forest Carbon Loss over the Tropics during the Early Twenty-First Century." *Nature Sustainability* 5, no. 5 (2022): 444–451. https://doi.org/10.1038/s41893-022-00854-3.

Fletcher, Charles H., Bradley M. Romine, Ayesha S. Genz, Matthew M. Barbee, Matthew Dyer, Tiffany R. Anderson, S. Chyn Lim, Sean Vitousek, Christopher Bochicchio, and Bruce M. Richmond. "National Assessment of Shoreline Change: Historical Shoreline Change in the Hawaiian Islands." U.S. Geological Survey Open-File Report 2011–1051, 2012. https://pubs.usgs.gov/of/2011/1051.

Fortini, Lucas, Jonathan Price, James Jacobi, Adam Vorsino, Jeff Burgett, Kevin Brinck, Fred Amidon, et al. "A Landscape-Based Assessment of Climate Change Vulnerability for All Native Hawaiian Plants." Hawai'i Cooperative Studies Unit. University of Hawai'i at Hilo. Technical Report HCSU-044, 2013.

Fortini, L. B., A. E. Vorsino, F. A. Amidon, Eben H. Paxton, and J. J, Jacobi. "Large-Scale Range Collapse of HI Forest Birds under CC and the

Need 21st Century Conservation Options." *PLoS ONE* 10, no. 10 (2015): e0140389. https://doi.org/10.1371%2Fjournal.pone.0140389.

Frazier, A. G., and T. W. Giambelluca. "Spatial Trend Analysis of HI Rainfall from 1920 to 2012." *International Journal of Climatology* 37, no. 5 (2017): 2522–2531. https://doi.org/10.1002/joc.4862.

Gaupp, F., Jim Hall, Dann Mitchell, and Simon Dadson. "Increasing Risks of Multiple Breadbasket Failure under 1.5 and 2°C Global Warming." *Agricultural Systems* 175 (2019): 34–45.

Hansen, J., Makiko Sato, Paul Hearty, Reto Ruedy, Maxwell Kelley, Valerie Masson-Delmotte, Gary Russell, et al. "Ice Melt, Sea Level Rise and Superstorms: Evidence from Paleoclimate Data, Climate Modeling, and Modern Observations That 2°C Global Warming Could Be Dangerous." *Atmospheric Chemistry and Physics* 16, no. 6 (2016): 3761–3812. https://doi.org/10.5194/acp-16-3761-2016.

Hausfather, Zeke, and Glen P. Peters. "Emissions—the 'Business as Usual' Story Is Misleading." *Nature* 577, no. 7792 (2020): 618–620. https://doi.org/10.1038/d41586-020-00177-3.

Hawai'i Climate Change Mitigation and Adaptation Commission. "Hawai'i Sea Level Rise Vulnerability and Adaptation Report." Prepared by Tetra Tech, Inc. and the State of Hawai'i Department of Land and Natural Resources, Office of Conservation and Coastal Lands, under the State of Hawai'i Department of Land and Natural Resources Contract No: 64064, 2017.

Intergovernmental Panel on Climate Change (IPCC). "Summary for Policymakers." In *Climate Change 2021: The Physical Science Basis. Working Group I Contribution to the Sixth Assessment Report of the Intergovernmental Panel on Climate Change,* 3–32. Cambridge: Cambridge University Press, 2023. https://doi.org/10.1017/9781009157896.001.

Intergovernmental Panel on Climate Change (IPCC). "Summary for Policymakers." In *Climate Change 2022: Impacts, Adaptation and Vulnerability: Working Group II Contribution to the Sixth Assessment Report of the Intergovernmental Panel on Climate Change,* 3–34. Cambridge: Cambridge University Press, 2023. https://doi.org/10.1017/9781009157896.001.

Intergovernmental Panel on Climate Change (IPCC). "Summary for Policymakers." In *Global Warming of 1.5°C: IPCC Special Report on Impacts of Global Warming of 1.5°C above Pre-industrial Levels in Context of Strengthening Response to Climate Change, Sustainable Development, and Efforts to Eradicate Poverty,* 1–24. Cambridge: Cambridge University Press, 2022. https://doi.org/10.1017/9781009157940.001.

International Energy Agency (IEA). "World Energy Investment 2023." May 2023. https://www.iea.org/reports/world-energy-investment-2023.

Jacobi, J. D., J. P. Price, L. B. Fortini, S. M. Gon III, and P. Berkowitz. "Baseline Land Cover." In *Baseline and Projected Future Carbon Storage and Carbon Fluxes in Ecosystems of Hawai'i,* edited by P. C. Selmants, C. P. Giardina, J. D. Jacobi, and Z. Zhu, 9–20. U.S. Geological Survey Professional Paper 1834. Reston, VA: U.S. Department of the Interior, 2017.

Jones, Dave. 2023. "European Electricity Review 2023." *Ember,* January 31, 2023. https://ember-climate.org/insights/research/european-electricity-review-2023/.

Kruk, M. C., Andrew Lorrey, Georgina M. Griffiths, Mark Lander, Ethan J. Gibney, Howard J. Diamond, and John J. Marra. "On the State of the Knowledge of Rainfall Extremes in the Western and Northern Pacific Basin." *International Journal of Climatology* 35, no. 3 (March 2014): 321–336. http://dx.doi.org/10.1002/joc.3990.

Lan, Xin, Sourish Basu, Stefan Schwietzke, Lori MP Bruhwiler, Edward J. Dlugokencky, Sylvia Englund Michel, Owen A. Sherwood, et al. "Improved Constraints on Global Methane Emissions and Sinks Using δ13C-CH4." *Global Biogeochemical Cycles* 35, no. 6 (2021): e2021GB007000. https://doi.org/10.1029/2021GB007000.

Lenton, Timothy M., Johan Rockström, Owen Gaffney, Stefan Rahmstorf, Katherine Richardson, Will Steffen, and Hans Joachim Schellnhuber. "Climate Tipping Points—Too Risky to Bet Against." *Nature* 575 (2019): 592–595. https://www.nature.com/articles/d41586-019-03595-0.

Lindsey, Rebeca, and Luann Dahlman. "Climate Change: Global Temperature." Climate.gov, January 18, 2024. https://www.climate.gov/news-features/understanding-climate/climate-change-global-temperature.

Marra, J. J., and M. C. Kruk. "State of Environmental Conditions in Hawai'i and the U.S. Affiliated Pacific Islands under a Changing Climate: 2017." https://coralreefwatch.noaa.gov/satellite/publications/state_of_the_environment_2017_hawaii-usapi_noaa-nesdis-ncei_oct2017.pdf.

McKay, D. A., A. Staal, J. F. Abrams, R. Winkelmann, B. Sakschewski, S. Loriani, I. Fetzer, et al. "Exceeding 1.5°C Global Warming Could Trigger Multiple Climate Tipping Points." *Science* 377, no. 6611 (2022). https://doi.org/10.1126/science.abn7950.

Miles, A. "If We Get Food Right, We Get Everything Right: Rethinking the Food System in Post-COVID-19 Hawai'i." Position Paper, 2020. http://hdl.handle.net/10790/5248.

Mooney, Chris, Juliet Eilperin, Desmond Butler, John Muyskens, A. Narayanswamy, and N. Ahmed. "Countries' Climate Pledges Built on Flawed Data, Post Investigation Finds." *Washington Post,* November 7, 2021.

Mora, C., Bénédicte Dousset, Iain R. Caldwell, Farrah E. Powell, Rollan C. Geronimo, Coral R. Bielecki, Chelsie W. W. Counsell, et al. "Global Risk of Deadly Heat." *Nature Climate Change* 7 (2017): 501–506. https://doi.org/10.1038/nclimate3322.

Murakami, Hiroyuki, Bin Wang, Tim Li, and Akio Kitoh. "Projected Increase in Tropical Cyclones Near Hawaii." *Nature Climate Change* 3, no. 8 (2013): 749–754.

National Oceanic and Atmospheric Administration (NOAA). "Carbon Dioxide Now More than 50% Higher than Pre-industrial Levels." News release, June 3, 2022. https://www.noaa.gov/news-release/carbon-dioxide-now-more-than-50-higher-than-pre-industrial-levels.

National Oceanic and Atmospheric Administration (NOAA). "Increase in Atmospheric Methane Set Another Record in 2021." News release, April 7, 2022. https://www.noaa.gov/news-release/increase-in-atmospheric-methane-set-another-record-during-2021.

Nolan, Connor, Jonathan T. Overpeck, Judy R. M. Allen, Patricia M. Anderson, Julio L. Betancourt, Heather A. Binney, Simon Brewer, et al. "Past and Future Global Transformation of Terrestrial Ecosystems under Climate Change." *Science* 361, no. 6405 (2018): 920–923. https://doi.org/10.1126/science.aan5360.

Orr, David W. *Earth in Mind: On Education, Environment, and the Human Prospect.* Washington, DC: Island Press, 2004.

Paxton, E. H., Richard J. Camp, P. Marcos Gorresen, Lisa H. Crampton, David L. Leonard Jr., and Eric A. VanderWerf. "Collapsing Avian Community on a Hawaiian Island." *Science Advances* 2, no. 9 (September 2016): e1600029. https://doi.org/10.1126%2Fsciadv.1600029.

Pokhrel, Yadu, Farshid Felfelani, Yusuke Satoh, Julien Boulange, Peter Burek, Anne Gädeke, Dieter Gerten, et al. "Global Terrestrial Water Storage and Drought Severity under Climate Change." *Nature Climate Change* 11, no. 3 (2021): 226–233. https://doi.org/10.1038/s41558-020-00972-w.

Rice, K., and Herman, J. "Acidification of Earth: An Assessment across Mechanisms and Scales." *Applied Geochemistry* 27, no. 1 (January 2012): 1–14. http://dx.doi.org/10.1016/j.apgeochem.2011.09.001.

Ripple, W. J., Christopher Wolf, Timothy M. Lenton, Jillian W. Gregg, Susan M. Natali, Philip B. Duffy, Johan Rockström, et al. "Many Risky Feedback Loops Amplify the Need for Climate Action." *One Earth* 6, no. 2 (February 2023): 86–91. https://doi.org/10.1016/j.oneear.2023.01.004.

Ripple, William J., Christopher Wolf, Thomas M. Newsome, Mauro Galetti, Mohammed Alamgir, Eileen Crist, Mahmoud I. Mahmoud, William F. Laurance, et al. "World Scientists' Warning to Humanity: A Second

Notice." *BioScience* 67, no, 12 (December 2017): 1026–1028. https://doi.org/10.1093/biosci/bix125.

Sproat, D. K. "An Indigenous People's Right to Environmental Self-Determination: Native Hawaiians and the Struggle Against Climate Change Devastation." *Stanford Environmental Law Journal* 35, no. 2 (2016).

Stevens, L. E., R. Frankson, K. E. Kunkel, P.-S. Chu, and W. Sweet. "Hawaiʻi State Climate Summary 2022." NOAA Technical Report NESDIS 150-HI. Silver Spring, MD: NOAA/NESDIS, 2022.

Still, Christopher J., Gerald Page, Bharat Rastogi, Daniel M. Griffith, Donald M. Aubrecht, Youngil Kim, Sean P. Burns, et al. "No Evidence of Canopy-Scale Leaf Thermoregulation to Cool Leaves below Air Temperature across a Range of Forest Ecosystems." *Proceedings of the National Academy of Sciences* 119, no. 38 (2022): e2205682119. https://doi.org/10.1073/pnas.2205682119.

Summers, Alisha, Charles H. Fletcher, Daniele Spirandelli, Kristian McDonald, Jin-Si Over, Tiffany Anderson, Matthew Barbee, and Bradley M. Romine. "Failure to Protect Beaches under Slowly Rising Sea Level." *Climatic Change* 151 (2018): 427–443. https://doi.org/10.1007/s10584-018-2327-7.

Tertzakian, Peter. *A Thousand Barrels a Second: The Coming Oil Break Point and the Challenges Facing an Energy Dependent World.* New York: McGraw Hill, 2006.

Thompson, P. R., Matthew J. Widlansky, Mark A. Merrifield, Janet M. Becker, and John J. Marra. "A Statistical Model for Frequency of Coastal Flooding in Honolulu, Hawaii, during the 21st Century." *Journal Geophysical Research-Oceans* 124, no. 4 (April 2019): 2787–2802. https://doi.org/10.1029/2018JC014741.

Tollefson, Jeff. "Scientists Raise Alarm over 'Dangerously Fast' Growth in Atmospheric Methane." *Nature,* February 8, 2022. https://doi.org/10.1038/d41586-022-00312-2.

United Nations Environment Programme. *Emissions Gap Report 2022: The Closing Window—Climate Crisis Calls for Rapid Transformation of Societies.* Nairobi, 2022. https://www.unep.org/emissions-gap-report-2022.

U.S. Department of Agriculture (USDA). "USDA Announces Streamlined Disaster Designation Process with Lower Emergency Loan Rates and Greater CRP Flexibility in Disaster Areas." Press release, July 11, 2012. https://www.usda.gov/media/press-releases/2012/07/11/usda-announces-streamlined-disaster-designation-process-lower.

Van Hooidonk, R., Jeffrey Allen Maynard, Derek Manzello, and Serge Planeset. "Opposite Latitudinal Gradients in Projected Ocean

Acidification and Bleaching Impacts on Coral Reefs." *Global Change Biology* 20, no. 1 (January 2014): 103–112. https://doi.org/10.1111/gcb.12394.

Wiens, J. J. "Climate-Related Local Extinctions Are Already Widespread among Plant and Animal Species." *PLOS Biology* 14, no. 12 (2016): e2001104. https://doi.org/10.1371/journal.pbio.2001104.

Williams, A. P., B. I. Cook, and E. E. Smerdon. "Rapid Intensification of the Emerging Southwestern North American Megadrought in 2020–2021." *Nature Climate Change* 12 (2022): 232–234. https://doi.org/10.1038/s41558-022-01290-z.

World Meteorological Organization. "State of the Global Climate." WMO-No. 1290, 2021. https://public.wmo.int/en/our-mandate/climate/wmo-statement-state-of-global-climate.

Xu, Chi, Timothy A. Kohler, Timothy M. Lenton, Jens-Christian Svenning, and Marten Scheffer. "Future of the Human Climate Niche." *Proceedings of the National Academy of Sciences* 117, no. 21 (2020): 11350–11355. https://doi.org/10.1073/pnas.1910114117.

Yeo, Sophie. "5 Ways Climate Change Harms Indigenous People." Climate Home News, July 28, 2014. http://www.climatechangenews.com/2014/07/28/five-ways-climate-change-harms-indigenous-people/.

Zeebe, Richard E., Andy Ridgwell, and James C. Zachos. "Anthropogenic Carbon Release Rate Unprecedented during the Past 66 Million Years." *Nature Geoscience* 9, no. 4 (2016): 325–329. https://doi.org/10.1038/ngeo2681.

CHAPTER 8

# Red Hill Underground Fuel Storage Facility and the Potential Impact to O'ahu's Water Resources

*O'ahu Board of Water Supply*

*This chapter was written prior to the events of November 2021, when there was a catastrophic spill at the Red Hill Bulk Storage Fuel Facility. Please read chapter 9 by Wayne Tanaka and Sharde Freitas on the aftermath of the spill.*

Ka Wai Ola—water is life.

Water is life—without it, we could not grow food, wash our bodies, keep ourselves from being thirsty. Water, along with the air we breathe, sustains us and gives life.

Ka Wai Ola has been the Board of Water Supply's (BWS's) motto for nearly twenty years, and is the driving force behind everything that we do. BWS is the steward of our water resources on the island of O'ahu. We are charged by our City Charter to manage our water system, a kuleana (responsibility) we take seriously. Our mission is to supply safe, dependable, and affordable water to our customers on the island of O'ahu. To accomplish this, we need to ensure that our groundwater resources, the source of the wai (water), are protected. This stewardship also extends to potential threats to our water resources.

We believe that Red Hill is a potential threat to these vital resources due to ongoing leaks from the fuel tanks situated at the site. BWS has taken a firm stand on the issues surrounding future potential leaks and possible resulting remediation of these leaks and the impacts to a major groundwater resource. Contamination of this water resource could negatively impact the water system for Honolulu and have a broader impact. This potential threat of contamination should concern all of us—as it could permanently destroy a critical water resource for our urban area.

### The Cultural Significance of Water

The value of water to our island cannot be separated from its cultural significance to the Hawaiian people.[1] Ka Wai Ola a Kāne—The Life-Giving Water of Kāne—is a term that reflects the special connection between the divine and all life forms. Hawaiians saw, and continue to see, gods everywhere in their world—in clouds, trees, stones, and all other parts of the sky, land, and sea.

In precontact times, fresh water was the key to life and prosperity. Early Hawaiians settled by perennial streams and springs where water was plentiful and reliable. The abundance allowed the Hawaiians to develop an extensive agricultural system and a sophisticated aquaculture structure. It is not a coincidence that there are so many words in the Hawaiian language for water in all its varied forms and uses.

Strict rules governed the use of water, and the Hawaiians developed a sophisticated land use system based on the ahupua'a (land division). The ali'i 'ai moku (district chiefs) were the trustees of water and exercised control over it as an instrument of the gods. They established and enforced regulations over water use in upland areas so that a pure flow was always available to those who lived at lower elevations.

This respect for water and ensuring its availability for everyone who lives on today as Native Hawaiians fight for the return of streamflows back to their natural courses. There have been several significant court cases on this issue, including the landmark Waiāhole Ditch Case, which established three public trust water uses: water in its natural state, water for traditional and customary practices, and domestic water use.

As the steward of this public trust for domestic water use, BWS continues with long-term planning and more efficient management of the water system. BWS wants to avail the people of O'ahu with ample drinking water, while providing safe, dependable, and affordable water overall. The BWS remains committed to maintaining policies and regulating and protecting watersheds and activities over the island's basal aquifers.

There are many threats to our precious water resources that we address daily in our work: climate change, lack of rainfall, overpumping.

While some we cannot control, there are others that require our attention and action in a timely manner. One of these threats is the underground fuel tanks at Red Hill, which we will cover in greater detail below. To better understand the concerns of potential contamination, a quick overview of our island's geology is in order.

## O'ahu Geology

The Hawaiian Islands are part of a long chain of volcanoes rising up from the seafloor. Eruptions over hundreds of thousands of years built an island that finally emerged above sea level. As the Pacific Plate moved to the northwest, new islands emerged above a stationary hot spot. O'ahu was created by two volcanoes, the remnants of which are the Wai'anae and Ko'olau mountains. Over time, the two volcanoes joined to form a single island that was further shaped by erosion and volcanic eruptions. Erosion covered the coastal reefs and built the high, fertile plateau of Central O'ahu. The volcanic landmass and high, rain-catching ridges gave the island the tools to create an efficient and dependable water source. The Native Hawaiians drew their water supplies from freshwater springs, lakes, streams, and wells. When the first artesian wells were developed in the late 1800s, residents of O'ahu increasingly came to rely on groundwater sources.

## Groundwater Sources

Groundwater sources are fed by rainfall and healthy watersheds that collect and recharge the underground aquifers;[2] they are dependent on three essential natural elements:

- ❑ Tradewinds from the northeast drive clouds, hydrated by evaporation from ocean waters, inland toward the Ko'olau Mountains.
- ❑ The Ko'olau Mountain range captures and forces the moisture-laden clouds to higher elevations, resulting in condensation and rainfall on the land below. The direction of the prevailing winds causes the windward (northeast) side of O'ahu to be generally wetter than the leeward (southwest) side, with most of the rainfall occurring over the Ko'olau range and Central O'ahu. Annual precipitation ranges from less than 25 inches in 'Ewa and leeward coastal areas to 240 inches along the northern end of the Ko'olau Mountains.

- ❑ As rainwater slowly percolates into the earth, the water is naturally filtered by volcanic soils and rocks and stored in groundwater aquifers. These aquifers are a natural freshwater reservoir in the permeable and porous volcanic rock from which the BWS eventually extracts groundwater to supply their customers. Figure 8.1 shows the aquifers on the island.

In the higher elevations, where rainfall is concentrated, groundwater is restrained by impermeable vertical rock structures called "dikes" formed by lava flows that intrude into existing, permeable rocks. In the valleys and middle elevations, groundwater exists as extensive reservoirs of fresh water that float on seawater under much of the southern and northern portions of the island. Where the fresh and salt waters merge, a brackish water mixing zone forms. Caprock formations, comprised of sediments and corals deposited along O'ahu's coastline, effectively prevent the fresh water from freely discharging to the ocean by acting as a

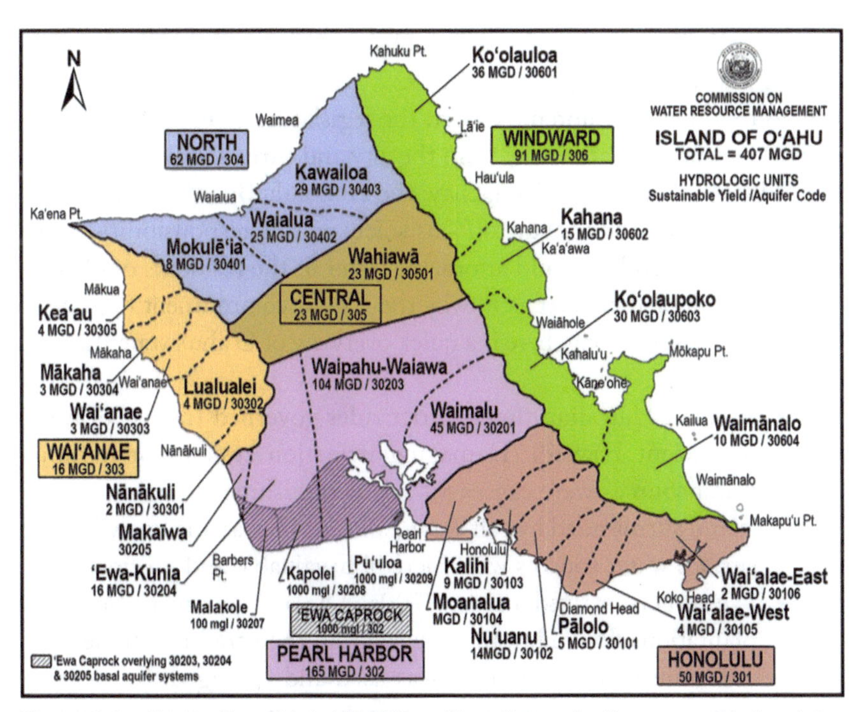


**Figure 8.1** O'ahu aquifers and their yields. MGD = million gallons per day. Map courtesy of the Commission on Water Resource Management, State of Hawai'i, August 2008.

dam. This in turn creates a deep pocket of fresh water floating on the salt water. The water under the caprock is generally brackish and supplies many small wells used for irrigation.

The average volume of rainfall falling on O'ahu's watersheds is approximately 1.4 billion gallons per day. Of this total, approximately one-third is lost through evapotranspiration (loss of water through evaporation and plant transpiration), one-third recharges the groundwater aquifers, and one-third becomes runoff and streamflow. As water moves from the mountains to the sea in streams, replenishment to groundwater aquifers can also occur from seepage through streambeds.

When factoring in climate change, the dry areas are expected to get drier, causing more frequent and severe droughts, and chloride levels in 'Ewa, Kunia, and Wai'anae sources are expected to increase. The windward areas could get wetter. This discrepancy in water availability will put a stress on the current system and will require BWS to develop alternative water sources to address this issue.

## BWS History

The Board of Water Supply was established to ensure a well-run water system that was above politics.[3] In 1929, after a battle that went all the way to the Hawai'i Supreme Court, the legislature took unilateral control of the island's water away from the city and turned it over to a newly created semiautonomous city agency: the Honolulu Board of Water Supply. This allowed the agency to focus solely on one responsibility—providing safe, dependable, and affordable water to the residents of O'ahu.

For additional context as to why the BWS's commitment to protecting the water is so strong, here is a quick overview of O'ahu's water history:

- Ancient Hawaiian times: Strict rules governed the use of water in ancient Hawai'i. Damaging irrigation systems or harming water sources were causes for severe punishment. Living on this island surrounded by salt water, ancient Hawaiians knew limited surface water sources were incredibly valuable and precious.
- Westerners arrived in the 1700s. Disease and recruitment by plantations drew Hawaiians away from their traditional lifestyles and transformed how land was used and managed. Cattle were brought in, and pineapple growers began to use millions of gallons of water.

- The 1800s brought the discovery of wells, but by the turn of the century, O'ahu suffered a water panic, with wells salting up and water levels dropping. The water system had grown too much, too fast, and too haphazardly; there was a complete lack of long-range planning. Fire protection was minimal and the threat of disease in the water was constant. The 1891 and 1894 droughts brought the formulation and installation of the first pumping station at Beretania and Alapai Streets.
- After Hawai'i became a Territory, a department under the Superintendent of Public Works of the Territory of Hawai'i managed the water system. In 1917, the department appointed a commission, which reported that the most urgent need was to protect the Nu'uanu water supply and recommended establishing a separate body to take charge of all the water works.
- By 1925, Hawai'i's legislature gave the governor power to appoint the Honolulu Sewer and Water Commission, which in turn reported that the city's system was antiquated and was not serving the needs of the city.
- In order to remove politics from water issues, the Territorial Legislature, by Act 96, created the Board of Water Supply (BWS) in 1929.
- Since 1925, the BWS has modernized the system, capped wasteful wells, installed consumer water meters, billed consumers at fixed rates, stabilized water tables, and made the water scare go away.
- Today, mountain watersheds are protected as forest reserves and forests are planted and replanted. The forest reserves you see today are a monument to an early understanding that the well-being of the entire island is tied to the health of areas with high rainfall.

### O'ahu's Aquifers

The sustainable yield from O'ahu's groundwater aquifers, as adopted by the Commission on Water Resource Management (CWRM), is 407 million gallons per day (mgd). Of this estimated sustainable yield, less than half was used in 2010.[4] However, demand is not always colocated with available supply, so supplies can be stressed in areas of high population density and high water use. Fortunately, water users on O'ahu have access to multiple sources of water to meet their needs: surface, groundwater, recycled, and brackish water, depending on use.

The BWS is the largest user of groundwater on the island, accounting for 64 percent (146 mgd) of the groundwater use in 2010. Our drinking water is drawn entirely from groundwater. Other major users include the military (23 mgd), agriculture (23 mgd), and landscape irrigation (17 mgd). Private domestic and industrial users account for 0.5 mgd and 20 mgd, respectively.[5]

Figure 8.1 summarizes the sustainable yield for each of the major freshwater aquifers as well as the sub-aquifers within each area:

- The colored boxes show the name of the aquifer, the daily permitted yield, and the aquifer ID number.
- Within each aquifer, the dotted lines delineate the sub-aquifers and their specific daily permitted yield.
- The map also shows the ʻEwa Caprock area in the hash lines.
- Approximately 294 mgd of the total sustainable yield of 407 mgd is permitted for use by the BWS and other users.

The CWRM monitors the sustainable yields and permitted usage closely, and all water users are to adhere to CWRM's set standards. If the permitted usage needs to increase, BWS or other applicants would need to petition CWRM on why the permitted usage needs to increase. BWS prefers to work within the allowable limits and preserve our water resources through a combination of water conservation, appropriate usage rates for the various customer rate classes, and the use of alternative resources such as non-potable and recycled water where appropriate. BWS believes this approach will provide greater results and better overall protection of our resources in the long term.

## Water Quality

The fresh groundwater aquifer that the BWS relies on for most of its water supply sits above a mass of denser, more saline water. For most of the island, the water that is pumped for our drinking supply is very pure, with small amounts of trace minerals. BWS is charged by the State of Hawaiʻi Department of Health (DOH) to ensure that all city, state, and federal regulations for maintaining the water quality for drinking purposes are followed. BWS also works closely with CWRM and provides data on the pumpage and salinity (i.e., chlorides). Updates on potential threats to our aquifers are also shared with CWRM as needed.

Water quality can be affected by a number of factors: salinity, chemical contaminants, biological contaminants. BWS adds small amounts of chlorine to prevent biological contaminants in the distribution system and to ensure the purity of the water. However, over time, we may see impacts such as our experience in Central O'ahu. In Central O'ahu, the water is also filtered through granular activated carbon (GAC) to remove leftover chemicals from the pineapple and sugar plantation eras.

The groundwater levels in the different aquifers used for our drinking water are monitored carefully using a series of groundwater index wells. If the levels fall into the critical area where they could be affected by salinity, a series of actions could be taken that include promotion of voluntary reductions in irrigation, mandatory restrictions in water use, or progressive actions to reduce water use through increased rates, reduced allocations, flow restrictors on meters, or civil actions. While agricultural practices have changed, the impacts of different chemicals, application methods, and best management practices (to the extent implemented) are not always known, and the same goes for the impacts of treatment and potential application of genetically modified organism crops.

Commercial activities can lead to the presence of solvents, degreasers, and refrigerants. Cesspools do exist, and while the BWS has mitigated effects near drinking water sources, monitoring continues. Leaking underground storage tanks, whether small-scale (e.g., gas stations) or large-scale (e.g., Red Hill), also pose a risk to water quality.

## Red Hill Bulk Storage Fuel Facility

The preceding information lays the framework for the concerns that BWS has regarding the Red Hill Bulk Storage Fuel Facility located near Pu'uloa or Pearl Harbor.

### *History of the Area*

An abundance of fresh surface water as well as artesian springs has made Ke Awalau o Pu'uloa into one of plenty for Hawaiians.[6] It was home to many types of fish and seafood, and the geography was favorable to the building of fishponds. In 1901, the area supported 24 loko lo'i yielding nearly 82,000 pounds of fish annually. In addition, taro, yams, and bananas flourished.

The mo'olelo of the area describe how Kāne and Kanaloa blessed a simple fisherman for his faithful devotion to the two gods by ensuring the water and land would provide for him and his family. The mo'olelo provides a sense of the resources and place-names that have long been forgotten. The area from Hālawa to Honouliuli is said to be associated with Kānekua'ana, the kia'i (protector) of the area. She was also associated with environmental stewardship and reciprocity, two key concepts when living on an island of limited natural resources.

In 1899, after the overthrow of the Hawaiian Kingdom, the U.S. Navy began to dredge the area now called Pearl Harbor and established a base. The name Red Hill, or Kapūkākī, refers to the 535-foot ridge in the area where fuel storage tanks are buried. While a brief look into how it got its name has not yielded any results, it could be inferred that the name was given due to the red color of the soil.

### *History of the Facility*

> As world tensions increased leading up to World War II, the decision was made to build a new facility that would store more fuel and in a location that would not be exposed to an enemy attack. Constructed from December 1940 through 1943, the facility is located under a volcanic mountain ridge near Honolulu, Hawaii, east of Pearl Harbor. Originally designed as a series of horizontal tunnels, the plans for the facility underwent a last-minute revision to construct 20 vertical tanks that would be installed inside cavities mined in the mountain ridge. This design change greatly reduced the amount of time it would take to construct the facility. Construction continued nonstop except for the time of the attack on Pearl Harbor on December 7, 1941.[7]

It would appear that there was no opposition to these facilities. Given the political climate and concerns about Japan during this period, it is unlikely that any protests would have been taken seriously or even considered, though a review of papers of the period could possibly turn up community concerns. In addition, these facilities were constructed under a veil of secrecy to ensure the safety of the fuel being stored there.[8]

The Red Hill fuel tanks are made of 1/4" and 1/2" thick steel encased in concrete. Each tank at the Red Hill Bulk Fuel Storage Facility is as tall as Aloha Tower and sits 100 feet above our groundwater—the same height as our State Capitol. Each tank can hold up to 12,500,000 gallons of fuel, and there are 20 of these tanks at the facility (see figure 8.2). More than 187,000,000 gallons of fuel are stored today over the drinking water quality groundwater aquifer.

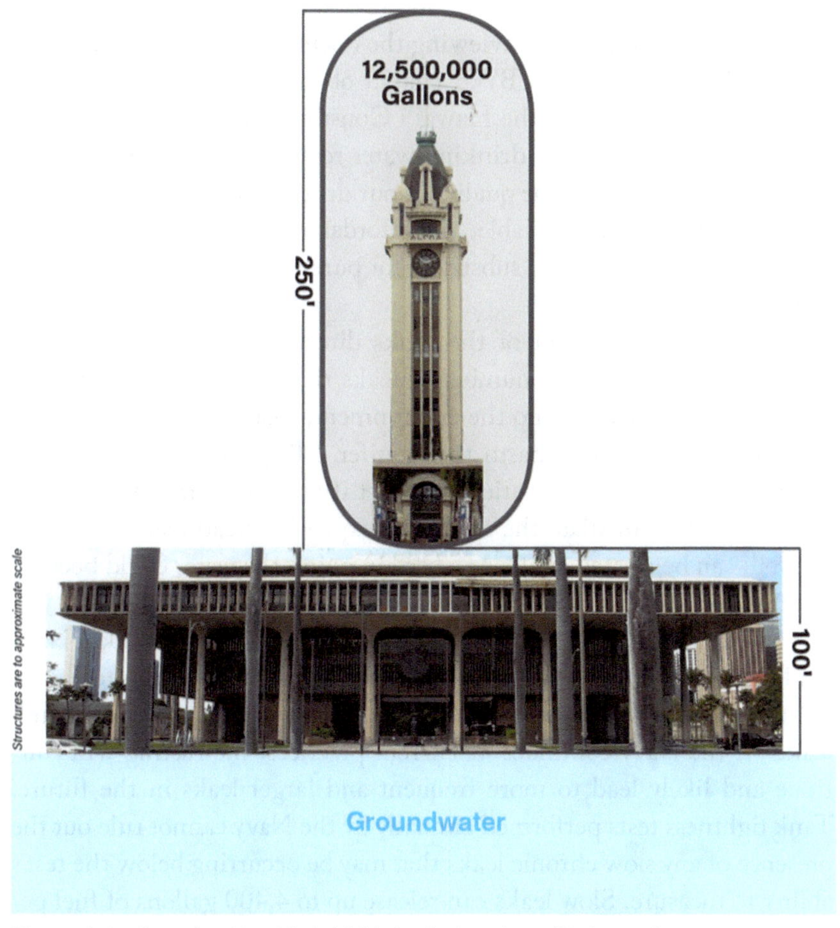


**Figure 8.2** Size and position of the Red Hill fuel tanks above the Pearl Harbor aquifer.
*Source:* Honolulu Board of Water Supply.

### *History of Leaks*

The tanks were completed by 1943, and leaks were detected starting in 1947. Between 1947 and the mid-1980s, an estimated 206,190 gallons of jet fuel may have leaked into the ground underneath the tanks.[9] The most recent spill occurred in January 2014, when 27,000 gallons of jet fuel leaked into the groundwater. This was from a tank that had just been repaired.

### *Why BWS Cares*

The question will arise, why does BWS concern itself with this issue? Why stay concerned about the issue since the 2014 leak? Why the staff expense of technical consultants and other subject matter experts when the EPA and DOH are all reviewing the documents? The answer is very simple—it is our kuleana! BWS has an obligation to protect groundwater in accordance with the Hawai'i Constitution, article XI, sections 1 and 7. Federal and state drinking water regulations have charged the BWS with maintaining the quality of our drinking water—it is our mission to ensure safe, dependable, and affordable drinking water now and into the future. There is no substitute for pure water. All life depends on water.

Given the placement of the tanks directly over a large drinking water quality aquifer, the number of leaks that have occurred, and the amount of fuel released into the environment, a catastrophic failure of a tank could permanently harm this aquifer. Mitigation measures could include mandatory conservation for most if not the entire island, along with trying to remediate the spill with expensive treatment—that is, if the spill can be treated at all. At a certain point, the water could become so saturated with fuel that no amount of treatment would work.

The Navy continues to do patchwork repairs on the tanks. Navy studies also show rusting is occurring on the back side of the steel plate of the tanks, which can lead to through-wall corrosion, producing a tank leak.[10] If the Navy continues its current practices, the rusting will continue and likely lead to more frequent and larger leaks in the future. Tank tightness tests performed annually by the Navy cannot rule out the presence of any slow chronic leaks that may be occurring below the test's ability to measure. Slow leaks can release up to 4,400 gallons of fuel per year per tank undetected.

Groundwater testing conducted by the Navy since 2005 continues to show petroleum contaminants present in the groundwater underneath Red Hill at levels that, in one case, exceeded the State of Hawai'i Department of Health environmental action limits. This contamination could have been caused by the 2014 reported leak, earlier leaks, or more recent unreported leaks.[11]

Because groundwater is always moving, fuel from the tanks that leak into the groundwater could eventually spread to neighboring wells. The amount of fuel contamination in the aquifer and how quickly it spreads will depend on the volume of fuel released into the groundwater. A large volume of fuel released due to pipe or tank failure will contaminate the groundwater much faster, and over a larger area compared to fuel that is slowly leaking from the tanks.

The existing situation at Red Hill poses a threat to the BWS wells that are presently not contaminated. If these wells were ever contaminated by a catastrophic, large-volume release from Red Hill, then BWS would be placed in a situation where the need for immediate funding for the installation of treatment systems to remove the contaminants from the water would be warranted. It is also possible that the release could be so large that the cost to treat it could become prohibitively expensive, thus rendering the wells and aquifer unusable for decades to come. A third scenario could arise where the release is so large that the water becomes untreatable.

The impact to Honolulu and 'Ewa would be felt immediately: Honolulu would lose a major water resource serving residents from Moanalua to Hawai'i Kai; 'Ewa's water would need to be rerouted through the system to assist the loss to Honolulu. While no one would be without water in the short term, this would negatively impact aquifer levels and increase the current sustainable yield beyond what has been set by the Water Commission in the long term.

For the community, the lack of access to fresh water would be felt immediately. Any widespread contamination would require a shutdown of the system as BWS worked on a solution to reroute water. However, any future water available would be far less than urban Honolulu is currently used to receiving. Mandatory conservation measures would be necessary to preserve the remaining water resources, and the impact would be felt by businesses as well as residents.

Whether the water resource is lost or contaminated but treatable, the financial impact would be considerable. This would affect water rates for all ratepayers as additional revenue would be needed to pay for alternatives to the groundwater loss. The impact would also be felt by businesses that cater to our largest industry, tourism. The threatened water source provides 25 percent of all the water for metro Honolulu, from Moanalua to Hawai'i Kai. Water rate increases will ultimately find their way into increased rates for rooms, restaurants, and other activities. That would be unavoidable as the cost for treatment and alternative water resources would need to be spread across all rate classes. And while the Navy has assured the community that they would pay the cost for any remediation efforts, any compensation to BWS for treatment facilities would need approval from Congress and the president. Remediation efforts would need to begin before any payment was received.

This is one possible scenario; due to the age and the ongoing corrosion of the steel plates, the structural integrity of the tanks is a major concern. A catastrophic fuel release could also occur as the result of an earthquake. This could result in millions of gallons of fuel being released into the groundwater and potentially several million gallons into Hālawa Stream and Pearl Harbor. Any release as described above could pollute the aquifer and our water supply for many years to come, and its impact would be felt for generations.

## Contaminants

Federal and state drinking water regulations require that all public water supply systems test for a variety of contaminants in drinking water. For many of the contaminants, the U.S. Environmental Protection Agency (EPA) has set enforceable standards for the maximum amount allowed in drinking water. These standards are based on possible health effects of consuming the water. The standards are known as maximum contaminant levels (MCLs), and are published as national primary drinking water regulations (NPDWR).

The MCLs are set at levels at which no significant health effects would occur if water was consumed for an entire lifetime. The MCLs are set by the EPA, and then adopted by the state. The state must adopt MCL standards at least as stringent as the EPA's, but may adopt more stringent standards. Although standards have been set for many contaminants, some contaminants that may occur in drinking water have not yet had NPDWR MCLs standards established. The EPA still requires systems to test for some of these unregulated contaminants, and report on their testing.

Even though MCLs have not been set for the unregulated compounds, many of them have been studied by the EPA for both their acute health effects and in some cases for their long-term effects, which may involve carcinogenic or noncarcinogenic effects. Petroleum hydrocarbons and various related chemicals such as total petroleum hydrocarbons as diesel (also called TPH-d), naphthalene, 1-methylnaphthalene, 2-methylnaphthalene, toluene, benzene, and lead have been found in the groundwater under the Red Hill tanks. In 2016, BWS contracted toxicology experts to determine the health significance of low-level petroleum chemicals in drinking water and their health effects. The study determined that levels below 100 parts per billion (ppb) (same as micrograms per liter) of total petroleum hydrocarbons as diesel (TPH-d) in water are safe and protective of public health. Consuming water containing TPH-d levels at or below 100 ppb is not expected to produce toxic effects. Research studies show that high levels of TPH-d can cause changes in red blood cell counts and noncancerous liver and kidney changes.[12]

Based on these results, BWS, as a subject matter expert, has provided input to the Administrative Order on Consent (AOC)[13] participants regarding the cleanup of the groundwater underneath and surrounding the Red Hill tanks, and the use of 100 ppb as the minimum cleanup level. This 100 ppb level is also a taste and odor threshold and environmental action level (EAL) established by DOH.

## Is the Water Safe to Drink?

All water served by the Board of Water Supply is tested to meet the requirements set by the Hawai'i State Department of Health (DOH) and the EPA in accordance with NPDWR. In addition to that, the BWS also conducts routine examinations of our waters for saltwater intrusion. This program manages O'ahu's natural drinking water resources by protecting them from overpumpage. The DOH's enforcement of the NPDWR testing enables both the DOH and BWS to cover a wide range of drinking water issues. BWS conducts regular testing of its wells in the Red Hill area. To date, no petroleum contaminants have been detected in BWS water sources, and the water is safe to drink.

## Impacts on the Water System

To prevent contamination to the water, there are two clear options:

1. remove the tanks, or
2. install secondary containment.

### *Remove the Tanks*

Removing and relocating the tanks from Red Hill would eliminate the possibility of continued contamination to the aquifer. A new location would necessitate new tanks with a structure to ensure against leaks. In Washington State, the Navy will be relocating the underground tanks at the Manchester Navy Fuel Depot to a new aboveground facility with a secondary containment structure to ensure any leaked jet fuel does not seep into the environment.[14] Double-walled or secondary containment is the industry standard for underground storage tanks and is mandatory across the U.S.

### *Secondary Containment*

The next reasonable measure would be to replace all the old tanks with new updated, secondary containment tanks. These are considered state-of-the-art and, as mentioned above, are now required by law for underground fuel storage. Any potential fuel leaks are trapped between the two walls of the tank and not let out into the environment. Secondary containment is by far the safest course (though not the least expensive) to ensure long-term safety of our water supply.

If a catastrophic leak were to occur, it is expected that BWS would have to immediately shut down Hālawa Shaft and the Moanalua Wells in order to try to keep the leak from moving into the water distribution system. The BWS would then look at instituting water conservation measures in metro Honolulu that may be extended into other areas of the island. The Navy's Red Hill Shaft would have to be shut down, creating a water shortage for Pearl Harbor. An excessively large fuel leak could render the groundwater aquifer unfit for drinking for decades, as treatment alternatives in such a scenario would be ineffective and costly. The best current treatment options are:

1. pump and treat, or
2. alternative water resources.

### *Pump and Treat*

This would require a significant investment in a treatment plant at the site of the BWS pumping facilities. The water would be treated as it is pumped from the wells. Treatment would be determined by the amount and type of contamination. However, this solution is only feasible if the contamination does not overwhelm the groundwater system to the point where treatment is not able to make the water potable.

### *Alternative Water Resources*

If too much fuel has leaked, then BWS may have to abandon the Hālawa Shaft and Moanalua Wells facilities and start looking for options to compensate for the loss of these wells and continue providing water to metro Honolulu. All of these options are expensive.

These alternative resources could include desalination, new wells in other parts of the island, and increasing the amount of available recycled water to replace existing uses and free up potable water to make up for the water sources contaminated by the fuel leak.

While not mandated by any federal or state regulation, the BWS, in an effort to be proactive and ready if the need arises, has begun the work to look at treatment and alternative water resource options and will be preparing cost feasibility studies and planning and design drawings. None of these options will be easily implementable or low cost, but the BWS believes it is critical to understand what challenges will be faced and look for solutions before the need is urgent.

## BWS's Role

The Board of Water Supply remains steadfast in our commitment to providing safe, dependable, and affordable water for O‘ahu's needs today and in the future. This includes taking a strong stance when protecting our groundwater aquifers.

The Hawai‘i State Constitution states in article XI, section 7: "The State has an obligation to protect, control and regulate the use of Hawaii's water resources for the benefit of its people." Protecting these valuable water resources from contamination honors the public trust as stated above and protects future generations. The BWS hopes that all appropriate government agencies at every level will work together to continue to protect this vital resource.

As the stewards of O'ahu's water supply, it is our duty to call out concerns where and when we see them; not doing so would be abdicating our responsibility. If the water in the aquifer were to become contaminated due to a major fuel leak from the facility, then the public will look to BWS to solve the problem, and because it is ultimately BWS's responsibility, we need to be proactive.

While we currently do not detect fuel in the water from our wells closest to the Navy fuel tanks under Red Hill, we will continue to test the water; no entity can guarantee that these seventy-four-year-old underground fuel tanks won't leak in the future or aren't leaking now at amounts that are currently undetectable. As stated previously, treatment would be very expensive, if it is even possible. If the water has some fuel but not enough to abandon the resource, we would look at installing costly treatment systems at the source to remove the fuel.

When the Navy states that the wells at Hālawa Shaft and Moanalua are not contaminated, they are downplaying the seriousness of the leaks from the fuel tanks at Red Hill and the potential for future leaks. It has been nearly three years since the AOC was signed, and the BWS is concerned that the longer it takes to implement the AOC, the longer it will take to implement a permanent solution that will ensure the safety of our water.

The BWS is not a party to the AOC, as we are not a regulator of underground storage tanks. We are committed to encouraging greater transparency on this important issue for our community. We remain as a subject matter expert and have strived to comment on all documents and letters produced by the Navy, EPA, and DOH.

In December 2017, the Navy released a tank upgrade alternative report that reviewed six tank upgrade options. The BWS supports the secondary containment option as the most effective in containing and detecting leaks from the tanks. The Navy needs to work toward ensuring their fuel tanks do not contaminate the aquifer. If the source becomes contaminated, installation of appropriate treatment should be at the Navy's expense.[15]

There are federal and state regulations that apply to all underground storage tanks. However, Red Hill, a field-constructed underground tank system, has been exempt from many of the requirements that must be met by smaller facilities. In 2015, the EPA revised the underground storage tank regulations. But, the revised rules still exempt field-constructed tanks like Red Hill from the regulatory requirements that must be met by all other underground storage tanks.

The BWS will continue to encourage the Navy to retrofit the fuel facility with double-walled tanks and to remediate past fuel leaks. BWS will continue to ask the community to stay involved and ask questions of the Navy, EPA, and DOH about the ongoing tank upgrades, their viability, and the long-term plans to ensure our water resources stay clean. We understand that this chapter presents only a snapshot in time of this important situation; however, the concerns about protecting this vital resource do not diminish or disappear at any time.

We cannot do it alone and will need every voice to be heard. Ka Wai Ola.

## Notes

1. Board of Water Supply, "Water For Life," April 2018, https://www.boardofwatersupply.com/bws/media/files/publication-water-for-life.pdf.
2. Board of Water Supply, "2016 Water Master Plan," section 8.2.1, p. 8–3, October 2016.
3. Board of Water Supply, "Water For Life," April 2018.
4. Board of Water Supply, "2016 Water Master Plan," section 8.2.2, p. 8–6, October 2016.
5. Ibid., section 8.3.1, p. 8–8, October 2016.
6. Kyle Kajihiro, "Becoming 'Pearl Harbor': A 'Lost Geography' of American Empire" (Master's thesis, University of Hawai'i at Mānoa, 2014).
7. William Cole, "Red Hill Hides WWII Engineering Wonder," *Honolulu Advertiser*, June 13, 2007, http://the.honoluluadvertiser.com/article/2007/Jun/13/ln/FP706130406.html.
8. Environmental Protection Agency, "What Is the Red Hill Bulk Fuel Storage Facility?," www.epa.gov/red-hill/what-red-hill-bulk-fuel-storage-facility, accessed January 19, 2017.
9. "Red Hill Bulk Fuel Storage Facility Final Groundwater Protection Plan, Pearl Harbor, Hawaii," prepared for Department of the Navy, Commander, Naval Facilities Engineering Command, Pacific, January 2008.
10. Board of Water Supply, Presentation to Board of Water Supply Board of Directors, September 2018, http://www.boardofwatersupply.com/getattachment/a5378073–4ebb-4823-a78a-e05e42381e60/board-meeting-minutes-2018–09–24.pdf.aspx.

11. Department of the Navy, "Work Plan/Scope of Work, Investigation and Remediation of Releases and Groundwater Protection and Evaluation, Revision 02, Red Hill Bulk Fuel Storage Facility, Joint Base Pearl Harbor-Hickam, Oahu, Hawaii, Administrative Order on Consent," January 4, 2017, https://www.epa.gov/sites/production/files/2017–01/documents/revised_section_6–7_scope_of_work_4_january_2017.pdf.
12. Board of Water Supply, Letter to Virginia Pressler, Director, State of Hawai'i Department of Health, December 13, 2016.
13. The AOC is an enforceable agreement between the U.S. Navy, the Defense Logistics Agency (DFA), the Environmental Protection Agency (EPA), and the State of Hawai'i Department of Health (DOH) on specific actions to be taken by the Navy and DLA to address fuel releases and implement infrastructure improvements necessary to protect the health of the community and the environment. All actions are subject to EPA and DOH approvals.
14. Julianne Stanford, "Navy Seeks to Upgrade Manchester Fuel Depot Storage Tanks," Kitsapsun, June 19, 2018, https://www.kitsapsun.com/story/news/local/2018/06/19/navy-proposes-upgrading-manchester-fuel-depots-storage-tanks/710706002/.
15. "Tank Grade Alternatives at Red Hill," Environmental Protection Agency, December 2017, https://www.epa.gov/red-hill/tank-upgrade-alternatives-red-hill.

## Bibliography

"2015 Red Hill AOC: Tank Grade Alternatives at Red Hill." Environmental Protection Agency, December 2017. https://www.epa.gov/red-hill/tank-upgrade-alternatives-red-hill.

Board of Water Supply. "2016 Water Master Plan." April 2018. https://www.boardofwatersupply.com/bws/media/files/water-master-plan-final-2016–10.pdf.

Board of Water Supply. Letter to Virginia Pressler, Director, State of Hawai'i Department of Health, December 13, 2016.

Board of Water Supply. Presentation to Board of Water Supply Board of Directors, September 2018.

Board of Water Supply. "Water for Life." April 2018. https://www.boardofwatersupply.com/bws/media/files/publication-water-for-life.pdf.

Cole, William. "Red Hill Hides WWII Engineering Wonder." *Honolulu Advertiser*, June 13, 2007. http://the.honoluluadvertiser.com/article/2007/Jun/13/ln/FP706130406.html.

Department of the Navy. "Work Plan/Scope of Work, Investigation and Remediation of Releases and Groundwater Protection and Evaluation,

Revision 02, Red Hill Bulk Fuel Storage Facility, Joint Base Pearl Harbor-Hickam, Oahu, Hawaii, Administrative Order on Consent," January 4, 2017. https://www.epa.gov/sites/production/files/2017–01/documents/revised_section_6–7_scope_of_work_4_january_2017.pdf.

Kajihiro, Kyle. "Becoming 'Pearl Harbor': A 'Lost Geography' of American Empire." Master's thesis, University of Hawai'i at Manoa, 2014.

"Red Hill Bulk Fuel Storage Facility Final Groundwater Protection Plan, Pearl Harbor, Hawai'i." Prepared for the Department of the Navy, Commander, Naval Facilities Engineering Command, Pacific. January 2008. https://health.hawaii.gov/ust/files/2014/08/2008-Final-Groundwater-Protection-Plan.pdf.

Stanford, Julianne. "Navy Seeks to Upgrade Manchester Fuel Depot Storage Tanks." Kitsapsun, June 19, 2018. https://www.kitsapsun.com/story/news/local/2018/06/19/navy-proposes-upgrading-manchester-fuel-depots-storage-tanks/710706002/.

U.S. Environmental Protection Facility. "What Is the Red Hill Bulk Fuel Storage Facility?" Environmental Protection Agency. www.epa.gov/red-hill/what-red-hill-bulk-fuel-storage-facility. Accessed January 19, 2017.

CHAPTER 9

# Uē ka Lani, Pouli ka Hōnua

## The Community Fight for Water and Life

*Wayne Tanaka and Sharde Mersberg Freitas*

### Uē

*Authors' note: This chapter was written after the 2021 Red Hill water crisis and paints a picture of the community response to this unprecedented and ongoing (as of this writing) nightmare that has upended the lives of tens of thousands of people; contaminated the southern O'ahu EPA Region IX Sole Source Aquifer, a once incredibly pure and ancestral source of water, and life, for the island; and left the million residents who call O'ahu home in a state of existential limbo.*

*With unknown amounts of forever chemicals and over one hundred million gallons of fuel still perched just a hundred feet above the island's principal drinking water supply, the U.S. Navy's Red Hill Bulk Fuel Storage Facility has left Hawai'i's current generation at risk of witnessing an unprecedented disaster—one that can only be described as "the death of an island."*

*The following interviews seek to highlight a cross section of community members and leaders whose actions have impacted and continue to shape the ongoing fight to protect O'ahu's water—one that has transcended social, cultural, and geographic boundaries, and that has tied the island's fate and our future with that of communities across Oceania, and beyond.*

**November 20, 2021.** Seven years after the January 2014 fuel leak incident, the U.S. Navy faced some formidable challenges in its pursuit of a Department of Health operating permit for the Red Hill Bulk Fuel Storage Facility: two additional leaks had occurred amid legal proceedings for its permit application;[1] whistleblowers had asserted that Navy officials had withheld critical information about the Red Hill Facility during those proceedings;[2] the Department of Health had issued a $325,000 notice of violation to the Navy for safety violations during a "routine inspection" of the facility;[3] the Navy had failed to complete numerous actions promised under a 2015 administrative order on consent after 27,000 gallons of jet fuel were released from one of the facility's twenty 25-story-tall underground storage tanks (including the completion of a full risk assessment[4] and the production of working groundwater and contaminant fate and transport models);[5] and the Navy's own reports had established that it had not inspected eight of its massive underground storage tanks for structural weaknesses for decades[6]—and that its tank inspection and repair process could not accurately assess the structural integrity of the tanks that had been inspected.[7]

Alarmingly, the portion of the risk assessment that the Navy did complete had also concluded that up to 5,000 gallons of fuel was "expected" to be chronically released into the environment per year, and that there was an 80 percent chance of a "sudden" release of up to 30,000 gallons in any five-year period.[8] These statistics did not account for the risk of an earthquake, fire, or human error or malice.

Nonetheless, the Navy had maintained that the Red Hill facility's safety systems "meet and exceed industry and regulatory standards,"[9] and despite finding the facility to be "inherently dangerous," a Department of Health hearings officer had just recommended the issuance of the underground storage permit, subject to inspection and repair requirements.[10]

Moreover, Hawai'i's congressional delegation and all but a handful of state elected officials had little to nothing to say about the pending permit, or about the continued storage of over 100 million gallons of jet fuel directly above O'ahu's primary drinking water source—and within a mile of municipal and Navy wells serving half a million island residents.

Then the nightmare began.

## The Nightmare Begins

*Jamie Williams*

*Jamie Williams moved into the Army's Red Hill Mauka housing complex along with her Coast Guard spouse in August 2020, shortly after the COVID-19 pandemic hit Hawai'i's shores. After fifteen years working in the medical research field, she decided to enroll at the William S. Richardson School of Law, beginning her first semester in the fall of 2021.*

*A first-time military housing resident, Jamie quickly came to appreciate the unique nature of military communities, where families are often stationed far away from any of their former social or support networks, and where civilian spouses must learn to make do as their significant others are deployed for months at a time. "There's this almost instant understanding that at some point, you're going to have to ask a stranger for a favor," she explains. "You join the community Facebook page, and it's like—I could have been like, 'I need an onion!' And somebody would have had it on my front doorstep in ten minutes—and vice versa."*

*As the Red Hill crisis unfolded, Jamie became a community connector, an advocate, a source for credible information and (thanks to her medical background) reliable analysis, and would later serve as an administrator on the Facebook group JBPHH Water Contamination Support.[11] The following account follows her journey from an affected resident to a community advocate and water protector in the Shut Down Red Hill campaign.*

I knew something was wrong starting in July [of 2021]. I didn't put pieces together until November, but July was when I started having health issues. [My husband] was away. I'm having cognitive problems, blacking out midday—not drinking, not using drugs. I'll come downstairs and there are cabinets open in the kitchen and a snack half made and I'm like, okay, that's weird. I actually have text messages between myself and a friend, a neighbor. I'm saying, "I'm terrified." I don't know what's going on. I'm forgetting things. This isn't right. And then I had a menstrual cycle that lasted from August through November and my doctor was like, "Well, now, you started law school so. . . ." Well guess what resolved itself as soon as I stopped drinking the water?

I didn't realize so many of us had these issues, but how often do you talk to your neighbors about their periods? In hindsight, speaking with those neighbors, almost every single woman on my street experienced menstrual problems in that time frame, which is mind-blowing.

*Authors' note: In May of 2021, a series of operator and safety system failures led to a pipe explosion at the Red Hill Facility, releasing what was then estimated to be approximately 1,000 gallons of fuel. Navy officials initially asserted that "there appears to be no release into the environment,"*[12] *despite soil vapor monitoring data indicating spikes in volatile organic compounds from ten to over a thousand times pre-spill levels.*[13] *A Navy investigation released in October 2021 later concluded that only 38 gallons of fuel out of an estimated 1,618 gallons had not been recovered.*[14] *After the November 2021 crisis began, the Navy then admitted that approximately 20,000 gallons of fuel was likely released in the May incident,*[15] *and groundwater data from the Navy water system's Red Hill shaft showed that the well itself had been contaminated with petroleum at levels exceeding the Department of Health's environmental action levels beginning as early as July 2021.*[16]

*In November, 19,000 gallons of fuel that had gone "missing" in May were suddenly released from a fire suppression drain line after a trolley cart collided with a low-hanging valve. The fuel cascaded onto the facility's floor for over twenty-four hours, sending one worker to the hospital and eventually ending up in the Navy's Red Hill drinking water shaft—and into the water system for 93,000 residents.*[17]

In November—I want to say [it] was the 20th or 21st of November, [my husband] was in the garage and I walked outside to go ask him something. I was like, "Oh, wow, it really smells like gas out here. . . . did you spill motor oil while you were working on the motorcycle or something?" He said, "No, that's really weird." And then we saw three ladder trucks, fire trucks go by. We met working in emergency medicine and we're like, wow, that's a really big call—you don't send three ladder trucks for burnt dinner. We kind of walked to the end of the road and didn't see any evidence of a house fire, so we headed back home.

I think we saw in total probably five or six fire trucks that evening. We were like, something's really not right. We did not connect the fuel smell with the fire trucks until after Thanksgiving, when knowledge that fuel had spilled became public.

The night of November 27, I started seeing some posts on Facebook from neighbors asking, "Does anyone else's water smell?" I got up out of bed and went to our primary bathroom. I was sniffing the water and [my husband] was just like, "What are you doing?" I said, "I don't know. People say the water smells like fuel," and I immediately connected where that would be coming from, because I knew about the Red Hill fuel facility. But I didn't smell anything.

The next morning I was about to make a pot of coffee. Right before I added the water, I thought to fill a pint glass with water to see if there was an odor. I sniffed it and my heart sank—it smelled like tires to me. I set it on the counter. I took [the dog's] water dish away. I texted [my husband]: "I need you to pick up cases of water. This is going to be a big deal. And if what I think happened happened, we're in trouble here."

*Authors' note: As word of fuel-like fumes and health symptoms associated with Navy tap water spread, bottled water quickly sold out in area stores. Navy officials denied any evidence of water contamination,[18] leading groups such as the Armed Forces Housing Advocates, the Sierra Club of Hawai'i, and the O'ahu Water Protectors to begin purchasing and distributing water to neighborhoods on the Navy's water lines. Menehune Water Co., a local water bottling company, donated a truckload of five-gallon canisters to the community water distribution effort. It would be days, and in some cases weeks, before affected residents would be provided water from military sources.*

On November 29, I contacted the EPA and Department of Health. I reported the odor to both agencies and got an email from the Department of Health's Safe Drinking Water Branch saying that until the "unknown is identified by testing," we should find an alternative source of water for all uses. My first thought was: people need to know as fast as possible. I got on the Facebook neighborhood group and started trying to disseminate that information to as many people as I could.

On December 2, I heard that there was going to be an emergency meeting with Army officials over at the AMR [Aliamanu Military Reserve] Community Center. I arrived and there were several military families gathered and a very chaotic scene. Busses were pulling up with soldiers running off, heavy equipment was being delivered, and temporary structures were being erected. There was a definite sense of emergency permeating the entire area, which made things feel really apocalyptic.

We ended up surprising a military official as he exited the auditorium and he was like, "This meeting was supposed to be at three. How did you people know about this?" I told him, "Listen, do not underestimate military spouses."

We were told we would have the opportunity to be moved to hotels. Despite being a Coast Guard family, because we lived in Army housing, we would be able to go, which was an immense relief. At this point it had been four days since we had access to a shower or laundry.

I tried my best to provide as many updates to the community as possible: passing information on meetings and town halls, official announcements, medical information, that sort of thing. Early on I set up a primitive public health survey for residents living on Navy water lines to see if we could quickly narrow down the most prominent health concerns. The survey was very basic and didn't collect identifying info, but it was easy to see anecdotally that even early on, people were really sick and struggling.

I think something that is so unique and proved to be an incredible advantage in this situation is that military families are often living in areas where you don't have a family support system. There is a collaborative connection among the military community. With common shared experiences, you don't always need to go through the standard pleasantries of getting to know your neighbor. It doesn't matter where someone is from or what their value system is, if they need help, you help. When we as a community became aware the water was contaminated, it was immediately like, "Do folks have enough water?" "Hey, I know someone who just had a baby six weeks ago. They're at home and their husband is gone, do they have enough water for bottles?"

From there, several residents started to advocate for community needs, telling military officials things like "Hey, what you are doing is really great, but we need more. We need your soldiers to go door to door and get water to people because a nursing mom doesn't have time to go fill jugs at a tanker truck." The few of us that felt strongly or had time kind of inserted ourselves in this way, to advocate for those who were not able.

I remember clearly having a moment very early on where I knew it was going to take getting loud to get things done, because that's just how the military is built. We were being treated like children: "Oh, don't worry your pretty little head about it—sit down, shut up, drink the water, you'll be fine." In my experience dealing with the military, unless you publicly embarrass officials, they don't do the right thing, unfortunately.

It was a four-second calculation for me to recognize my immense personal privilege. [My husband] has been in the military twenty years. I had a successful career before going to law school. God forbid the worst happens and he gets kicked out of the military due to my actions, at least we can eat. There are young military members that have families they need to support, who do not have that luxury. I think there was a small contingent of us that came together, saying that hell or high water, we were going to continue speaking out until the military did the right thing. It wasn't even a choice.

For example, a lot of people were really concerned about reimbursements [for alternative lodging expenses]. The Coast Guard initially authorized alternate lodging stipends once the Army stopped paying, but informed members that it wasn't clear when or if funding would be appropriated to cover the cost. You know, that's fine, for me. I have a savings account. But you can't tell that to a junior enlisted member who technically qualifies for food stamps and maybe doesn't even have a $500 credit limit. At that point I began to try and generate some public pressure on the commandant [highest ranking member] of the Coast Guard via Twitter. Eventually the funds came through, but by the time we were paid back, we were owed over $20,000.

Communication wasn't the best organized at the beginning and there was a little bit of division because every branch does things differently. I actually had a lot of people tell me that it was inappropriate for me to even question the Army or Coast Guard about anything that was going on because they had provided me with a hotel.

In mid-December, day-to-day life started to stabilize. As a community, things certainly were not ideal, but we had a place to stay and an alternate water source. I started to think—this event is so much bigger than affected residents on the Navy's water line. I wanted to connect with the local community, which at that point had spent years mounting an impressively organized response to the threat lurking at Red Hill. I reached out to [my professor], and I was like, I don't know necessarily what I need to be doing. But here's what I feel like I need to be moving towards, and he put me in touch with [the Sierra Club of Hawai'i].

Then in January, Mikey Inouye got wind of the video I had of the army and EPA officials speaking in my front yard, and I got involved with [Oʻahu Water Protectors] shortly after that, because they released the video as well.

*Authors' note: After acknowledging the fuel contamination crisis, the Navy and Army began sending teams to assess and "flush" affected houses' water systems, sometimes repeatedly. One team, unusual in its inclusion of an Army colonel and EPA staff, visited Jamie's home, where a home security camera automatically recorded them in animated conversation on her front porch. An EPA official gestured across the yard, warning that "whatever is in here . . . is probably not fuel. It's likely worse. . . . This is bad. This is like, high hazard. . . ." Fearing for her community, including the neighborhood children, Jamie shared the video on social media, as well as with the Oʻahu Water Protectors.*[19] *The video went viral, and the Army's later explanation—that they were simply reacting to the presence of unknown irrigation lines—was far from convincing.*

I think I made it maybe a month into my second semester of law school before it was like okay, there's no way I'm going to be able to eke out another semester with a decent GPA and still continue organizing and passing information and basically doing surveillance on the military, in my own neighborhood, as crazy as that sounds. So I made the choice to take a leave of absence. . . . It just felt more important to be focusing on Red Hill and those issues than to be getting a law degree at that point.

I still get a lot of flak for being outspoken, you know, people saying "You're a traitor!" But once I got involved with the Oʻahu Water Protectors and some folks that have been doing this a long time—you start to see the obvious issues. I mean, I knew about the military's pollution of Vieques, I knew about Guam, I knew about Okinawa—but certainly not in the intimate way I do now. When you start to comprehend the intersectionality of environmentalism, race, and militarism intertwined with their collective violence, you can't unlearn what you've learned. There's not a way to unsee that in my mind.

*Authors' note: To this day, Jamie continues to advocate for the protection of Oʻahu's water, for help and relief for those directly harmed by the poisoning of the Navy's water lines, and for military accountability to the communities and environment it has harmed.*

I've been very happy to have the opportunity to meet a few military-affiliated individuals who really get the bigger picture regarding Red Hill and understand the layer cake of cultural and social issues, generational traumas, all rolled into one.

There are still a few affected families working together even though we have relocated away from O'ahu. I'm not as active as I was, but I'm still a member of Shut Down Red Hill Mutual Aid. I'm still keeping an eye on things and definitely talk to people involved in the movement regularly. I think we're all struggling acutely with a huge amount of PTSD because we were so deeply involved, going to every meeting, watching every hearing, all while trying to move forward with our lives, plan moves, survive deployments.

## They Knew All Along

*Marti Townsend*

*Marti Townsend was raised in Central O'ahu, graduated from Moanalua High School, and eventually earned her juris doctorate from the William S. Richardson School of Law. She now lives in Kalihi Valley with her family. As a law student, she worked as a volunteer to stop the further expansion of the military's presence in the islands, fighting against the U.S. Army's plans to station its newly formed Stryker Brigade at Schofield Barracks. Seeking to combine her professional career with her lifelong passion for "sticking it to the man," Marti joined KAHEA: The Hawaiian Environmental Alliance, then later became the executive director of the Outdoor Circle, before taking the helm of the Sierra Club of Hawai'i in 2015. Under her leadership, the Sierra Club of Hawai'i would soon be recognized as the primary organization advocating around the dangers of the U.S. Navy's Red Hill Bulk Fuel Storage Facility.*

By the time I had joined the Sierra Club, the 2014 leak was old news in many ways. The fuel had leaked, pollutants were detected in the aquifer, but nobody got arrested, and nobody seemed sick, so life just went on with 100 million gallons of jet fuel precariously perched over our water supply.

I had been in enough bouts over military expansion in Hawai'i to know that you just can't win against the U.S. military. It's the world's largest, richest war machine, and there is no telling them no. But I also thought, logically it makes strategic sense to protect the primary drinking water source for soldiers stationed in the middle of a very salty sea. I thought the defense contractors would probably make as much if not more money from building a new state-of-the-art fuel storage system than from tinkering around on these janky old ones. Plus this was the water my mother drinks, my kids drink. I really wanted it shut down. So in a perverse kind of way, I saw the retiring of the tanks at Red Hill as a kind of win-win situation for all of us—even the military. Yeah, I was so wrong.

The first "action" I worked on in the campaign to shut down Red Hill was a letter signed by more than two-thirds of the state senate to the newly elected Governor David Ige, asking him to not sign the [2015 administrative order on consent, or AOC] because it gave the military over twenty years to "fix up" the tanks, took authority away from local officials to protect Oʻahu's water supply, and gave final say on the future of this "inherently dangerous" facility to the EPA.

Water is fundamentally a local asset that can really only be managed well by those who are actively relying on it. DC politicians and bureaucrats are not here. The EPA is not here. So they should not have the final say in the fate of our water supply. Plus, we know from past experience that the EPA is impotent when it comes to holding the military accountable. . . .

At the end of the day, the military has much more influence over federal agencies and federal decision-makers than local ones. So, we knew it was a bad idea to give them the final say on anything regarding our water. Governor Ige ignored all of us, signed the AOC, and thus began our seven-year rollercoaster ride to the worst-case scenario becoming real.

*Authors' note: Many believe that the 2015 AOC contributed to and exacerbated the current Oʻahu water crisis. Due to the AOC's general lack of deadlines or enforcement provisions, many of the Navy's commitments, as of February 2023, still remain unfulfilled. For example, the Navy has failed to produce: a complete risk assessment for the Red Hill Facility; a tank upgrade alternatives report with factually supported conclusions; a working groundwater flow model; and a contaminant fate and transport model. In addition, a required analysis demonstrated that the Navy's nondestructive tank inspection technique was inaccurate 40 percent of*

> *the time, and compiled records revealed that eight of the fourteen tanks in active operation had not been inspected in decades.*[20] *Nonetheless, the Navy was allowed to continue operating the Red Hill Facility, up until the 2021 water contamination crisis.*
>
> *Limited transparency mechanisms and mandatory nondisclosure agreements signed by the parties also allowed the Navy to use the 2015 AOC to downplay community concerns and thereby defer protective actions: as late as February 2021, the Navy released a press statement claiming that "under the Administrative Order on Consent, the Navy, Defense Logistics Agency, [Department of Health] and EPA work closely 'to protect drinking water, natural resources, human health, and the environment.'"*[21]

The AOC was crap because it gave the military considerable room to disorient, delay, and distract from the goal of shutting down the tanks. . . . The military would publish very long, impressive-looking studies required by the AOC, but the documents would be late or incomplete or not what was required. I can't say whether the military was just incompetent or willfully negligent. On some level I guess it doesn't matter. They weren't getting the job done.

Of course you didn't know this until you read it all, which for a slow reader like me was a bigger challenge than I thought I could handle. And there was all kinds of math involved. It was a double whammy for me, absolutely awful. Luckily there was a hui of nerds helping us make heads or tails of these reports, and figure out how to respond.

Most of their reports were followed by comment letters from the Sierra Club, BWS [Board of Water Supply], and many others explaining to regulators—which meant the EPA, with the DOH [Department of Health] in their pocket—why each report was bulls*%!. Sometimes it seemed pointless, but sometimes there would be a news story about the military's piss-poor work product, or even better, a few times the regulators rejected a report outright and required the military to do it again, *immediately*.

Woohoo! We used to celebrate those victories. It took us a few times to finally learn that "immediately" actually had no actual deadline, and the regulators had no intention of forcing the military to do anything it did not want to do.

As disappointing as the AOC was, I did have hope that the 2019 risk assessment it required the military to publish would change people's minds about keeping the Red Hill tanks. This report by an independent third party was condemning. It found that there was an 80 percent chance of an acute leak over the next five years—acute leaks are between 1,000 and 30,000 gallons. The study also revealed chronic leaks at the facility that could release upwards of 5,800 gallons of fuel into the environment annually. These estimates were based on the assumption that the Navy was keeping up with inspections and repairs (which it most certainly was not), and did not account for unexpected natural disasters, like earthquakes. Knowing all this, I thought they would surely listen to us now!

The Sierra Club undertook a major public education campaign to ensure that every government official knew that these rusty old fuel tanks were too dangerous to leave perched over our water supply for any length of time. We made presentations and went to every neighborhood board on O'ahu that would have us, sometimes with [the BWS], to educate people on the risk assessment and the permitting process that was underway. Many of these neighborhood boards passed resolutions renewing their call for the tanks to be retired. . . . Even at the height of COVID restrictions, we met with federal, state, and county officials via Zoom, we launched a new online petition, started socially distanced sign waving, and had several stories written in the local news.

Unfortunately, that was not enough to convince our elected and appointed leaders. With the exception of Congressman Kaiali'i Kahele, all of Hawai'i's congresspeople said the risk assessment was insufficient to compel closing the facility sooner than the military's 2045 date. Sadly, the story was much the same for most of our local lawmakers. Except for a few key champions—with little influence inside the Capitol power structure—our state lawmakers just did not have the juice to hold the military accountable.

The bright exception was the Honolulu City Council, which had consistently and unanimously called for the closure of Red Hill since 2014 and backed BWS when the powers that be tried to silence them.

As much as we celebrated the wins at the City Council, there was no denying that the lack of significant political movement by those in power after reading this report was heartbreaking. How could our own elected leaders knowingly expose us all to so much risk? I mean, I knew that the military doesn't care about us. But for our own people in elected office to sacrifice us like this destroyed my faith in representational democracy. Who were they representing? Whose interests were they protecting?

One common thing I have noticed between 2014 and 2021 is the litter of informational briefings, task forces, and special committees that the government produces after a leak. Some of them are a mix of state and federal officials, others a mix of government officials and community members. None of them have any authority to make the military do anything actually useful.

The most useful thing to come from these meetings in my opinion was exposing the collusion between the regulators and the military. For example, the Health Department had a decades-long practice of exempting the Red Hill tanks from the basic permitting and oversight that every corner gas station must follow. They had no authority for this exemption but would not change this practice until the Sierra Club sued them in state court and won.

At one point, the then-director for the environmental division of the state Health Department lectured us about how it was impossible to protect the environment without first ensuring U.S. national security. And yes, he admitted, that may mean that some places are sacrificed, but it is for the collective good: "You wouldn't want some foreign government destroying what you are trying to protect, would you?" I was dumbfounded. I mean, that is literally what is happening right now. The U.S. occupies Hawai'i as a war-making outpost and has no problem destroying the one thing we *all* need to survive—our water.

After two lawsuits and a lot of advocacy, we were finally able to pull the state Health Department to the public's side. A new health director and a new deputy director for the environmental division were appointed in February 2020, and the whole dynamic with the regulators shifted. The new deputy director had staff conduct unannounced inspections at Red Hill, issue violations, and impose fines. For me, the lesson here is to never assume that government officials are working to protect the public's best interest. It is important to make them prove it regularly with their actions.

*Authors' note: After prevailing in 2017 and 2019 in legal interventions to force the Health Department to require a permit for the continued operation of the Red Hill Facility, the Sierra Club would then challenge the permit in an administrative proceeding, with the Board of Water Supply joining as a co-intervenor.*

That is the context for the public hearing on the military's request for a permit to operate the Red Hill tanks. The Health Department had proposed a permit for the facility that was not awesome, but not awful. Both the Sierra Club and the Board of Water Supply challenged it in an administrative procedure called a "contested case hearing." We worked together to make the most of this bureaucratic hopscotch.

At that point, we thought it was highly unlikely that we could get their permit application denied outright. So, our goals were more modest. We wanted 1) to impose as many restrictions, requirements, oversight as possible to help ensure we knew everything that was going on at the facility, and 2) to make a strong record that the future could rely on to pin the military down.

The Sierra Club, represented by the annoyingly brilliant David Kimo Frankel, helped to make an excellent record, and to keep the public in the loop on an otherwise esoteric process. To this day I am amazed at the amount of information Kimo can hold in his head. While my office resembled a rabbit warren of consultant reports, he functioned in a high-tech, well-organized mind palace.

This was the hearing where Naval officials lied under oath about leaks from the Red Hill Facility. One of the military's main talking points at the time was that the 2014 leak from tank #5 was a fluke. It was faulty repairs, human error. In January of 2021 the military said, "There has been no release of fuel from Red Hill since 1988, except for the 2014 leak."

However, by the summer it was well-known that in fact there were two active leaks at the time of that hearing. A whistleblower released internal military emails showing that officials chose to not report the leaks to the Health Department for fear that people like me would use a revelation like that to sink their permit. By October, the Health Department, Sierra Club, and BWS were advocating for the permit hearing to be reopened so that the facts about these two leaks could be put on the record. Of course, that [hearing] never happened because a few weeks later jet fuel started pouring from people's faucets.

*Authors' note: The Pentagon's March 2022 decision to close the Red Hill Facility eventually led to the Navy's withdrawal of its permit application before the Hawai'i Department of Health.*

A big obstacle when challenging the military is their master manipulation of mainstream media. We worked hard to keep Red Hill near the top of people's minds for years after the 2014 leak. In some ways it was easy because the military is so negligent, careless, and incompetent that they were all the time screwing up, and thanks to the Sierra Club team, those screw-ups regularly made the news. But for all of their incompetence in fuel management, the military is quite skilled in propaganda. The military has a limitless budget, a well-oiled shibai machine, and spiffy white uniforms that they used to manipulate us into complacency. And they are still doing it.

A good recent example is the announcement the secretary of defense made in March 2022 that the Red Hill fuel tanks weren't that strategically important after all, and so will be closed as soon as possible. That announcement got a lot of attention. People celebrated. "Woohoo! Red Hill is closed!" Local politicians used that announcement to win votes. And people went back to their daily lives thinking the crisis was being handled.

But what does it mean that the military suddenly changed its mind on the strategic importance of the Red Hill Bulk Fuel Storage Facility? It means our water was sacrificed for nothing but the military's convenient access to gas. It means the military seriously injured those they said they were protecting because it is simply too lazy, too dumb, and too heartless to find a better way to do what it wanted. It means the military considers all of us—civilian and soldiers alike—and all our resources to be expendable in the name of war.

What that announcement does not mean is any substantive change in our situation. Those tanks are still there, still holding 100 million gallons of fuel directly above our water, and still have a very high risk of leaking again. And the military is still sneaking in provisions here and there—like in the 2023 Defense Appropriations bill—to allow them to reopen the tanks. Please note that they are doing this at the same time that they are beating the war drum around China.

The lesson here is: try not to fall for it. The military will try its best to distract us with shiny announcements that have no substance because they know most people cannot keep up with all the stories they tell. No matter what they announce or promise, we must remain committed to ending the military's oppression in Hawaiʻi.

## The Water Protector

*Ernie Lau*

*Ernie Lau, PE, is the manager and chief engineer of the Honolulu Board of Water Supply. Raised in Waipahu and 'Ewa, Ernie spent his youth working on his uncle's farms, fishing in Pu'uloa, and attending school at Waipahu Elementary, Waipahu Intermediate, and Waipahu High School.*

*Ernie obtained his bachelor of science and master of science degrees in civil engineering at the University of Hawai'i at Mānoa, and would go on to work for the Board of Water Supply, the Kaua'i Department of Water, the State Commission on Water Resource Management, and the Public Works Division in the Department of Accounting and General Services, before returning to the Honolulu Board of Water Supply in 2012, where he was appointed chief engineer and manager.*

Just kind of looking back at my childhood, little did I know. The first three letters in Waipahu are W-A-I, and all the places that I actually hung around, fishing in Māmala, Waipahu—all those affected areas [in the 2021 water contamination crisis]. I never knew that I would be involved in a water crisis that affected those areas.

I have this memory that sticks in my mind, of an elementary school field trip to the Board of Water Supply, where my office is located now. There is a pipe in front of our yard, it's kind of rusty right now—I gotta ask the guys to clean it up and paint it. But that's where classes came in for field trips, and they would turn off the pumps and then open up the valve and water would gush out of this vertical pipe under artesian pressure—natural pressure. So, I just have this image in my mind of water gushing out of that pipe. And that's stuck in my mind for almost my whole life.

In my office there's a faded photo of that very pipe, with the artesian water gushing out of it. It's kind of faded, but that photo was taken here at the Board of Water Supply.

So I never knew. I never knew I'd be involved in this Red Hill issue that impacts the very area I grew up in. But it's kind of starting to make sense.

*Authors' note: In January of 2014, just two years after Ernie's appointment as the Honolulu Board of Water Supply manager, the Navy reported a leak of an estimated 27,000 gallons of fuel from the Red Hill Bulk Fuel Storage Facility.*[22] *Media coverage of this event was the first time that many O'ahu residents learned about the facility's existence, and its location just 100 feet above the island's sole source aquifer.*

I still remember the day our office got a phone call from the Department of Health—it was around noon or so on January 13. They let us know that there was a leak of a fuel tank at Red Hill.

My first thought, I still remember, was "What fuel tank? Where is it located? How close is it to our wells?" And I just had so many questions trying to understand more about this facility. I had heard about these underground storage tanks that were in the Moanalua area, but never really knew anything in detail about them. My staff was very proactive—even without me telling them, they immediately shut down five wells closest to Red Hill.

I have this memory of a legislative hearing at the state legislature later. I think it was maybe March or April. There were six committees there—I think even current Governor Green was there when he was in the Senate. Gary Gill, deputy director at the time for the Department of Health, was there also. Captain Williamson—he was the person in charge of the facility at the time.

Gary and Captain Williamson went up and shared what had happened. It almost sounded like it was under control. It was being dealt with. And then I was the last person to speak. I remember going up there thinking. "What am I gonna say? The other two guys in front of me downplayed the seriousness of the situation." But when I went up there, something—my heart just said I gotta tell it like it is. So, I just said—the words I used were something to the effect of, "This is a disaster waiting to happen. All this fuel right over our aquifer. It's a disaster waiting to happen."

We started to ask the Navy questions to get more information about what was happening at the facility, but weren't getting a whole lot of answers.

The Sierra Club at some point, not too long afterwards, under Marti Townsend, became very active. Marti did a lot of advocacy about this situation at Red Hill, and the threat to the water. Then in 2015, the Department of Health, the EPA, the Navy, and the Defense Logistics

Agency signed on to an administrative order on consent—the 2015 AOC. I still remember the day they signed it—Governor Ige called me on my cell phone and wanted to give me the good news that they had signed the agreement.

I was polite, but in my mind, I was thinking, "This isn't good. This ain't good news!" We had submitted almost forty pages of comments on why this [AOC] wasn't a good idea to do, because the whole AOC was structured to actually allow the Navy to continue to operate the Red Hill Facility for twenty more years.

*Authors' note: Due to AOC's inability to hold the Navy accountable to its commitments, the Navy and regulators today continue to remain in the dark in many respects: in understanding and balancing the risks at hand, in assessing the integrity and vulnerability of the fourteen underground storage tanks still containing over 100 million gallons of fuel above Oʻahu's drinking water aquifer, in developing contamination plume recovery and groundwater monitoring strategies, in developing proactive remediation options, and in prioritizing defueling actions.*[23] *In December of 2022, the EPA announced that it had negotiated a new consent order with the Department of Defense, containing eerily similar limitations as the 2015 AOC.*

That's why we decided to send the letter to all BWS customers—over 170,000 letters across the island, alerting people about this proposed consent order in 2023, and raising some of our concerns. My concern is that the 2015 AOC didn't prevent the disaster in 2021, and [the Navy and regulators] didn't really make much progress from 2015 to 2021. A lot of back-and-forth letters, draft reports, but no real progress on groundwater modeling, on contaminant fate and transport modeling. . . .

The Navy was kind of ignoring the regulators, and no one called them to the carpet. I remember one letter sent by the EPA and [Department of Health] around June 2016, they sent this letter that even attached the Board of Water Supply's letter commenting on the groundwater [modeling] scope of work. And I thought, "Wow, this is hopeful—they're gonna finally tell the Navy to stop putting out the baloney that the water only flows from mauka to makai and not across Hālawa Valley [toward our wells]."

But after that, nothing happened. They convened a groundwater modeling task force or working group, they had meetings that would last like all day long. [BWS Deputy Manager] Erwin went to those meetings with our technical team, and yet, still no progress, we still have no groundwater model.

On the risk and vulnerability assessment, when the conclusions were not good, the Navy then turned around and tried to refute or discredit their own report that they submitted to the regulators, that they paid their consultant to develop. Which is insane. That's the risk and vulnerability assessment that indicates a 27 percent chance of a release of up to 30,000 gallons from this facility every year. So, the AOC of 2015 to me was a failure, yeah. It didn't really make substantial progress. It didn't prevent the disaster.

With the current [consent order] we will have to really see—I would prefer that they don't do an AOC or a consent order because fundamentally it would be a settlement agreement. And why do we have to settle with the Navy right now? Why can't the regulators just unilaterally order the Navy to take action and tell them what to do and hold them completely liable for this situation and not give them any concessions?

If it was any private company or even the Board of Water Supply, and not the DoD, I don't think the EPA would hesitate to slam them. But when it comes to another federal agency like the DoD, I don't know why they can't be treated the same way they treat the rest of us.

*Authors' note: After the November 2021 crisis began, and the Red Hill Facility was suspected and eventually confirmed as the cause of the widespread illnesses and contaminated water, the Board of Water Supply became the primary source of agency information on the safety of the island's water supply, the inherent dangers of the Red Hill Bulk Fuel Facility, and the ongoing threat to Oʻahu's aquifer. Ernie's continuing willingness to provide information and assurances as well as his pleas for Navy and regulator action would become an inspiration for residents across Oʻahu to do their part to protect the island's water.*

[In November of 2021,] our 24/7 operation center started to get calls from people complaining about fuel in their water. Initially, we didn't know if it was our customers. Later we kind of figured out it was the Navy people living at the base.

Typically, in a situation like this, you would have crisis communications where the Department of Health and maybe the utility operator would come out and hold press conferences. It's important to come out and be transparent, let people know what you know and what you don't know and what you're trying to do to address it. Let people ask questions and also share their concerns.

But the Department of Health, the Solid Waste Branch that regulates underground storage tanks and the Safe Drinking Water Branch, we didn't see them coming out to provide updates and information, and the Navy kind of did a pretty—unfortunately, not a very good job. So, we expected that we would start to get questions and we started to get media inquiries left and right. That first week and the next week I was holding or giving five or more media interviews a day. By the end of the first week, it was like thirty or forty interviews, and it continued the next week and then it slowly kind of died down.

We still had no or very little complete or accurate information coming from the Navy, and the Department of Health was kind of silent, absent. And that wasn't good. So, Kathleen and I figured we've got to share what we're doing, what the risks are to our customers, you know, because we're getting calls already from our customers concerned about the situation.

So, I reached out to the reporters, the television stations and radio stations. I needed their help in getting information out to the community. This is a crisis situation. And if we don't get information out, fear will just run out of control and people will start to really panic.

We took a lot of time during those early weeks and months. If a reporter didn't quite understand, and if they had a lot of questions, I basically set up an hour or two in a virtual meeting to brief them and answer any questions that they had. I said we want you to become educated about this issue, because you can help get the message out to the public. And we did that for quite a few of the reporters.

We called our own press conferences at times. We would invite the DOH to join us but they never showed up, but we still held press conferences about shutting down Hālawa shaft and all those things, and then more recently, it was on the AFFF concentrate spill.

The latest is that we have plans to try to drill monitoring wells in Moanalua Valley, at the Moanalua Golf Course. And also more wells in Hālawa Valley. And the Navy has plans to add another twenty-two wells. But the problem is—this is now over a year [later], and we still don't have a good, coordinated plan to investigate the aquifer. To determine the proper placement of monitor wells. A year later, we haven't progressed from where we were before the crisis. And while the Navy is busily trying to drill wells, they are going to easy spots on their property.

## Water Protectors Rising

*Healani Sonoda-Pale*

*Healani Sonoda-Pale is a Kanaka Maoli woman and mother, a citizen of Ka Lāhui Hawai'i, a founding member of Pu'uhonua o Wailupe, and a veteran advocate and organizer who was mentored by the late, and legendary, Dr. Haunani Kay-Trask. Her three decades of work in "the Movement" include campaigns to restore Native Hawaiian governance, defend Native Hawaiian intellectual property from commercial exploitation, uphold Native Hawaiian land and water rights, and protect iwi kūpuna and cultural sites from desecration and destruction. Most recently, she has joined the O'ahu Water Protectors as a key organizer and thought leader in the Shut Down Red Hill Campaign.*

When I found out what happened [in November 2021], my first reaction was total alarm and shock. It was like someone punched me in the gut. I was fearful for the future of O'ahu, and I realized immediately that this was now the most important issue we faced. I knew, as a Kanaka Maoli, the value of water, and its importance for the sustaining of life. Without water, there is no life. There is no future without water.

My ancestors, who flourished on this island since time immemorial, knew that water was sacred, knew that water was important for life. They protected the wai, *revered* the wai—and now we're in a situation where that wai that has sustained life on this island for 2,000 years is in danger of being contaminated to the point where it would be undrinkable, and unable to sustain life any longer. Knowing that the future of my children and my grandchildren could be very bleak—that put the fire in me to do something, to stand up, and join the O'ahu Water Protectors.

*Authors' note: Prior to the November 2021 spill, the O'ahu Water Protectors—an ad hoc group of individuals formed in September of that year—had already been working to educate the broader public about the dangers of the Red Hill Fuel Facility, and to increase the pressure on government decision-makers to shut the facility down.*

I thought it was super important to join the O'ahu Water Protectors, because they were on the front lines. They were the ones stepping up, saying the things that needed to be said, and demanding that the U.S. Navy shut down the Red Hill complex permanently, and to defuel those fuel tanks immediately. When we first started, it just seemed like we had jumped on this canoe, and pushed the canoe out to the ocean, not knowing where it was going to end up, but knowing that we had to do it—we had to make that journey, not for ourselves, but for our families and our communities and everyone who lives here. Without hesitation, without fear, we jumped on that canoe and went on that journey to challenge the most powerful military in the world—with basically no money, no supplies, with just our knowledge and our 'ike and our trust in each other.

And what I'm most proud of is that by March, the secretary of defense announced that he was going to shut down the Red Hill Facility. That was a big one, because from the time we started, from before November to March, everyone else kept telling us, "There's no way they're gonna shut that facility down." But guess what? Our group of community advocates and water protectors, we were able to force the hand of the most powerful military in the world.

We did it with just our voices, and with our people power, and in a peaceful manner, using nonviolent direct actions, organizing, and working in the community. Calling for truth, calling for transparency. We forced their hand to shut down the facility—or at least say they were going to shut it down.

It really showed me the power of people, how powerful our voices are when we come together. And it wasn't just Kānaka in this group. There were all different types of cultures, ethnicities, genders, ages, a real diversity of thought and experiences and people. And we were able to come together and fight for the protection of our water.

*Authors' note: The O'ahu Water Protectors' efforts to maintain pressure and attention on the U.S. Navy have been well documented in local and national media, with notable actions including rallies in front of the Board of Water Supply and the Prince Kūhiō Federal Building, a "die in" at the Hawai'i State Capitol, a "Lie-Aversary" event at the Pearl Harbor Visitors Center, an "Eviction" action at Pacific Base Command headquarters, and a massive "Walk for Wai" march to the Navy's Honolulu NAVFAC headquarters.*

My work on previous campaigns taught me about the basic building blocks of community organizing: You gotta have your facts down. Research and education are super important. You gotta keep the issue alive. You need to know how to write a press release. You have to build a relationship with the press—Dr. Haunani Kay-Trask, she always said that. You need to know how to organize events. You gotta be a good friend, a supporter of your comrades. Sometimes, you have to cook. Sometimes, you have to farm, so you can have food to cook for events. And you have to know how to move people to action. That can be the hardest part.

I also brought my perspective, as a woman, and as a mother. Women's voices have been super important in this movement to protect water, including the mothers whose children were poisoned.

*Authors' note: On November 29, 2022, one year after Oʻahu's aquifer was conclusively contaminated by jet fuel, approximately 1,300 gallons of PFAS-based firefighting foam concentrate were released in the Red Hill Facility, spilling out across the outside landscape.*[24]

*This caused immediate alarm for those familiar with PFAS, or forever chemicals. Some of these extremely toxic compounds, including those found in firefighting foam, are considered dangerous at concentrations equivalent to one drop in a thousand Olympic-sized swimming pools.*[25] *Exposure to PFAS has been associated with increased risks of cancer, reproductive issues, lowered immune responses, and thyroid disease, among other health effects. Most alarmingly, these compounds do not break down naturally, and may persist for centuries or longer.*

*Rainy weather on the night of the November 29 spill and over the following days confounded the Navy's efforts to contain the spill, leading to a decision to pave over the contaminated area without soil testing data to show whether all of the foam concentrate had been excavated.*[26] *As a result, these forever chemicals may slowly—or quickly—migrate throughout the environment, for centuries or longer.*

*In a televised press conference after the spill, Honolulu Board of Water Supply Chief Engineer Ernie Lau broke down in tears, demanding that no more harm be inflicted on Oʻahu's ʻāina and wai. His candid display of sadness and frustration spurred Oʻahu residents to take action, with the Shimanchu Wai Protectors organizing a rally outside of the Department of Health within a matter of days.*

*Another community member, Malia Marquez, felt that a larger, mass mobilization action was needed. "For years, Ernie Lau has done everything in his power to keep our water safe, to keep our children, our keiki safe from this decrepit Navy facility, and I felt deep down that we had to answer his call," she explained. "The U.S. Navy made this dedicated public servant, this humble, quiet, selfless man cry with their callous attitude, their horrifying negligence. That was a big, big mistake."*[27]

The Walk for Wai was the brainchild of a community member. As community organizers with the water protectors, we were able to support her. And we brought in other community members, all different types of people jumping in. Everybody brought their own superpower, their own expertise to channel the community outrage and help bring to life this beautiful march. It all happened in a short time period, so it was a little bit chaotic in the background, in the beginning, but it ended up being very, very good.

*Authors' note: The "Walk for Wai" saw upwards of 1,500 people—doctors, lawyers, construction workers, teachers, students, kūpuna and infants, military spouses and demilitarization activists, individuals from all walks of life—march to the Navy's NAVFAC headquarters, with Ernie Lau leading the way.*[28]

The biggest challenge is burnout. It seems like there is an issue, an emergency every week, where the U.S. military does something really bad, and we have to respond. It just never seems to end. All we're trying to get to is that point where those fuel tanks are emptied and the facility is permanently shut down. It's going to take a while to get there. So we're gonna have to continue voicing our concerns and speaking truth to power to get there. But, how are we going to sustain ourselves on that long and arduous road to that goal? We're not a corporation with endless resources. We're not the U.S. military. We are not getting paid. This is all volunteer work.

Looking back, it's been over a year, and it's been amazing what the O'ahu Water Protectors achieved working at a very grassroots community level. But it's not a fight that is going to end tomorrow. Sustaining ourselves, keeping ourselves motivated and moving forward and trying not to get burned out in this whole process, is the challenge. We're motivated by our love for this island, for our children and the future generations to come. The military industrial complex has no heart, has no feelings. Their motivation is insidious. Even though they say they're here for national security, they're really not here to protect us. They're here to protect America's imperialist interests in the Pacific. We're here to fight for our life. This island's ability to sustain life. And to me, that's what makes this movement so different, so sacred.

## The Same Old Playbook

*Susan Gorman-Chang*

*Susan Gorman-Chang was born and raised in Munster, Indiana, and later moved to Van Nuys, California, where she met her husband. They raised their son in their home near Porter Ranch and would later become involved in a local environmental campaign that would make the later crisis at Kapūkakī feel all too familiar. After moving to O'ahu in 2019, Susan would quickly find herself involved in the Shut Down Red Hill campaign, advocating and organizing actions through her role as the chair of Faith Action for Community Equity's Environmental Justice Task Force, and as a member of the Shut Down Red Hill coalition.*

In July 2014, two of our neighbors in Porter Ranch, California, let us know that Termo, an oil company, was seeking permission to construct twelve new oil wells in the unincorporated hills behind our housing development. We, along with some neighbors, created the nonprofit Save Porter Ranch, and, assisted by Food & Water Watch, we demanded that an environmental impact report be conducted for the proposed oil wells.

Then, on October 25, 2015, we came home after dinner one evening to an overwhelming smell of mercaptan—an odorant added to natural gas—enveloping our entire community. We learned over the coming months that the Aliso Canyon Gas Storage Facility stores 87 billion cubic feet of natural methane gas within porous rock, and SoCalGas uses old oil wells to pump the methane in and out of this facility. Some of these old oil wells dated back to the 1930s, and they were never designed to withstand this type of pressure. In addition, SoCalGas was utilizing these [wells] without any redundancy or any secondary hull to contain a leak or a breach. As a result, there was a massive blowout that began in October.

With SoCalGas downplaying the incident, an Environmental Defense Fund representative and a Save Porter Ranch director went up into the hills surrounding the Aliso Canyon Gas Storage Facility, and used a FLIR camera to record what our eyes could not see—methane gas pouring out of a blown-out well.

As an officer of Save Porter Ranch, I helped organize protests and educational workshops and met with several elected officials.

*Authors' note: The Alison Canyon Facility blowout would turn out to be the worst natural gas leak in U.S. history, with a carbon footprint larger than the Deepwater Horizon oil spill.*[29] *Nonetheless, to this day, the facility continues to operate.*

While renting a house in Kapolei, after moving to O'ahu in 2019, I opened our Board of Water Supply bill and saw a flier that gave a lot of information about the decrepit Red Hill [underground storage] tanks. It was eerily familiar as I read that the tanks were single hull, old, decaying, under the ground, and a danger to our water supply.

As I recall, the letter explained that the Navy was promoting the least-expensive upgrade option. I remember researching and coming across the Sierra Club Hawai'i's work on this issue as well. I wrote testimony advocating for what the Board of Water Supply and Sierra Club recommended. When I watched a recording of the November 2019 town hall, the Navy spokesman seemed to be using the same playbook as SoCalGas, carefully wording and couching his answers and evading questions. It was déjà vu.

*Authors' note: As part of the 2015 Administrative Order on Consent, the Navy agreed to complete a "Tank Upgrade Alternative and Leak Detection Alternatives" analysis to study and make recommendations on structural upgrade options for the Red Hill Facility's underground storage tanks. The document, submitted in September of 2019, was met with tremendous public opposition, including during a November 2019 Environmental Protection Agency and Hawai'i Department of Health Town Hall;*[30] *the regulators would later reject the Navy's analysis and chosen tank upgrade option, finding that it lacked "detail, clarity, rationale and justification" and that there was "no clear nexus between the proposed decision and protection to drinking water aquifer."*[31]

Meanwhile, I was searching for a church and bounced around until I found Trinity United Methodist Church in Pearl City. Pastor Amy's sermons revealed a deep commitment to social and environmental justice, which fit my view of faith reflected in action. In our church bulletin, I saw the announcement that Faith Action for Community Equity was starting an Environmental Justice Task Force, which I joined. Right after the November 2021 leak disaster, we turned our attention to Red Hill. SoCalGas and the Navy still seemed to be using the same playbook: Deny, get caught, admit but downplay any health or environmental danger, and gaslight. More déjà vu for me. So, one of the first things we did was join the Shut Down Red Hill Coalition, which formed around that time.

One of the most memorable and moving protests we took part in was the December 10, 2021, Die-In. After we "died" and rose up, it was very touching to hear [Heolimeleikalani Osorio] sing "Aloha 'Oe." Then Nani and Makaio's chant, "Aia i hea ka Wai a Kāne"—"Where Are the Waters of Kāne"—was so incredibly moving and poignant. Another memorable event for me was the Lie-Aversary in October 2022 at the Pearl Harbor Visitor Center. I felt the funeral attire, signs, the "Top 10 Lies," black umbrellas, and speakers made it such a powerful event.

In contrast to my church in California—which would not even sign a letter to shut down Aliso Canyon or attend any of our protests—my church on O'ahu was and remains in full support of the Shut Down Red Hill movement. Twice now after church services we have held signs protesting Red Hill and calling for quicker defueling. Pastor Amy Wake was the first name on the Interfaith Statement of Support for the Permanent Shutdown of Kapūkakī, which took up an entire page of the Honolulu Star Advertiser when it was published on Sunday, February 2, 2022.

One of my fellow church members also started a petition to Shut Down Red Hill and permanently decommission it. We will submit that to the EPA site seeking comments on the [2023] Consent Order. One gentleman in particular who cannot physically attend protests anymore still has a lot of social connections with activists, and using his email and his phone, he pressed more folks to sign on to that petition as well.

*Authors' note: The consent order announced in December 2022 between the Environmental Protection Agency and the U.S. Department of Defense was immediately panned by the Board of Water Supply, O'ahu Water Protectors, the Sierra Club of Hawai'i, citing its lack of deadlines, meaningful penalties, or mechanisms for meaningful transparency, among many other shortcomings. A community town hall hosted by the EPA and Navy elicited a "barrage of criticism" before public testimony was cut off.*[32]

*As of this writing, nearly 2,000 public comments, nearly uniformly critical, have been submitted on the EPA's consent order.*

Because of what happened to me and my family in Porter Ranch, I know how isolating and demoralizing it feels to be gaslighted and ignored by the local, state, and federal government elected officials and agencies that are supposed to protect us.

I know how hard it is to speak out and to be heard and how terrifying it is to wonder if by speaking out, harm will come to you or your family. I also know how important it is to have advocates who have not necessarily themselves been impacted, but are nevertheless willing to stand up and help, and utilize networks of other environmental organizations to come together for actions and events.

I am so thankful and inspired by the military families who are speaking up regarding what they have been through and the devastating health impacts that have been inflicted on themselves and their children. That cannot be easy. I am really proud of the way Native Hawaiian organizations, environmental organizations, neighborhood boards, and some churches have come together for this cause, and most importantly have stuck with it. Seeing all of these organizations come together to fight and shut down Red Hill has been truly inspiring.

## Okage Sama De

### *Reverend David Nakamoto*

*Rev. David Nakamoto grew up in a family of five children, to a father who immigrated to Hawai'i from Japan, and a mother who was born in Hawai'i and raised in Japan before returning to the islands. A graduate of 'Iolani High School, David earned his bachelor's and master's degrees in social work at the University of Hawai'i, and worked at the Hawai'i State Hospital and later for the Queen Lili'uokalani Children's Center for thirty-five years. David and his wife, Irene, became ordained Buddhist ministers after their retirements at the Honpa Hongwanji Mission of Hawai'i.*

*After moving into a retirement facility at 15 Craigside, David would continue to serve as an engaged member of his new community. This would eventually lead him, along with his close friend and fellow 15 Craigside resident Irene Zane, to organize actions to save O'ahu's aquifer and protect Hawai'i's keiki and future generations from the existential threat of the Red Hill Facility. David and Irene's work has since inspired other kūpuna as well as younger generations, to join in the campaign to Shut Down Red Hill.*

I grew up in the Makiki area, where my parents lived and worked running a floral business. I have fond memories of growing up in the community—of friendly neighbors and childhood friendships, and of attending school as well as Japanese language classes.

My sense of community deepened as I served as a social worker on the Wai'anae Coast for over twenty years—a sense of connection to a network of people concerned with the preciousness of what was important to them. This became clearly apparent during the most challenging times. The rallying together of collective energy to meet these challenges brought people closer together, and thereby further strengthened the community.

Initially my view of community was a diverse entity of people and organizations with varying concerns, but a clearer view emerged over time. Each community is unique in its ability to sustain itself effectively.

Upon entering our retirement complex, I was able to connect to a wide range of retirees and discovered, as we became further acquainted, the unique and varying experiences they each possess. They come with tremendous skills, expertise, and much wisdom. The intended goal, for many, is to retire in a comfortable setting with fellow retirees and enjoy their retirement—a well-deserved time in their lives. There are retirees, however, who continue to serve in various capacities within the immediate retirement complex, serving in leadership positions to guide and oversee the overall operations of the complex. Others serve in the broader community in various capacities, such as on community boards, and through volunteer work.

I heard about the Red Hill crisis by attending a meeting at a friend's home. The informative and thought-provoking discussion led by [Sierra Club executive director] Wayne Tanaka made clear the urgency of the issue, and its relevance to the health and safety of people. It was then that I decided to join in on an organized effort within my community to effect change.

*Authors' note: David and Irene circulated and gathered signatures on a heartfelt and thoughtful letter to President Joe Biden and top Department of Defense officials. The delivery of this letter, as part of a broader letter- and postcard-writing campaign to the Pentagon and White House, would be followed just a few weeks later by an announcement by Secretary of Defense Lloyd Austin that the Red Hill Facility was no longer essential to national security—and that it would cease operations permanently.*

It was during that initial meeting with Wayne and others when I raised a question about a letter-writing campaign and how this may have an impact on the issue at hand. When the response was "Go for it!," it inspired within me an impetus to carry on what I believed could make a difference.

This effort would involve several organizing steps within our retirement complex. Irene Zane, a fellow resident, and I took on a number of tasks to successfully bring about the letter-writing campaign. Providing further information on the issue was the initial step, and we involved the top leadership at the retirement complex to gain support for the initiative. With their support and assistance, we developed and distributed a survey to all residents at the complex 1) providing information on the issue and 2) enlisting their support through a questionnaire. After receiving a positive response from the returned survey instrument, we were ready for the petition letter to be acted upon.

Within a span of about three days, we were able to receive 200 signatures from residents, as well as some staff, in support of the letter to President Joe Biden, with copies sent to congressional leadership and all pertinent governmental officials.

When Wayne was contacted about our successful efforts, he came down personally to the complex to receive the signed petition letter. This became part of a strategy to demonstrate the involvement of concerned seniors, who seriously sought to advocate for the health and safety of our citizens, as well as for the future well-being of our beloved Hawai'i. Later communication to us from Wayne indicated that this involvement of concerned seniors had a clear impact in inspiring other community organizations in their advocacy efforts.

*Authors' note: Despite the March 2022 Pentagon decision to shut down the Red Hill Facility, the situation was—and remains—far from resolved. The subsequent Navy claim that it would take until July 2024 just to remove the over 100 million gallons of fuel still stored over O'ahu's sole source aquifer was quickly and roundly criticized by the Board of Water Supply, Department of Health, and broader Hawai'i community. It soon became clear to many that, despite its unimaginably vast resources, the Department of Defense, along with its leadership, would not consider the need to defuel and close the Red Hill Facility as a true emergency.*[33]

The Red Hill crisis at one point turned from orders to shut down the facility to an attempt by the Navy to drastically slow down the defueling process. It was at this point that our effort united a drive at our retirement complex to press forward with a second letter-writing campaign. Rallying our seniors to respond to this pressing concern, the second campaign began, and within the span of one day, over 200 signatures were secured and passed on to Wayne. The willingness to respond was clearly evident.

*Authors' note: This second letter-writing campaign would also expand to other retirement communities, becoming part of a monthlong series of community protests and activism to expedite the defueling and closure of the Red Hill Facility.*[34]

Basically I feel, as well as many seniors in Hawai'i, that our forefathers before us have done well in their effort to establish themselves here. They faced many challenges and difficult hardships to pave the way for economic, social, and educational opportunities for future generations. It is the "okage sama de" spirit ("because of you, I am able to live the life that we now live") that is part of our experience today. The gratitude for this support may be manifested in our effort to support the keiki of today as well as future generations to come. The Red Hill crisis is clearly one that can have a major impact on their future.

## A Light in the Darkness

### *Dr. Aurora Kealohilani Kagawa-Viviani*

*Dr. Aurora Kealohilani Kagawa-Viviani was born and raised in urban Honolulu. Her mother was a soil chemist at University of Hawaiʻi at Mānoa, leading Aurora to spend much of her childhood familiarizing herself with science and the outdoors. A graduate of Kamehameha Schools, Aurora earned an environmental engineering degree at the Massachusetts Institute of Technology, before returning home to pursue graduate research on watershed management. After working for a few years on a traditional agriculture restoration and research site at Puanui, Kohala, she returned to Honolulu, where she eventually earned a PhD in geography and the environment at the University of Hawaiʻi at Mānoa.*

*Aurora is currently a mother of two young children, the newest member of the State Commission on Water Resource Management, and works as an assistant professor at the University of Hawaiʻi at Mānoa's Water Resources Research Center and the Department of Geography and Environment. She is also a member of the University's Red Hill Task Force, an ad hoc, volunteer, cross-disciplinary initiative that has shined a light (quite literally) on the most pressing, and vexing, questions in the Red Hill crisis.*

I was applying for what is now my current job, sitting in wall-to-wall interviews at the end of November 2021. The evening of the 30th after the interviews wrapped up, my partner and I were having a celebratory dinner when we got a message from his colleague at Leeward Community College [LCC]—a student who wanted help testing her water.

The first thing I thought was, "Okay, we're at the university, we have mass specs, we can do this." My partner forwarded the message to a mentee who is a grad student in chemistry, asking, "Any idea who might be interested in examining water samples from homes that seem to be affected by the Red Hill leak?"

I reached out to [Water Resources Research Center, or WRCC] staff about in-house capacity and learned that the Navy had reached out earlier inquiring whether the WRRC lab could analyze total petroleum hydrocarbons—TPH—at a detection limit of 400–500 ppb. The lab manager's reply: "Short answer, most likely no."

We learned that at the WRRC instruments once used to quantify petroleum in water had not been used for many years. I had this sort of jaw-dropping realization that if the Navy—the institution you'd expect to have this capacity—is reaching out to UH, and we haven't done this since the 1980s . . . we're in a pickle.

At this point, a few of us decided to convene an ad hoc working group meeting with our network of contacts from across the campuses—our goal was to beat the thirty-day turnaround time for sample analysis. There was a flurry to identify what on-campus resources could be mobilized. Our WRRC director, Tom Giambelluca, called a broader all-hands-on-deck meeting. We set up a listserv and have been meeting regularly ever since. Long story short: "testing" at the standards required by regulators was more difficult than we anticipated.

*Authors' note: As of this writing, there is still no EPA-certified water testing lab within the islands, with the Navy, Department of Health, and Board of Water Supply still sending their water samples to commercial labs across the continent for analysis. The turnaround time of weeks to months associated with extensive EPA quality assurance/quality control requirements continues to confound efforts to identify and track contamination issues both in the Navy's water system and in O'ahu's groundwater aquifer.*

That first week, the WRRC lab analyzed the student's water samples for non-purgeable organic carbon (a nonspecific water quality indicator that is relatively simple to measure), and requests for help kept coming in. Frustrated community members began dropping their bottles of water off at LCC. We were reaching out to all the UH Mānoa labs to see who had instruments, and if they would like to try running a tap sample, and we hit a lot of dead ends.

In searching for sample storage space and bottles at UHM, we ended up connecting serendipitously to an oceanography lab led by Craig Nelson that used fluorescence to characterize organic carbon compounds in water. That was a pivotal moment.

Nelson Lab grad student Sean Swift, who's this really unassuming guy, ended up running samples over the winter break at the crazy end of the semester. The lab had an instrument and workflow that could evaluate samples quickly, with little cost. They could basically tell if there is a strong positive fluorescence, versus a non-detect, or a possible detect.

There are also background organics that also fluoresce and can complicate things. In this way the method is not definitive testing, but a rapid screening method to flag suspicious samples since clean drinking water should not fluoresce.

The WRRC Lab and the Nelson Lab have continued to collaborate closely to figure out how this new method compares to the standard EPA methods that the Department of Health, Board of Water Supply, and Navy all are much more familiar with. The EPA methods are the current benchmark, but other researchers have established that they don't capture everything, and the methods are costly and time-consuming.

In the crisis, members of the UH community were able to mobilize very quickly. Faculty, staff, and students had more flexibility and bandwidth than Department of Health staff who were getting bombarded. Everyone was shocked by the crisis, and wanted to help, and it was a matter of trying to figure out, well, what can we do? What tools do we have? What teams can we form?

A grad student who does completely unrelated research helped me keep notes and reach out to new contacts in the early days; faculty who had experience working with the Navy advised us on navigating politics. The issue is terribly complicated, but at the same time, you can't just sit by and not do anything. Our anxiety about having information that was contradicting what the Navy was publicly saying—that was really stressful. Since then, the atmosphere has changed and there is more openness now.

The situation catalyzed so many people even at the crazy end of semester, and folks showed up and have stayed. Our listserv has over eighty members, our meetings have about twenty regulars. We've had different key volunteers come and go—students graduate, community volunteers have moved. We have shifted out of crisis response and are settling in to strategize and prioritize research and educational initiatives that play to our strengths and can be helpful to the public.

Our strongest team is the individuals doing the chemistry and fluorescence work, leveraging the freshwater testing capacity of the Water Resources Research Center with the expertise of the oceanographers who are looking at dissolved organic carbon in the environment. Our early results from tap samples had multiple agencies very concerned, asking, "If it's a detect, what does it mean?" The skepticism could be seen as naysaying, but it's also a challenge to improve the method, improve our articulation, improve our understanding of what the method picks up and what it doesn't.

At present, we still work at the pace of volunteers, which means sometimes it's slow. Early on we were reallocated some existing defense funding, which had initially supported corrosion research on the [Red Hill] tanks and was redirected toward water testing supplies. That gave us some concern about accepting Navy money, although we've begun to understand that there are different "flavors" of [Department of Defense] money. Some come with strings, and some don't. So an additional challenge is walking that line in maintaining the trust of the community while funding the work that is necessary to support public interest, where the agency best funded to support it is the Navy.

Early on, a water crisis specialist giving us some guidance on Red Hill, tweeted that "researchers are helpful in a crisis . . . because we're used to not knowing the answer." That's really stuck with me. And I think it speaks to why academics were not deterred, and we were able to ask questions. Not to intellectualize the harm experienced. But, we could channel our efforts into asking questions. And so that's a lot of how I have processed my response to Red Hill.

## We Don't Not Show Up

*Noel Kaleikalaunuoka'oiai'o Shaw*

*Noel Kaleikalaunuoka'oiai'o Shaw is a mother, fourth-generation Hawaiian homestead lessee, member of O'ahu Water Protectors and Hui Aloha 'Āina 'o O'ahu, and part of the kaiāulu kaiapuni (Hawaiian immersion community). Noel participated in various leadership trainings as a youth and has since used her experiences as a school counselor, and also as part of an environmental justice compensation team under the Radiation Exposure Compensation Act, to organize around aloha 'āina. Noel moved back home during the COVID-19 pandemic and gave birth to her youngest child—one month before the November 2022 spill at Kapūkakī.*

All the waters that flow through me come from Hawai'i. From Mauna Kea, Hawai'i on my mom's side and Kawaikini, Kauai on my dad's side.

So of course I know to protect, to represent, to kūʻē Hawaiʻi. Of course I know the waters of Hawaiʻi to be my life source. Those waters then become the waters of the keiki I have, and they also inherit the waters of whom I partner with to have them. They inherit from us relationships to places, people, and just an unwavering connection to ʻāina-feeding spaces. Our eldest, the Kokoʻula rains of our family homestead in Waimea and traveling across waters to connect me and my kāne. Haukea, our middle, named for the tenacity of wāhine coming at a time when that was demonstrated at by mana wāhine leadership at Mauna Kea and by the Black women who used their grief for their lost children at the hands of police violence to awaken a nation. Then you birth this little baby back home on Oʻahu, where you lived since you were eleven years old, twelve minutes from this place that you also grew up in. Your grandparents lived there in Aliamanu, and you lived there for a couple of years. Your grandfather served in the Air National Guard as a radio man. He would go on base, he'd hide you so you could get on because they're like, "Oh, no kids," he's trying to take you to the good swim spots.

You bring this baby, and you've been working around water your whole life because even when you're really young, you're indoctrinated to know about the water cycle. You come from a family who on one side of the family never had water running through their home for a while, and another side of the family used a water catchment system. So your whole life you grew up with five-minute showers, don't let nothing run. We got to save water, always saving water. And water more than money, or any other thing, water is the most important thing.

So when Kapūkakī happens, you're like, "This is literally the worst thing that could ever happen." I was freaking out; I still freak out about it.

When I was an undergrad, I had done an internship with the Department of Justice under the Radiation Exposure Compensation Act. Essentially, we were trying to get Native communities compensated for being exposed to nuclear waste in the 1940s, because they were the ones who weren't enrolling the most for the compensation. I just happen to be going on this internship, and I'm the only Native Hawaiian, of course. And there's all these Navajo Diné people, and I'm sitting at these tables [and] these people around me are saying, "That is not enough money. Do you know what my family had to go through? I can't even hear this history from you because you're telling me that they knew that whole time that it was bad. And now I have to go out and try and get my, my análí, my grandparents to fill out this form, and they literally can't breathe."

So I sat on the compensation side of an environmental injustice, and I saw the struggle of what it meant to try to obtain some sort of justice for wronged people. And they're still having to compensate people sixty, seventy, eighty years down the road. People die during that time, people develop horrible illnesses. And we live on an island. It's not like the continent where people can just disperse like other places. Where are we supposed to go?

When I found out about [Kapūkakī], I think my grandma called me, and I was like, "Well, what do you think Papa would do?" She's like, "He would leave. He would leave right now. And then he would find out who was in charge and hold them accountable." And I was like, "Man, you're right."

*Authors' note: As a leader in her homestead community and with deep kuleana to Hawai'i, Noel took quick action to try to support impacted families, with her newborn and keiki in tow. After collecting and distributing supplies and gifts to impacted families with keiki, Noel's instinctive motherly abilities to care for others led her to explore the creation of "story time" spaces for keiki to understand and, along with their entire 'ohana, become involved in the burgeoning movement to care for the wai that we all need to survive.*

One of the things that we first organized was trying to take care of all the military families who were displaced. It was Christmas time and they got poisoned from our water. No one should ever come to Hawai'i and get poisoned from our water. So the first thing that we organized was making little coloring book kits. We reached out to the Board of Water Supply. We got coloring books, some other stuff. Then this other mākua, Kate Righter, made a coloring sheet with Red Hill in the shape of a mama carrying a baby and crying with tanks leaking and with the bold words "You are worthy of protecting" that we stuck in front of the kits.

Simultaneously, I had to teach our own keiki what was happening, because they deserved to know what I was doing making these kits and why I was asking them to help—and I just needed to prepare them for what organizing to protect meant for our family. Part of how I taught our own keiki was I read a puke [book], *We Are the Water Protectors.* I explained what had happened in an age-appropriate way and had given them this color sheet. "You see what the mama is holding? We care about you, and you matter, you're worthy of being protected."

We crowdfunded to put all of these makana together, and I had a tita who was doing ʻukulele time at the hotels that [directly impacted and displaced families] were staying at, and she distributed it all to the families. That clicked off to another organization, Pūnana Leo, who made 500 goodie bags of homey things like snacks and they had some books, cricket kits, stuff like that. And we were distributing it to all the impacted communities, not just the military, but also where they do the water distribution, at Kapilina Beach Homes.

And that's kind of how it sparked off—we had done it at home, it was an effective way to engage the keiki and prepare them for what they were going to be a part of, because . . . we [parents] just don't have capacity to leave them out of the fight.

And in many of these organizing spaces that are frontline, there isn't space for keiki for them to have programming. To have someone who listens actively to engage them in the conversation or the movement. Being a mākua is a political thing. There's a power dynamic. It's a power play to have a child. It's creating a legacy and you have to participate in life on a greater level because there's literally someone to make it better for along the way. I think that's also because I want mākua to feel like they have space to participate. They hold frontlines every single day.

*Authors' note: "I'm just a mom" has been Noel's main refrain in her movement work. Doing whatever she can to contribute however she can in this "wā pēpē" (childrearing years where our keiki are more dependent on mākua) has included the strategic use of social media to educate, call others into the movement, and raise critical questions to spark further interest and curiosity.*

Sometimes it's really an opportunity to educate people. With education, there's always ea. Everybody's on different pages and everyone has drunk different kinds of water with different kinds of flavoring. Colonization and assimilation and occupation exist very differently in everyone's bodies. So how do you get everyone to drink the same waters again? Sometimes we think about Kapūkakī, and what's happening with the Navy as the answer to that. But still people hold back. They haven't drunk the jet fuel. They haven't built empathy because they need to actually feel it. They actually need to be the same or have the family feel it.

So I guess that goes back to social media and just you do what you can. You go work the stream by your house, one day you're gonna show people the eggs from the chickens you raise and you demonstrate community building 'cause people forgot—shoot, you forgot too. Maybe that's how, so that's what community water looks like in our community.

## Shimanchu Solidarity

### *Janice Toma Shiira*

*Janice Toma Shiira is a lifelong Uchinanchu resident of O'ahu. She grew up in Kalihi and Liliha before her working-class family was displaced to 'Ewa. A graduate of James Campbell High School, she earned her degrees in public administration and political science from the University of Hawai'i West O'ahu and worked in the financial sector for forty years until her retirement in 2021. A mother of two daughters, Janice now resides in Mo'ili'ili.*

*Her volunteer work involves assisting with various community events in 'Ewa Beach; canvassing and working on political campaigns for elected officials; volunteering at Ukwanshin Kabudan events; and creating and leading the Shimakutuba (island language) exhibit at the Hawai'i United Okinawa Association's annual Okinawan Festival. Most recently, Janice created the Shimanchu Wai Protectors, a grassroots group of Loochooan residents of Hawai'i, dedicated to defending O'ahu's life-giving wai from the Red Hill Facility.*

I first heard about the Red Hill crisis [in 2021] via the news. I was angry that the Navy covered up their mismanagement of the Red Hill Fuel Tank Facility and did not let the public know about the leak in a timely manner. The truth is they lied about it. I was really sad and continue to feel that the Navy has not been a protector for their own military service members and families, as well as all the civilian community members that live in housing on the Navy's water line.

As a third-generation Okinawan settler and a water drinker, I felt a sense of responsibility to get involved, as a protector of my birth land. I believe my Okinawan background helps me to understand the similarities between the Indigenous Kānaka Maoli fight to protect their environment and the Indigenous Loochooan community that is also fighting the same issues with the military. [My background] helps me to stand strong with my birth land's community as we fight for environmental justice here.

*Authors' note: The main island of the former Ryūkyū Kingdom, Okinawa has been subject to U.S. military occupation for nearly eighty years, since the 1945 Battle of Okinawa. Currently, thirty-one U.S. bases, or 70 percent of the U.S. military's installations in Japan, are located in Okinawa. Decades of associated environmental contamination, economic depression, public health impacts, and social issues—including hundreds of documented sex crimes by U.S. military personnel and contractors—have sparked years of protests by Uchinanchu and other residents of Okinawa against the U.S. military's occupation of their island.*[35]

I wanted to get my friends within the Okinawan community [in Hawai'i] involved in the Red Hill crisis, by reminding them that in our ancestral homeland, they also have issues with water and land contamination created by the military. Thankfully, those that we contacted to join in our efforts to stand in solidarity were excited to know that in some small way they could help bring awareness to the crisis. The members of our group felt it was our sukubun—in the Okinawan language, this word roughly translates to "responsibility." In English, the word "responsibility" often carries negative connotations—a chore, something you must do or suffer consequences. Sukubun is different, in that it reflects opportunity and even honor, to be able to carry out this "responsibility" for oneself, one's family, or community. Therefore, I and others in our group personally felt our sukubun to take action against injustices committed by the military against the people and lands of our home.

While most of the members of our group have Okinawan ancestry, one of our members does come from one of the other island groups in the Loochooan Islands. So it would be erroneous to say we are all Okinawan. In Uchinaaguchi [the indigenous Okinawan language], the word "shima" is often oversimplified to mean "island." In the traditional usage, shima refers to one's home village, or even a small area within a village. In many diasporic communities, the term "shimanchu" has evolved into a broader idea, including the Loochoo archipelago, which consists of several island groups, each with its own unique language and culture. Therefore I felt it fitting to call ourselves Shimanchu Wai Protectors, rather than Okinawan Wai Protectors.

The Shimanchu Wai Protectors have since banded together with all the water protector organizations to establish a united front, educate the public, and keep pressure on the Navy, Department of Health, and Environmental Protection Agency. We have held sign-waving events and submitted testimonies and attended in-person meetings. We did a water distribution at Pearl City Peninsula for the Navy housing and loved engaging with the affected family members as they thanked us for the water. Other memorable moments included walking with the other wai protectors to deliver the eviction notice [to the U.S. Navy] and going to the [Navy Exchange Mall] to pass out ice cream and information about the Red Hill fuel tank leak. Also memorable was participating in the Walk for Wai and seeing all the energy of the people coming together to continue our fight for clean water. I would like to see more of the Loochooan diaspora get involved as many are residents on Oʻahu. I am hoping that they will feel a sense of responsibility to our ʻāina and stand in solidarity too.

## In a Colonizer's World

*Mai Kapuaoʻihilani Hall*

*Mai Kapuaoʻihilani Mei-Lin Hall was born and raised on her one hānau of Kalihi, Oʻahu. She received both her bachelor's and master's degrees in education from the University of Hawaiʻi at Mānoa, with an emphasis in early childhood education and Hawaiian culture. A former preschool teacher, public school teacher, Hawaiian culture kumu, and nonprofit program manager, Mai also served for seven years on the Executive Board for the Hawaiʻi affiliate chapter of the National Association for the Education of Young Children. She currently works as a children's health coordinator for the Hawaiʻi Children's Action Network, a local nonprofit. A parent of two children with disabilities, Mai is also a military spouse whose home was on the Navy's water system in November 2021.*

It was a Sunday and the neighborhood was out having a party. We all noticed a smell right away in the morning. We talked to each other, confirmed that we all smelled the same thing, and decided to call emergency maintenance. Maintenance did not pick up. I called the [Honolulu] Board of Water Supply and they said they didn't have jurisdiction over our water, and to call the Navy. The number they gave me didn't work. I didn't know whom else to call, or how to ask for help. We went out to buy bottled water to drink. We ordered food so we wouldn't need to wash any dishes. We still bathed in the water, but I told my son to not drink the water. This was all on Sunday.

Then my family and I began experiencing the following symptoms: gastrointestinal issues like sore stomachs, diarrhea, nausea, vomiting, bloody stools. We had headaches too. Our cats got sick and vomited as well. Yet daily life had to move on, as the adults in the household had full-time jobs and had to continue working. We spent money on takeout because we couldn't wash dishes. It put a lot of strain on my immediate and extended family as we relied on them to help us through this hardship. We had to drive to my mother's apartment to shower and do laundry.

Over time, other health issues became more apparent, with seemingly odd issues like brain fog, trouble concentrating, and attention issues, with more complaints from our son. Everyone just had a harder time focusing on work and school, and we had no rationale or explanation why. My daughter and I also had a few new medical diagnoses since the water incident and we have no way of knowing if the fuel caused those new symptoms.

Doctors at military treatment facilities didn't believe us when we said we were affected. It took months to get a doctor to put "water contamination" into my diagnosis. I'm not even sure if my children's records document that, even though I asked. I understood that our issues were not as complex as others and felt grateful that we had my mother and sisters here to help us. I know others were not as fortunate.

*Authors' note: As affected residents helped each other respond to and navigate the water crisis, community groups such as the Oʻahu Water Protectors and the Sierra Club of Hawaiʻi also increased their advocacy and outreach around the Red Hill Fuel Facility, and worked to provide resources and information to affected communities. Some affected residents, like Mai, would begin to join the larger community's actions, despite a prevailing fear of retaliation and the challenges of life without access to clean water.*

I started going out when I felt well enough to. I met others affected and we got along just fine. It was fine testifying at hearings or writing statements to our leaders and elected officials or making phone calls to our senators and representatives. I can easily do those things. I just found it difficult to show up in person at events because I had a job and a child who has lots of therapy in the evenings. I felt sad that I couldn't participate in everything I wanted to do. I miss the weekly pule at the koʻa.

As a Native Hawaiian, it is in my blood to protest. It's sad that we have to. I remember protesting Mākua Valley and linking arms with kākoʻo against the military. I remember marching at Kū i ka Pono, and Onipaʻa, and other events in red shirts, blue shirts, or whatever colors we chose for the year. We Hawaiians know how to mālama each other and to show up. We know how to kapu aloha and to show maluhia in every aspect of what we do. It is expected that we kapu aloha and to demonstrate that we understand this process well.

Unfortunately, currently, there is strain in the military community. Many residents have moved on and don't want to talk about the water. I have been asked to stop posting so many articles and links to community events on our neighborhood page because I was told "no one wants to hear that anymore." The military community just wants to forget the incident happened. I have been bullied for living here and protesting the way I did. I have been asked to move off my island since I complain so much about the military. To be asked to leave my own homelands by colonizers! The nerve!

I have been asked to take down my signs on my home and to not pass out my "Shut Down Red Hill" stickers to friends. So I just stopped for a while because the military community did not support me for what I was doing for them. I needed a break from advocating for my own sanity. It is a challenge to see racism, classism, and colonialism existing loud and proud. Being told to leave my kuʻu one hānau, having my husband told he should leave the military, and seeing the officer spouses receive far better treatment than I ever could, it all makes me severely disappointed and sick to my stomach.

Those who could afford to tackle this head on had the power and resources to be heard. They could figure out how to move off island and still manage a life for themselves. But for those who are tied to this ʻāina, what choice do I have but to fight to coexist with the military? I cannot tell the military to just pick up their ʻopala and leave, even though that is what I believe with my naʻau. I can't say that. I know it will never happen. Hawaiʻi is tied to the military and they will never, ever go away. So we have to compromise and bring to the table real solutions that maintain our independence and ʻāina.

I have learned enough to know that the military will never leave our islands. It makes me sad. But we must learn to coexist and to force the military to exercise caution, kuleana, and diligence in whatever they do. Those that do not should be the first ones to leave.

### He Pūnāwai No ke Ola

*Dr. Kalehua Krug*

*Dr. Kalehua Krug of Lualualei, Waiʻanae, Oʻahu, is a father of three, and has dedicated his life to education, primarily in Hawaiian language immersion. He is now the principal of a charter school in Nānākuli, Ka Waihona o Ka Naʻauao. He is also one of the leading members of Kaʻohewai, a coalition of Hawaiian organizations rising in defense of Kapūkaki and dedicated to the shutdown of the Navy's Red Hill Bulk Fuel Storage Facility.*

*At the end of 2021, shortly after the November fuel spill, Kaʻohewai led the building and dedication of a koʻa at the doorstep of the command headquarters of the U.S. Pacific Fleet. As explained by Andre Perez, another leader of Kaʻohewai: "It's a ceremonial place where our lāhui, and others who support the long-term vitality of Hawaiʻi's waters, can call upon our akua (gods), ʻaumākua (ancestral gods), and kūpuna (ancestors) to help restore life and full health back to Kapūkaki. It is also a place where we can commit ourselves to do the same."*

*"The koʻa will remain in place as a focal point for Hawaiian religious practices until the Navy removes its fuel tanks. Kaʻohewai is urging all those who care about our precious water resources to offer at the koʻa their prayers and wai from their own lands. Koʻa are ceremonial focal points that draw and multiply abundance, health, and ola (life). Koʻa can focus on fishes, birds, or in this case, life-giving water and all the akua associated with ola."*[36]

> *Beyond and as part of the community response building of the ko'a, the ko'a includes yet another opportunity to rethink the ideologies, methodologies, and frameworks being uplifted and honored with the protection of our wai for ourselves and generations to come.*

The ko'a is this resource magnet. . . . The kūpuna would build a ko'a to draw resources, to utilize spirit and our ancestors and the gods to enter into a support role as we ourselves try to do as well, for our ecosystems. Ko'a were built not just in times of famine and need. It was just a basic part of the Hawaiian response in the ecological landscape.

The [Ka'ohewai] ko'a started off to attract resources and, in this case, attract cleanliness, attract clean water, attract proper resourcing, attract people and advocates to be able to help move what is right forward. We were just going to build that ko'a as many others have been built in the water to protect the baby fish in fish ponds. . . . There are ko'a on the island of Kaho'olawe to attract rain. So [utilizing the ko'a], it's just following cultural precedence.

It's not an ahu because an ahu is a more permanent structure, more something that has to be given to a deity, and it's a given responsibility that you feed it, in the traditions. A ko'a can stand there as something for all of us. It doesn't just have to belong to one person or one organization. And so the ko'a belongs to the water, it belongs to our people. It belongs to the concepts of Kāneikawaiola, and life itself.

It was dedicated to Kāneikawaiola, and we put an image in there. It was serendipitous that Andre actually had an image of Kāne that we put in the ko'a. The stars kind of aligned in that moment to be able to utilize our spiritual practices and beliefs as our primary and initial step of advocacy.

I think in the space that we are in, that our gods are the characters of our stories, and our stories drive our behaviors. If we're not using our stories, if we're not using our gods, then we're actually not learning from our past. We're not actually utilizing methodologies of our kūpuna.

I think in the community response, there were mixed feelings. I think in the time frame that we live, most of our people are Christian now. By building [the ko'a], I don't know if our community at large was ready. Many of our community members feel like we left the time frame of our ancestors, and there's no reason to go back to that or to utilize those beliefs. So there was some conversational trouble, I think, for some people when it came to dealing with the koʻa.

But I do believe the significance of utilizing a traditional structure moves people of our lāhui a little bit differently. A part of changing the narrative and utilizing our gods and our stories, is again, a methodology for us to change the behaviors that got us to this point. I think the basic premise in the Indigenous realm is that our language and our practices and our culture is in a place of resurgence.

*Authors' note: after the building of the koʻa, Dr. Krug, together with his ʻohana, wrote and released "Ola i ka Wai," a song that seeks to question the western ideologies that have led us to where we are today, and how we can make an ideological return to a more sustainable existence in Hawaiʻi.*

**"Ola i ka Wai"**

Āhea ana lā e ʻike ai i ka waiwai
O ka ʻike nā kūpuna
He pūnāwai no ke ola
Hui:
Ola i kākou i ka wai
Ka wai ola a Kāne
Neʻe papa hoʻi i nā kānaka
No ka pono o ka ʻāina
Nā ka nuʻa o kānaka
Ka wai e mālama mai
He mālama ʻana ia
E lōkahi ai ʻo Hawaiʻi

[Our ancestors and our kūpuna] collected longitudinal data for thousands of years on what the oceans and the skies and the mountains and the plants and the animals, all of this place, needed from them. The beauty of what we offer here in Hawaiʻi is the methodology of how to embrace the ideology that humans are here to enhance the natural world in the way that the natural world teaches us to enhance it. That the world is not built for us, and we are not made to rule it.

And we have tangible evidence from our ancestors and lots and lots of newspaper articles and lots and lots of people still alive who embrace the values of our tradition and who were raised in such a way to be able to share that narrative with us in the present.

I think the hope of the song was to help whoever listens to it, now or in the future, in whatever way, to ask themselves the question: When will we stop seeing our antiquity as primitive? When will we start seeing where we come from as actual progress, and see the indicators in our world, especially like fuel leaks in [Oʻahu's] main aquifer, that our current ideology is wrong? That the core values that we currently carry are wrong? I don't think it's easy for people to look at what they may have built their entire lives upon as being potentially flawed. But that's what the song is asking.

It's asking to see that maybe we came from more wisdom. And we left. We are less now than we were. And that moving forward under the current understanding of progress, utilizing fuel and fuel lines and water pumps, and digging intrusively into the ground to suck out the blood of our island, that maybe that's not the best thing. [It's asking us] to question our behaviors. That was the main idea behind the song. And, to remind people that we need water to live.

Awareness is built through education, and the words and the lyrics of these songs are the vehicles of those educational words. The beats become those catchy tunes that help us to build these devices in our memory and record this in our subconscious as well. So when we do walk through life, and we do speak to our children, the lyrics of these songs, because they become ingrained into our subconscious, they become a part of the way that we describe our experience.

I think it's important for all of us to have a description of our experience and have it tied to an emotional response. I think that these songs do a good job of getting people to that emotional response.

*Authors' note: During the Summer of 2022, during the Rim of the Pacific Exercise or RIMPAC, the koʻa was desecrated. In response to this desecration, an act that many Hawaiians must continue to endure in their homelands, Dr. Krug led a campaign to once again call for unity with an "ʻAnahulu at the Koʻa." This campaign was a ten-day stand at the site of the koʻa, featuring music, activities, sharing of cultural learning, and education as a way to come together to plan for future strategies to protect our wai. The ʻAnahulu called upon others to reflect upon our ideologies and assumptions and become activated in the protection of our wai to prevent us from witnessing the "the death of an island."*

There are existential truths that we're avoiding, and we avoid them under American rule and capitalism because American law is already perceived to be untarnishable or infallible. When you see [American] law in that way, you're gonna run into a brick wall because it means that something written in 1776 should never be changed. Or that you must walk through these processes to "change" it as a voice of the people, when the people don't actually have a voice.

Another really good example is housing. We're not gonna be able to keep building, we're not gonna be able to keep taking the water. Cause the water will go. Then as the placement of dominion that we think we have over the islands, we'll just do desalination plants. And we're not gonna ask ourselves, "Well, how are the trees and the grasses and the birds and the estuary gonna survive?" Cause we don't care, we believe that the world was made for man and man was made to rule it. That's the existential problem that we're here with, and we're just not talking about it because American law doesn't allow us to.

Recorded data, the data that we have available to us, show that our ways kept our islands better than they are now. I think that's the value of our culture, and in not just educational spaces. Hopefully all human beings who dwell in this geographic vicinity can look to those teachings as a guide for all behaviors so that we can continue the process of building more stable ecosystems and not dismantling them and tearing them down like it seems like we're doing now. And that's embedded in hula, mele, ʻōlelo Hawaiʻi, kākau, fishing, everything. It's embedded in everything.

## Nānā i ke Kumu

### *Dani Espiritu*

*Dani Espiritu grew up in Kāneʻohe and Heʻeia, in the moku of Koʻolaupoko, on the island of Oʻahu. A graduate of Kamehameha Schools, Dani attended college in Oregon and then became more involved with Native Hawaiian community issues after her return to Oʻahu. A teacher by profession, Dani has worked to heal ʻāina and restore loʻi kalo for the last decade, and works within the faith community to advance justice issues. She has dedicated much of her recent life to the restoration of ʻāina and kaiāulu at Kaʻōnohi, a kīpuka of Native Hawaiian agriculture and land stewardship in the current concrete jungle of Kalauao.*

*Notably, Ka'ōnohi's lo'i kalo—along with the adjacent Sumida watercress fields—are fed by the aquifer underlying Kapūkakī. The eventual contamination of O'ahu's groundwater would soon lead to Dani's heavy involvement as a member, spokesperson, and organizer with the O'ahu Water Protectors.*

When I moved home [from college] it was a lot more like, you show up to the rally, get your friends to the rally; if there's petitions, we would do that. A lot of it also was through my teaching, trying to subtly infiltrate the [Department of Education], which is hard, but good work. Then when Mauna Kea happened—well not the first round, but when things picked up again—I had left the classroom already, and I had the flexibility to be able to fly back and forth.

A couple of my best friends basically raised their boys on the Mauna when that second round happened, and so I tried to be there as much as possible, to help with the kids and hold space. I was working at a farm and going to school and lecturing at [the University of Hawai'i], so I would try to stack all my stuff within a few days, and then have a week-and-a-half or two-week gap to go to the Mauna, then come back and teach, and then go fly back. That was a big chunk of what life looked like back in 2019, going into 2020. And then that was when all hell broke loose.

There were those Department of Health testimonies for Red Hill—that was the first time I testified in person, which is interesting that it was about this issue. Then some of my closest friends got arrested at Hunānāniho—which was a trip because I'm, like, teaching keiki from Mālama Honua [Public Charter School] from Waimānalo at the lo'i, while the parents and chaperones are watching their kumu and our friends get arrested at Hunānāniho for protecting iwi kūpuna. I had done legal observing on the Mauna, and because of that we then did legal observing here at Kalaeloa, related to the windmills.

So everything just kind of exploded at that time and then we went straight into COVID. We really didn't have time to process—it went straight from that into the pandemic, and then George Floyd got murdered, and then the Red Hill hemorrhage. We haven't been able to process it all.

I do think going through those things and especially around the Mauna—there is something about the Mauna that was so galvanizing, that drew in all kinds of folks, from the Hawaiian community specifically but allies as well, and in a way that I don't know that I've seen in my lifetime. It also equipped so many of us with the language and skills to be able to engage. Like, it was the first time that we're learning about bail bonds and what do you do when you get arrested and what are legal observers, those kinds of things.

*Authors' note: In December 2021, Hilo artist Makaiwa Kanui visited O'ahu, where she learned more about Kapūkakī and the water crisis facing the island. In reflection, she wrote: "It's undeniable that as Kanaka 'Ōiwi we are tied to our 'āina. When one part of the body suffers, we all suffer. It grieves us in ways we don't fully understand. Our bodies reflect the bodies of 'āina and wai that sustain us. Our spirit laments, as our minds do our best to understand, adapt, resist, and fight for those we love. We do not lack injustices, rampant in our homelands. We are tired. Yet we push, and push, and do what we can to hold on to what we have left."[37] Her subsequent artwork Can Our Tears Clean the Aquifer? (figure 9.1) was dedicated to the Shut Down Red Hill campaign.*

When November [2021] happened, I feel like I was depressed for like five months. And I still kind of ebb and flow in that. It's still an added component of the work—the grieving. There's grieving, then action—the coalition building, the organizing is our outlet to put our energy somewhere, and then to try to effect change. And then there is also a spiritual component.

You saw Makaiwa's drawing. Basically she happened to be in town in the first week of December [2021] for something, and we just lay around and cried together, and drove through different places in 'Ewa and Pu'uloa and I would just share mo'olelo about spaces. I feel like for me and my process, and even like engaging in this work, whenever something new happens, like when the PFAS spilled, there's always like a component of a spiritual, emotional, and physical process of grieving, in the midst of the doing. And that balance is human, and also very challenging, and also challenging because collectively we are going through all of that together and individually.

This is tied a bit to my dissertation—wai hū or pūnāwai can be a metaphor for resurgence for Kānaka and community. It came out of helping to restore lo'i kalo in Kapalai, in Maunawili with the Wilhelms—where it was completely overgrown, it was a dump site, completely covered with

**Figure 9.1** Makaiwa Kanui, *Can Our Tears Clean the Aquifer?*, 2022.

hau and cane grass, and fill—like a contractor had just dumped it in the area. I was able to learn from them and be a part of, over the last ten or so years, of clearing all of that, and then, as we cleared it, noticing that the wai would just start to come up from the ground, and then, as we built the kuāuna, and those pathways of connection that will then hold the ability to grow food for generations, the wai continued to flow.

And so it was like the wai—the wai was always pure. The wai was always ready to flow. It was just the toxicity of the world had clogged it and had made the ʻāina paʻa. But as soon as you take those things out, whether it be like settler colonial structures, or indoctrination, whatever those ʻopala are, as soon as you take those out the wai starts to flow and then it heals the ʻāina, and then it heals the people, and then it allows you to grow food that can then feed generations.

That has been my own process. What are the colonial mindsets that we've learned as we've been Americanized, as we've been indoctrinated, that we need to huki out? That then allows the ancestral memory to flow, and that then allows the healing to flow.

And I've seen that on an individual level, I've seen that physically in ʻāina, in seeing ʻāina restored, and I've seen that in ʻohana and in community. And so part of getting rid of all the ʻopala that's been packed on is the unlearning of that. And as we do that, then the remembering of moʻolelo comes up, and the practicing of our traditions comes up, and the restoring of ʻāina comes up. I give all of that context to say that what if the aquifer is contaminated and the water that's coming up is tainted? Then that whole paradigm is shaken.

I think that shook me in particular, because it's like, well, if the core of our island is poisoned, then what else is there? It's not just drinking water, and then we go tap someplace else, and then we have to just monitor chloride levels, which is what they're doing already. But really it's the death of the island from the core of it. The core of Oʻahu.

Feeling that emotionally, and spiritually, and physically, was significant, and then I think it's compounded because I had just started working at Kaʻōnohi [Farm]. Whatever chemicals in their facility which they've still not disclosed have spilled into the aquifer, to pollute and contaminate our springs and all our ahupuaʻa, between Kapūkakī and Kalauao. And so it's jeopardizing anyone moving forward from being able to connect—in the growing of food. The animals, the streams, the ocean. All of that.

*Authors' note: In 2015, decades of Native Hawaiian and environmental concerns regarding the mismanagement and overdevelopment of the sacred and ecologically fragile Mauna Kea summit*[38] *came to a head, after the state gave the go-ahead for the heavily contested Thirty Meter Telescope—an eighteen-story-tall observatory with a five-acre footprint—to begin construction. A partial détente after the arrest of over thirty Mauna Kea protectors engaged in civil disobedience—and after the success of procedural legal interventions*[39]*—ended in 2019, when attempts to resume construction were met by a sustained occupation involving tens of thousands of Native Hawaiians and others from across the islands and beyond. To this day, construction of the billion-dollar Thirty Meter Telescope remains halted.*

*Many have observed how the convergence of individuals, networks, and lessons from local and Indigenous movements in Hawaiʻi, the continent, and across the Pacific informed the Mauna Kea Movement, and how the Mauna Kea Movement has also informed Native Hawaiian and larger community efforts to push back against environmental and social injustices, and reclaim Kānaka ʻōiwi stewardship and governance over ʻāina and wahi pana across the islands.*

We were raised super American—the intensity of the colonization is real. So I say that because seeing how my family has engaged in this issue has been really transformative. It started with Mauna Kea, where in the beginning it was like, "Ah, what are they grumbling about now?" And then I was like, "Well, you know my friends, and well, you know their kids, and they're living on the Mauna. And you know me, and I'm going." So it helped to kind of soften that. And so one of the things that's been beautiful to see is the ways that my family, who are *not* political, have engaged on this issue.

I think one of the challenges that I've been trying to think through is like, we have a movement. But how do we continue to build the movement and the culture of the movement? I've been reflecting a lot about [this], especially because I sit in these different communities and circles. Once we're able to bring folks along together collectively, that's when I think things will shift and I think with the Mauna, we saw that. But with this issue, I don't know that we've seen the tipping point, and I think that's one of the challenges. And perhaps it's tied to the one before—where we went from a season of back-to-back intensity, where so many people just put their bodies on the line for a year.

I've always seen myself in a supporting role. But with this, I'm available, I'm not raising keiki right now. I have the capacity to be available, whereas others, who I would have looked to, are needing to take time for ʻohana, for other kuleana.

So I don't know if it's a challenge but it's just been something that I've been reflecting on over the last year, like how do we, as a community, as a lāhui, support ourselves as we ebb and flow, and then also keep everybody together. Unity is a constant challenge when you have so many different people.

### He ui, he ninau: What Is the Future of O'ahu?

*Wayne Tanaka*

"He ui, he ninau" is a query that starts the 'oli "Aia i Hea Ka Wai a Kāne," honoring the many lifeways and waterways where water gives us life. With O'ahu and Hawai'i still hanging in the balance, we asked these community members and leaders to reflect upon their hopes and fears for the future, as we collectively re-remember that there is no future without abundant, clean wai.

As you will read below, the Navy's modus operandi remains a source of daily fear: continual delay tactics, wielding its power to push a narrative that prioritizes its needs at the cost of all else, and evading all accountability for its actions, in Hawai'i and elsewhere. For Jamie Williams, who was a member of a directly impacted family, her fear hits much closer to home, with a lump in her breast that she is being evaluated for. The Hall 'ohana similarly fear unknown long-term health effects from their own exposure to contaminated water. The ultimate fear for those with the deepest connections to the 'āina is the death of O'ahu. But they also hang onto hope.

Their hope is not only for everyone to come together, rise up, and stand for our water, but to also raise consciousness to see the connections to other issues we should care about as well. Their hope is that we will think of the generations to come, and rethink Hawai'i's relationship with the military. Their hope is that we will remember the value of every person in our community, from keiki to kūpuna, and remember that water is life.

The continued fight for the protection of our water is not necessarily unique to Hawai'i. However, Hawai'i is unique in the ways that we have stood up for the protection of our wai, our 'āina, and the people and living things that make Hawai'i so special to us all. Answering the call to stand for our water, that too is the waiwai of Hawai'i.

Let us not forget that we are made up of water ourselves, connected by water, and that water is life. As you read these final thoughts, consider your own connections and relationships and attachments to your ʻāina, to your wai, and what you can do to build community, re-remember the preciousness of life, and protect your live-giving wai, and all that you love.

Aia i laila ka wai a Kāne.

### *Jamie Williams*

My hope is that the EPA steps up as a regulatory body. Why is there not a proactive stance? And if you can't be proactive, at least flex your regulatory muscle.

I also hope this puts a spotlight on what the military is doing in the mainstream media. Prior to the Red Hill leak, I didn't realize the systemic culture of passing the buck and the utter disregard for the environment and human life was as bad or as rampant as it is, worldwide.

I do so much posting about these issues on social media to try to raise awareness. The amount of people that were commenting on national news articles about Red Hill saying, "Oh yeah that happened in New Mexico," and "You know an even bigger spill happened here. . . ." We're normalizing this and that's not OK. My hope for the future is that enough of a spotlight is placed—because again, you have to publicly embarrass the military to get them to do anything—so that this violence—because that's what environmental harm is—stops.

My fears? Right now I have a lump in my breast that I'm being evaluated for and the thought does cross your mind: Is this the result of the chemicals that I ingested? To this day the Navy still has not disclosed all of the compounds that were in the fuel mixture. I worry about the future health of my family and the community of Oʻahu at large. A constant fear is, where do we go from here? We've screamed into the void for over a year. And I keep thinking, this is finally it—this piece of information is going to be what gets the regulators' attention, or the governor's going to care, or somebody's going to do something—and then nothing happens.

The wheels of justice move slowly through the courts, I know, but it's really disheartening to see the Navy accept culpability for all of this, and then ignore a federal tort claim. I worry that there may never be any consequence for the Navy and that similar events just perpetually occur. In Alaska, in every state in the U.S., and worldwide—we're the biggest polluter.

*Authors' note: Jamie Williams now resides in West Virginia. She hopes to one day return to Hawai'i to resume her law school career. Aia i laila ka wai a Kāne.*

## Marti Townsend

Remember the military correctly predicted the massive fuel leaks from the Red Hill tanks multiple times—in reports published decades ago, and most recently in 2019. They knew this was going to happen but did nothing about it. The military abuses us and our environment and calls it "maintaining the peace." This is not peace. We are not safe. The military's sole mission is war and there is nothing safe, peaceful, or protective about that. The sooner we recognize this reality, the sooner we can work together to collectively free ourselves from this cycle of abuse we are trapped in.

That is why resistance is so important. We have to make sure the whole world knows that we do not consent to the U.S. military contaminating our water, bombing our land, destroying what we love about Hawai'i nei. We make clear to every politician that we do not care how many jobs the military "creates" in Hawai'i; an economy that exploits, oppresses, and pollutes is no economy for people to live on. We link arms with people around the world who are suffering the injustice of U.S. military bases in their communities and say no more. We are dealing with a global bully. The response calls for a global intervention.

This is no small task. It means dismantling the whole system of exploitation that humans have managed to barely survive in all these years. It means pulling up the roots of our economy that feed off war, and planting the seeds of what we really need to thrive: clean water, local healthy food, respect, holistic education, reciprocity, innovation, art. It means dropping capitalism once and for all.

What gives me hope at this point is imagining a distant future in Hawai'i where everything is okay. It is better than okay, it's epic. It is a place where no one is thrown away and everyone makes a meaningful contribution to their community. I imagine a future Hawai'i where our economy is rooted in reciprocity, not war. Where each of us are the seeds of Hawai'i's epic comeback in the distant future. I am not talking about [getting rid of] plastic straws or riots—although I admit the logic is consistent. I am talking about speaking truth to power with honesty and respect. I am talking about being an antidote to the poison that oozes from a society that thrives off war and exploitation. I am talking about applying the lessons learned from the Kiaʻi Mauna movement to stand firm in Kapu Aloha for the better future we know is possible.

> *Authors' note: Marti Townsend continues to advocate for Hawaiʻi's ʻāina, wai, and people as the regional engagement specialist at the Earthjustice Mid-Pacific Office. Aia i laila ka wai a Kāne.*

### *Ernie Lau*

My hopes are to get the facility defueled, sooner than June of 2024. And I hope we can continue to pressure the Navy and the regulators to try to do it faster. Right now, there's momentum built up through everybody's efforts. But if the momentum dies down, the military, and the regulator, will also lose momentum. This could start to drag on longer than June. [The Navy is] gonna come up with a reason, and then the regulator is going to let them get away with it, and then it's going to be 2025 and we will still have 104 million gallons [of fuel above our aquifer]. I hope that never happens, because 104 million gallons is like a sword hanging over our heads. At least the EPA—a few times now they've told me they got over 1,700 comments from the community on the draft [2023 consent order]. They told me that to my face in an in-person meeting, and it appeared that they were a little nervous that they'd received so many comments. And I hope that they are now having to go back and look at changes to the consent order. The question is gonna be whether the Navy is gonna agree or push back. I don't trust the Navy. You know, when they talk about other beneficial uses, and they hired a consultant on "beneficial reuse" [of the Red Hill Facility]—that concerns me. The closure needs to be permanent, so it can never store any kind of fuel or chemical again. That's my concern.

The other one is the investigation or remediation cleanup of the aquifer. That's the other rail that's needed here. It amazes me that this story is repeated all over the world, especially the Pacific. So, I hope it's an opportunity maybe to create a network, across our country and maybe even to other countries, to share experiences because for us, what the Navy was telling us, all these nine years, was that what we were experiencing was unique to us. And that they couldn't replace the fuel tanks. This baloney, this propaganda that our experience here on Oʻahu is only a one-off—they weren't telling us the truth that they've got similar issues in Okinawa, Guam, and Saipan, and maybe even in Japan too, or other countries or other states. They are conveniently not telling us. They are not transparent. They're the big Navy, we're the locals, they have the ships and the weapons, and we are collateral damage that is not important.

*Authors' note: As the Board of Water Supply chief engineer, Ernie Lau, along with his dedicated team, continues to advocate for the protection of Oʻahu's water, while endeavoring to take on the precautionary measures needed to navigate the current contamination crisis. Aia i laila ka wai a Kāne.*

### *Healani Sonoda-Pale*

In this moment, it's constantly on my mind, and especially when I'm around my children—I always think about whether they'll have clean water in twenty years. Whether their children's children will have access to clean water in 100 years. As a mother, as a Kanaka Maoli woman, the question for me is whether our family, our community will be able to survive here on this island for the next 100 years, for the next 1,000 years.

Right now, the way that the military is moving and dragging its feet, the future looks pretty bleak. No one has been held accountable for what has happened, for thousands of people being poisoned in November 2021. Not one person. And so there is nothing to encourage the U.S. military to do better so they're going to continue doing what they're doing.

But when we look at all the things we've achieved. . . . Everybody said they would never agree to shut down Red Hill. But they did. And then you think, okay, there is hope. Right? There's always hope. But that's what is difficult to hold on to when you're in the trenches, fighting, exhausted, organizing, urging people to testify. So there is hope, but it is hard to hold on to that.

Something has to change. We can no longer allow the U.S. military to continue contaminating our water, our land. They're destroying our economy, too, driving up housing costs, destroying our sacred sites, continuing to sit on stolen Hawaiian lands. We should be calling for them to leave Hawaiʻi. It's time. It's very clear to everyone who lives here and who values life. The U.S. military is the greatest threat to life on this island.

A lot of people have come to that conclusion. Not just Kānaka Maoli, not just the activists who have been fighting for the demilitarization of our islands, but also now a wider group of people who otherwise would have never seen how the U.S. military is actually a danger to life here on this island. Even people in the military have come to the conclusion that the U.S. military doesn't care about life. They don't care about protecting even their own people, let alone residents, let alone Kānaka Maoli. The U.S. military needs to leave.

*Authors' note: Healani Sonoda-Pale continues to organize on multiple fronts to restore Kānaka Maoli stewardship and governance over 'āina, wai, and iwi kūpuna, and to seek a brighter and more just future for the lāhui and all of Hawai'i nei. Aia i laila ka wai a Kāne.*

### Susan Gorman-Chang

I hope the Red Hill Jet Fuel Storage Facility is defueled safely and earlier than 2024. I hope it is entirely decommissioned and the tanks filled with an inert substance so that they are never used and can never be used for anything ever again. My fear is that the Navy will be successful in stalling the defueling until the next presidential administration and I fear, if that administration does not value our lives and our aquifer, that the Navy will be allowed to continue using the tanks for fuel. The worst-case scenario in that case would be our aquifer getting poisoned, and O'ahu rendered no longer habitable.

*Authors' note: Susan Gorman-Chang remains active in her church and as the chair of Faith Action for Community Equity's Environmental Justice Task Force. She continues to spearhead a charge to get all of O'ahu's neighborhood boards—representing every community on an island of over one million residents—to pass a resolution urging the timely defueling and closure of the Red Hill Facility. As of this writing, twenty-eight of O'ahu's thirty-three neighborhood boards have passed such a resolution. Aia i laila ka wai a Kāne.*

### Rev. David Nakamoto

For me, as I reflect on life—especially now, in our so-called golden years—I see that opportunities are driven by our choices we make as we interface with the influences that come our way. With senior living, there may be limitations based on our individual unique circumstances, including physical capacity, but it is important to remember that a critical valuable resource for community enrichment lies in the senior community. My hope for the future is that community organizations, like the Sierra Club of Hawaiʻi, continue to value the potential for senior involvement in community advocacy efforts, using unique and creative approaches to tap into retirement communities.

*Rev. David Nakamoto and his wife, Irene, continue to reside and work with their fellow residents at 15 Craigside, to enrich their lives and the lives of others—and to pass on the spirit of "Okage sama de" in their work as kūpuna seeking to protect the future generations of Hawaiʻi nei. Aia i laila ka wai a Kāne.*

### *Aurora Kealohilani Kagawa-Viviani*

What gives me hope is—I've said this in multiple places—there's never been a time when so many people were riveted on water. We need to take advantage of that and train people to increase their water literacy and understanding about the many facets of wai. Including the water that we take for granted, coming out of our taps.

Out of sight out of mind—we don't think about our corroding infrastructure, or how climate change is going to affect our water supplies and our infrastructure and our delivery systems and our wastewater. And our contamination issues. We are a heavily urbanized island. And we have not given real thought to water quality—we've taken it for granted, for far too long. We have—you know how there's American exceptionalism? We have Hawaiʻi exceptionalism, and we think our water is so special, and we take it for granted. And we're not doing enough to steward it individually. So I think it's a challenge for everyone. It's a challenge for the [Water] Commission to better address water quality and quantity issues together. We really need to retool how our agencies work and communicate with each other and collaborate with each other. And we really do need all hands on deck. So I think arranging the relationships in ways we're all pulling together, instead of pulling at cross purposes, should be a top priority.

My biggest fear is realizing how much our aging infrastructure, plus climate change, plus legacy contamination—not just Red Hill but everywhere—is gonna put us in some trouble. We also have an aging, retiring group of workers in the water sector. I'm not sure who's going to replace them. So another fear is that we are so behind on training young people to work in water and ʻauamo kuleana in this space. But no better time than the present!

*Authors' note: Dr. Aurora Kealohilani Kagawa-Viviani continues to serve on the University of Hawaiʻi's Red Hill Task Force, aiding their search for knowledge that can be "the antidote to fear and anxiety linked to lack of information and transparency" that have accompanied the Red Hill crisis. Aia i laila ka wai a Kāne.*

## Noel Shaw

When I think of the future of Hawaiʻi, really if we can rid ourselves of the military, we would be so much better off. If we could shift Hawaiʻi back to he aliʻi ka ʻāina. . . . But not just Hawaiʻi, the whole continental United States, the whole world back to that, we would be a lot better very quickly again. The other day I was thinking: Imagine if the EPA had the Department of Defense budget? Imagine if we shift the whole perspective away from defending to protecting? Well, that would just be explosive, right? I don't go into my mothering every day defending. I'm there to protect, and that is a nourishing and sustaining relationship. If people ask: "What's your theory of change?" It's family, ʻohana is a theory of change. All of these collective things working together to support a huli to support generations to be able to plant the next generation, then the next generation. That's the whole point of being Native or Indigenous, you know somebody comes after you. For some reason, all these other people in the world forgot that the whole point of existing is you're here to make it better for those people coming after.

*Authors' note: Noel Kaleikalaunuokaʻoiaiʻo Shaw continues to be "just a mom," and is active in the continued care and protection of places and people and creating safe spaces for all to join and mobilize for justice. She continues to activate the lāhui in re-remembering the paths our ancestors already created for us with Hui Aloha ʻĀina ʻo Honolulu and pushing for ea (Hawaiian sovereignty—in its many forms, political, intellectual, and educational) and demilitarization. You can follow her on Instagram at @noelchu. Aia i laila ka wai a Kāne.*

### *Janice Toma Shiira*

What gives me hope is that people involved genuinely know that water is *life,* and we need to be in a united front in our efforts to reveal the ugly side of the Navy. We must not let the Navy disregard and disrespect the residents on Oʻahu. I also hope that we have a stronger state agency—that the Department of Health will hold the military accountable to clean up and protect our environment. My fear is that the Department of Defense will delay the defueling, and not consider this fuel leak crisis as a state of emergency. My greatest fears are of a catastrophic fuel leak, or the use of fire suppressant containing PFAS at the facility—the forever chemicals.

> *Authors' note: Janice Toma Shiira continues to organize within the Loochooan community on Oʻahu through the Shimanchu Wai Protectors and other community groups—pushing for the revitalization and perpetuation of Shimakutuba and Loochooan culture and identity, and for the U.S. military to be held accountable for its actions in Okinawa, Hawaiʻi, and throughout the Pacific. Aia i laila ka wai a Kāne.*

### *Mai Hall*

I really, truly hope that my family does NOT develop cancer from this. That's my biggest fear—for long-term health problems to occur. I want the military to contract out someone else to clean up their mess, since they seem to be incompetent in clean-up efforts. I do not trust the military. I never have, and never will.

> *Authors' note: Mai Hall currently serves as the co-chair of the Honolulu Community Children's Council and as a parent representative on the Special Education Advisory Council, advocating for children with disabilities and special health care needs. She also continues to advocate for the ʻāina, for military accountability, and for all who call Hawaiʻi home. Aia i laila ka wai a Kāne.*

### *Kalehua Krug*

My hope is that we can actually get to a place where the military is relinquished of their responsibility of caring for our land and our resources. I think there are a lot of good people in the military, but the military is constructed in such a way that orders come from people and places that have no bearing on our islands. So there's always the construct law of those systems that the orders will always take precedence over the actual observable needs of the natural world. My hope is that we can relinquish that responsibility for our people. And really, re-situate a Hawaiian worldview, as far as ecology is concerned, with land and resource management systems. I think my fear is that we will kill Oʻahu. And I don't know if Oʻahu is willing to sacrifice itself so that we can save the other islands and utilize what we learned in the death of this island to not do it again on the other islands. I hope that doesn't have to happen.

*Authors' note: Dr. Kalehua Krug continues to dedicate his activism, professional career, and personal life to an ideological shift that he believes we must adopt in Hawaiʻi, that recognizes how our continued existence is dependent on returning to the time-tested ideology and science of our kūpuna. Aia i laila ka wai a Kāne.*

### *Dani Espiritu*

I keep pointing back to the Mauna. Perhaps for me it's because it was a galvanizing thing. But I remember the first time that I came back home from there, driving on the freeway, and I just had this moment of *realizing*. It was like sadness and also like an "Aha!" moment. With the ʻaha three times a day on the Mauna, and the consciousness of everything that you're doing and producing—because it all needs to be hauled out, or our footprint will be lasting. We're conscious of what kind of chemicals we're using, what kind of toothpaste I'm using up there, you know, all that kind of stuff. And I think especially for us on Oʻahu, because there's so much development, we're just so numb to the fact that those things have already been paved over, and they have been already for several generations. But there was like a consciousness raising because of how we were expected to conduct ourselves on the Mauna, that reminded me of the importance of translating that elsewhere and connecting that to other issues.

I guess that's one of my hopes—that there is a consciousness raising for people so that it's not just "Oh those are just radical crazy Hawaiians," or, you know, like tree huggers, or whatever the stereotype is. But like my advisor talks about it, that there is a defamiliarizing of what has become so familiar, and realizing what is actually crazy. Like having to live off bottled water—*that* is crazy, that the water coming out of your faucet isn't healthy. That is crazy. We should have clean water. Period. And so I hope to defamiliarize with things that have become really familiar for us in a way that can create broader, collective awakening and change.

*Authors' note: Dani currently resides in ʻEwa and continues to spend much of her energy taking care of kūpuna and healing ʻāina with Hoʻōla Hou iā Kalauao at Kaʻōnohi Farm, and in educating, advocating, and organizing around the Shut Down Red Hill campaign. Aia i laila ka wai a Kāne.*

## Notes

1. Catherine Cruz and Sophia McCullough, "Navy Says Pearl Harbor Friday Fuel Leak Contained, Sheds Light on March 2020 Leak," Hawaii Public Radio, July 19, 2021.
2. Christina Jedra, "Amid 'Political Concerns,' Navy Kept Quiet about Red Hill Fuel Leak into Pearl Harbor," *Honolulu Civil Beat,* October 8, 2021; Christina Jedra, "Whistleblower Says the Navy Gave False Testimony about Red Hill Fuel Facility," *Honolulu Civil Beat,* November 9, 2021.
3. Christina Jedra, "State Fines Navy $325,000 for Environmental Violations at Red Hill Fuel Facility," *Honolulu Civil Beat,* October 27, 2021; Department of Health, Notice of Violation and Order No. 21-UST-EA-01, October 26, 2021.
4. The completed portion of the risk assessment, looking only to "internal events" and not external events like fire or flooding, found that over 5,000 gallons of fuel was "expected" to be chronically released from the facility each year, and that there was an 80 percent chance that up to 30,000 gallons of fuel could be suddenly released over the next five years. Honolulu Board of Water Supply, Board of Water Supply's Post Hearing Memorandum, Proposed Findings of Fact, Conclusions of Law, and Recommended Decision, July 13, 2021, p. 23, citing ABS Consulting, "Quantitative Risk and Vulnerability Assessment Phase 1 (Internal Events without Fire and Flooding)," 2018, ES-4.
5. EPA, 2015 Administrative Order on Consent.
6. Sierra Club of Hawaiʻi, Sierra Club's Proposed Findings of Fact and Conclusions of Law & Order, July 13, 2021, p. 7.
7. Honolulu Board of Water Supply, Board of Water Supply's Post Hearing Memorandum, pp. 26–27.
8. ABS Consulting, "Quantitative Risk and Vulnerability Analysis Phase 1," 5ES-4; see also Honolulu Board of Water Supply, Board of Water Supply's Post Hearing Memorandum, p. 12.
9. Commander, Navy Region Hawaii, "U.S. Navy Red Hill Facility Exceeds Industry and Regulatory Standards, Protecting the Environment, Water, and Health," press release no. 21–01, February 8, 2021.

10. Mark Ladao, "Approval of Red Hill Permit Recommended despite Risks," *Honolulu Star-Advertiser,* September 12, 2021.
11. JBPHH Water Contamination Support, Facebook group, https://www.facebook.com/groups/311603564169307.
12. William Cole, "Navy Confirms 1,000 Gallon Fuel Release at Red Hill," *Honolulu Star-Advertiser,* May 7, 2021.
13. Red Hill Bulk Fuel Storage Facility Notice of Interest 20210507–0852 JP-5 Spill That Occurred on 6 May 2021, Table 2.a. Soil Vapor Concentrations at Tanks 13–18 and 20.
14. Commander, Naval Supply Systems Command, FINAL ENDORSEMENT on [redacted] ltr 5830 of 15 Sep 21, Subj: Command Investigation into the Fuel Spill at the Red Hill Fuel Storage Facility on or about 6 May 2021, October 14, 2021.
15. Commander, U.S. Pacific Fleet, FINAL ENDORSEMENT on RDML Christopher J. Cavanaugh, USN, ltr 5830 of 14 Jan 22, Subj: Command Investigation into the 6 May 2021 and 20 November 2021 Incidents at the Red Hill Bulk Fuel Storage Facility, June 13, 2022.
16. Sophie Cocke, "Navy Water Sampling at Red Hill Well Showed Petroleum Contamination Months Ago," *Honolulu Star-Advertiser,* December 5, 2021.
17. June 13, 2022, Commander Endorsement, note 15; EPA, Drinking Water Incident Response at Joint Base Pearl Harbor-Hickam, Honolulu, Hawai'i [sic], November 2021–March 2022.
18. June 13, 2022 Commander Endorsement, note 15; Haley Britzky, "The Navy Told This Military Family They Were Safe from Toxic Water. Then Both Their Children Ended Up in the ER," *Task and Purpose,* January 6, 2022.
19. See @oahuwaterprotectors, Instagram post, February 2, 2022, https://www.instagram.com/tv/CZeCBLzJjhp/.
20. See EPA, 2015 Administrative Order on Consent.
21. Commander, Navy Region Hawaii, "U.S. Navy Red Hill Facility."
22. EPA, Red Hill Administrative Order on Consent in the Matter of Red Hill Bulk Fuel Storage Facility and Attachment A—Statement of Work, August 17, 2015, p. 5.
23. See, e.g., Christina Jedra, "Where Is the Red Hill Contamination Moving? Regulators Are 'Working Blind,'" *Honolulu Civil Beat,* April 4, 2022.
24. Bryce Moore, "Navy Hikes Toxic Leak Estimate: 'They're on Thin Ice,'" KHON2, December 4, 2022.
25. "Stricter Federal Guidelines on 'Forever Chemicals' in Drinking Water Pose Challenges," *Harvard T.H. Chan School of Public Health News,* June 22, 2022.
26. Sophie Cocke, "Report on Toxic Red Hill Spill Delayed," *Honolulu Star Advertiser,* January 4, 2023.
27. Sierra Club of Hawaii, "Hawaiians Rally behind the Honolulu Board of Water Supply, March to Navy," *Popular Resistance,* December 11, 2022.

28. Wayne Tanaka, "Protect What You Love," *Ka Wai Ola,* January 1, 2023.
29. Tim Walker, "California Methane Gas Leak More Damaging than Deepwater Horizon disaster," *The Independent,* January 10, 2016.
30. See Honolulu Board of Water Supply, "Joint EPA-DOH Meeting on Navy Proposals for Red Hill—November 19, 2019," video.
31. See Steven Linder and Roxanne Kwan, Letter to Gordie Meyer Dated Oct. 26, 2020, Re "Notice of Deficiency for the Tank Upgrade Alternatives Decision Document and New Release Detection Alternatives Decision Document, for Red Hill Administrative Order on Consent Statement of Work Sections 3.5 and 4.8."
32. Sophie Cocke, "Residents Critical of Red Hill Agreement," *Honolulu Star-Advertiser,* January 19, 2023.
33. Wayne Tanaka, "Column: Incompetent Leadership as Dangerous as Red Hill Tanks," *Honolulu Star-Advertiser,* November 17, 2022.
34. Wayne Tanaka, "Red Hill Time Bomb Ticks Away as Community Members Continue to Mobilize," *Mālama Monthly,* November 2022.
35. Thisanka Siripala, "50 Years after US Occupation, Okinawa Continues to Resist Military Bases," *The Diplomat,* May 24, 2022.
36. "Kaohewai Demands Shutdown of Kapūkaki (Red Hill) Fuel Tanks," Kanaeokana.
37. Makaiwa Kanui, "Can Our Tears Clean the Aquifer?," Sierra Club of Hawai'i.
38. See Sharde Mersberg Freitas, "Kanaka Maoli Health Does Not Matter, Kanaka Maoli Do Not Matter," *Honolulu Civil Beat,* July 14, 2015.
39. See *Mauna Kea Anaina Hou v. Board of Land and Nat'l Resources,* 323 P.3d 224 (2015).

## Bibliography

ABS Consulting. "Quantitative Risk and Vulnerability Analysis Phase 1 (Internal Events without Fire and Flooding)," ES-4, 2018. https://www.epa.gov/sites/default/files/2019–06/documents/red_hill_risk_assessment_report_redacted-2018–11–12.pdf.

Britzky, Haley. "The Navy Told This Military Family They Were Safe from Toxic Water. Then Both Their Children Ended Up in the ER." *Task and Purpose,* January 6, 2022. https://taskandpurpose.com/news/navy-water-contamination-crisis-hawaii/.

Cocke, Sophie. "Navy Water Sampling at Red Hill Well Showed Petroleum Contamination Months Ago." *Honolulu Star-Advertiser,* December 5, 2021. https://www.staradvertiser.com/2021/12/05/hawaii-news/navy-water-sampling-at-red-hill-well-detected-petroleum-contamination-months-ago/.

Cocke, Sophie. "Report on Toxic Red Hill Spill Delayed." *Honolulu Star Advertiser*, January 4, 2023. https://www.staradvertiser.com/2023/01/04/hawaii-news/report-on-toxic-red-hill-spill-delayed/.

Cocke, Sophie. "Residents Critical of Red Hill Agreement." *Honolulu Star-Advertiser*, January 19, 2023. https://www.staradvertiser.com/2023/01/19/hawaii-news/residents-critical-of-red-hill-agreement/.

Cole, William. "Navy Confirms 1,000 Gallon Fuel Release at Red Hill." *Honolulu Star-Advertiser*, May 7, 2021. https://www.staradvertiser.com/2021/05/07/breaking-news/navy-confirms-1000-gallon-fuel-release-at-red-hill/.

Commander, Navy Region Hawaii. "U.S. Navy Red Hill Facility Exceeds Industry and Regulatory Standards, Protecting the Environment, Water, and Health." Press release no. 21–01, February 8, 2021.

Commander, Naval Supply Systems Command. FINAL ENDORSEMENT on [redacted] ltr 5830 of 15 Sep 21, Subj: Command Investigation into the Fuel Spill at the Red Hill Fuel Storage Facility on or about 6 May 2021. October 14, 2021. https://npr.brightspotcdn.com/c9/c0/b9225ecc405dbfa0d83b88150735/6-may-21-jp5-spill-ci-final-report-redacted.pdf.

Commander, U.S. Pacific Fleet. FINAL ENDORSEMENT on RDML Christopher J. Cavanaugh, USN, ltr 5830 of 14 Jan 22, Subj: Command Investigation into the 6 May 2021 and 20 November 2021 Incidents at the Red Hill Bulk Fuel Storage Facility. June 13, 2022. https://www.epa.gov/system/files/documents/2022–07/FOIA-Release-Red%20Hill-CI-%28June%202022%29.pdf.

Cruz, Catherine, and Sophia McCullough. "Navy Says Pearl Harbor Friday Fuel Leak Contained, Sheds Light on March 2020 Leak." Hawaii Public Radio, July 19, 2021. https://www.hawaiipublicradio.org/the-conversation/2021–07–19/navy-says-pearl-harbor-friday-fuel-leak-contained-sheds-light-on-march-2020-leak.

Department of Health. Notice of Violation and Order No. 21-UST-EA-01, October 26, 2021. https://health.hawaii.gov/ust/files/2021/11/SHWB-NOVO.21-UST-EA-01-signed.pdf.

Environmental Protection Agency. 2015 Administrative Order on Consent. https://www.epa.gov/red-hill/2015-administrative-order-consent. Accessed February 5, 2023.

Environmental Protection Agency. Drinking Water Incident Response at Joint Base Pearl Harbor-Hickam, Honolulu, Hawai'i [sic]. November 2021–March 2022. https://www.epa.gov/red-hill/drinking-water-incident-response-joint-base-pearl-harbor-hickam-honolulu-hawaii-november. Accessed February 4, 2023.

Environmental Protection Agency. Red Hill Administrative Order on Consent in the Matter of Reed Hill Bulk Fuel Storage Facility and

Attachment A—Statement of Work 5. August 17, 2015. https://www.regulations.gov/document/EPA-R09-UST-2015–0441–0404.

Freitas, Sharde Mersberg. "Kanaka Maoli Health Does Not Matter, Kanaka Maoli Do Not Matter." *Honolulu Civil Beat,* July 14, 2015. https://web.archive.org/web/20201202130921/https://www.civilbeat.org/connections/kanaka-maoli-health-does-not-matter-kanaka-maoli-do-not-matter/.

Honolulu Board of Water Supply. Board of Water Supply's Post Hearing Memorandum, Proposed Findings of Fact, Conclusions of Law, and Recommended Decision, July 13, 2021. https://static1.squarespace.com/static/5e28fa5870afe4486a9e6a2d/t/60efb4cf4063550b157920ef/1626322130860/BWS+Post-Hearing+Memorandum%2C+Proposed+FOF%2C+COL+%26+RD.pdf.

Honolulu Board of Water Supply. "Joint EPA-DOH Meeting on Navy Proposals for Red Hill—November 19, 2019." Video. https://vimeo.com/374979821.

Jedra, Christina. "Amid 'Political Concerns,' Navy Kept Quiet about Red Hill Fuel Leak into Pearl Harbor." *Honolulu Civil Beat,* October 8, 2021. https://www.civilbeat.org/2021/10/amid-political-concerns-navy-kept-quiet-about-red-hill-pipeline-leaking-into-pearl-harbor/.

Jedra, Christina. "State Fines Navy $325,000 for Environmental Violations at Red Hill Fuel Facility." *Honolulu Civil Beat,* October 27, 2021. https://www.civilbeat.org/2021/10/state-fines-navy-325000-for-environmental-violations-at-red-hill-fuel-facility/.

Jedra, Christina. "Where Is the Red Hill Contamination Moving? Regulators Are 'Working Blind.'" *Honolulu Civil Beat,* April 4, 2022. https://www.civilbeat.org/2022/04/where-is-the-red-hill-contamination-moving-experts-are-working-blind/.

Jedra, Christina. "Whistleblower Says the Navy Gave False Testimony about Red Hill Fuel Facility." *Honolulu Civil Beat,* November 9, 2021. https://www.civilbeat.org/2021/11/whistleblower-says-the-navy-gave-false-testimony-about-red-hill-fuel-facility/.

Kanui, Makaiwa. "Can Our Tears Clean the Aquifer?" Sierra Club of Hawai'i. https://sierraclubhawaii.org/blog/rh-poster-23.

"Kaohewai Demands Shutdown of Kapūkaki (Red Hill) Fuel Tanks." Kanaeokana. https://kanaeokana.net/shutdownredhill.

Ladao, Mark. "Approval of Red Hill Permit Recommended despite Risks." *Honolulu Star-Advertiser,* September 12, 2021. https://www.staradvertiser.com/2021/09/12/hawaii-news/approval-of-red-hill-permit-recommended-despite-risks/.

Linder, Steven, and Roxanne Kwan. Letter to Gordie Meyer Dated October 26, 2020, Re "Notice of Deficiency for the Tank Upgrade Alternatives Decision Document and New Release Detection Alternatives

Decision Document, for Red Hill Administrative Order on Consent Statement of Work Sections 3.5 and 4.8." https://www.epa.gov/red-hill/2015-red-hill-aoc-tank-upgrade-alternatives.

*Mauna Kea Anaina Hou v. Board of Land and Nat'l Resources,* 323 P.3d 224 (2015).

Moore, Bryce. "Navy Hikes Toxic Leak Estimate: 'They're on Thin Ice.'" KHON2, December 4, 2022. https://www.newsbreak.com/news/2846063924875/navy-hikes-toxic-leak-estimate-they-re-on-thin-ice.

Red Hill Bulk Fuel Storage Facility Notice of Interest 20210507–0852 JP-5 Spill That Occurred on 6 May 2021. Table 2.a. Soil Vapor Concentrations at Tanks 13–18 and 20. https://health.hawaii.gov/shwb/files/2021/05/05–06–2021ReleaseMonitoringData-05–22–2021.pdf. Accessed February 5, 2023.

Sierra Club of Hawaii. "Hawaiians Rally Behind the Honolulu Board of Water Supply, March to Navy." *Popular Resistance,* December 11, 2022. https://popularresistance.org/hawaiʻians-rally-behind-the-honolulu-board-of-water-supply-march-to-navy/.

Sierra Club of Hawaiʻi. Sierra Club's Proposed Findings of Fact and Conclusions of Law & Order. July 13, 2021. https://static1.squarespace.com/static/5e28fa5870afe4486a9e6a2d/t/60efb3f0b3de5a36b123a37a/1626321909343/Proposed+FOF+COL+Sierra+Club.pdf.

Siripala, Thisanka. "50 Years after US Occupation, Okinawa Continues to Resist Military Bases." *The Diplomat,* May 24, 2022. https://thediplomat.com/2022/05/50-years-after-us-occupation-okinawa-continues-to-resist-military-bases/.

"Stricter Federal Guidelines on 'Forever Chemicals' in Drinking Water Pose Challenges." *Harvard T.H. Chan School of Public Health News,* June 22, 2022. https://www.hsph.harvard.edu/news/features/stricter-federal-guidelines-on-forever-chemicals-in-drinking-water-pose-challenges/.

Tanaka, Wayne. "Column: Incompetent Leadership as Dangerous as Red Hill Tanks." *Honolulu Star-Advertiser,* November 17, 2022. https://www.staradvertiser.com/2022/11/17/editorial/island-voices/column-uninformed-leadership-just-as-dangerous-as-red-hill-tanks/.

Tanaka, Wayne. "Protect What You Love." *Ka Wai Ola,* January 1, 2023. https://kawaiola.news/aina/protect-what-you-love/.

Tanaka, Wayne. "Red Hill Time Bomb Ticks Away as Community Members Continue to Mobilize." *Mālama Monthly,* November 2022. https://sierraclubhawaii.org/blog/red-hill-nov-2022.

Walker, Tim. "California Methane Gas Leak More Damaging than Deepwater Horizon Disaster." *The Independent,* January 10, 2016. https://www.independent.co.uk/news/world/americas/california-methane-gas-leak-more-damaging-than-deepwater-horizon-disaster-a6794251.html.

PART III

# FLUID RELATIONSHIPS

## Management and Stewardship

In part 3, "Fluid Relationships: Management and Stewardship," we are challenged to ground ourselves in pilina to place and space so that we may strengthen our ʻike wai (water knowledge) and become stewards skillfully equipped to care for our most precious resource. In powerful and reflective essays from Hawaiʻi hydrologists, we get an inside look into the thoughts and questions of those experienced in managing substantial portions of Hawaiʻi's waterways. What does the integration of cultural and ecological processes within an instream values assessment look like? How can we ensure that the integrity of wai is preserved within the entities and structures that currently manage the majority of this resource today?

CHAPTER 10

# Ma Ka Hana Ka ʻIke Wai

## Reflections and Lessons Learned in Stewardship

*T. Kāʻeo Duarte*

Both Hawaiian and Western repositories of historical and scientific knowledge inform our current understanding and stewardship of land and water resources in Hawaiʻi. It is important that one approaches each place or situation with an open mindset, and there is no replacement for spending time getting to know an ahupuaʻa (land division), kahawai (stream), punawai (spring), or related land and water systems. Assuming that the physical or ecological dynamics of a ʻāina-wai system will behave a certain way based on non–site-specific moʻolelo, scientific research, or legal precedent is often intellectually limiting, and will often handicap one's ability to fully appreciate the complexity of a given situation and to find optimal sustainable solutions. Much has changed and continues to change in these islands, and we must remain open to new realities, trends, innovations, and opportunities. Through a number of examples from the field, I share various observations, lessons learned, and unexpected results.

Ma ka hana ka ʻike is a widely used ʻōlelo noʻeau[1] that I translate as "knowledge and wisdom come from the application and doing of things." This chapter, "Ma Ka Hana Ka ʻIke Wai," is a reflection of my experiences as a hydrologist and kānaka Hawaiʻi (Indigenous Hawaiian) seeking to understand and relate to water and water issues in Hawaiʻi. It is a talk-story piece meant to stimulate reflection and dialogue. I use my own journey to contemplate the importance of multiple sources of knowledge—as well as the continuous application, testing, and improvement of knowledge—in order to gain some level of wisdom. To that end, my focus here is not exactness, and few sources are cited. It is my hope that those of us who work in the area of water (and ʻāina, land or that which feeds, generally) in Hawaiʻi will be more vulnerable, honest, and collaborative than we have been in the past. Scientists, government agencies, private landowners, farmers, cultural practitioners, and a myriad of other stakeholders need to keep striving to better understand each other's ways of "knowing and seeing" water and ʻāina.[2] As a collective, we must be courageous and humble so that we can try together, fail sometimes together, learn and improve together, and ultimately grow wiser together in stewarding our water ecosystems.

## The Decline of Kupaʻāina (Localized) Water Knowledge

I grew up in the ahupuaʻa of Holualoa, Kona, Moku o Keawe (Hawaiʻi Island). Near the top of our family property—a small farm and ranch—was a water reservoir that we frequented as kids. We made small canoes to play in it, and ate the green frogs (not toads) that lived in there. It was lined with black plastic and we used it to provide water to cattle troughs. The surrounding area was cool and shady, with tall ʻōhiʻa trees and some remnant understory of hāpuʻu and smaller ferns.

When I was older, my father and grandfather broadened my perspective on that water source by noting that it was originally a natural water hole and shallow spring, and that koloa (Hawaiian duck) and nēnē (Hawaiian goose) would even stop there in the old days—a sight you cannot see today and that I have never seen. They would see them flying from Kīholo or ʻAnaehoʻomalu in a V formation, and apparently they would stop at this and other water holes along the Kona coast. Adding to that ʻike wai (water knowledge), they pointed out that when the forest understory was intact prior to the introduction of cattle, the area was naturally marshy, to the point that it was difficult to traverse. In this same area are extensive lava tubes that our Hawaiian ancestors clearly used and frequented, and in which water seeps and gathers. Taken all together, a whole other landscape exists in my mind that contrasts with what my generation and subsequent generations have seen and will see.

I do, however, remember seeing an aukuʻu[3] at the mauka pond when I was a teenager and wondering what it was doing there at 2,000 feet above sea level. As I reflect on that now, that auku'u was probably the last of its 'ohana to have the kupa'āina memory of that punawai as part of its landscape. The bone memory of this pool of water far mauka away from the shoreline or waterways. A pool that is dried up and gone now.

Later in life, I would come across other such shallow-fed springs and punawai in Kona at high elevations. Many are relatively small, only a couple feet across, but the bigger ones could be many tens of feet across, like the one on our family 'āina. Though small, many of them always had water in them, even in periods of dry weather. Some folks keep fish in these water holes to this day. I am sure there are many others I don't know of, and equally sure there are many more that were simply bulldozed over or filled in as Kona developed over the last one hundred plus years.

These springs and pools would likely have been the critical, life-sustaining water sources for kā naka and wildlife in this district and, I am convinced, known and protected by the people of this region. The region is known today for being dry with no surface water, and we really only talk about (and fight over) deep groundwater, which would have been a relatively irrelevant conversation in old Hawai'i, especially at mid to higher elevations. The other primary sources of water are rainfall, water seepage in caves, and shallow wells along the coastline.

It is likely that the stream channels that are always dry nowadays flowed more in the Kona of the past. Just as the springs and pools have dried up due to changing rainfall and/or deforestation, so would the shallow hydrologic systems that fed the streams. While I have no data, I remember hearing the water rage in the small stream channel a quarter mile from my childhood home during large rain events. I have not heard it since I moved back home ten years ago. While a generally drier Kona climate may be partially to blame, I believe it to also be because that area has been heavily modified due to grading, grubbing, and a complete change in the vegetative landscape. The wet, marshy areas are dried up and have been replaced by pasture and cropland. (See discussion later in this chapter on the importance of a complex forest cover to shallow Hawaiian water systems.)

In the ahupua'a of Kahalu'u there is a large swath of traditional agricultural system that was not bulldozed. It is at the same elevation as the punawai on the family farm. In this field system we see what appears to be 'auwai, diversions for water, gathering pools, and other features indicating the use and manipulation of surface water in Kona agriculture. This expanse of Kona field system is where kalo and many other traditional crops were grown. It is amazing to see, and staggering to contemplate that this system stretched along miles and miles of coastline from North Kona to South Kona. Besides highlighting the productivity and ingenuity of my people in order to feed a large population, it reinforces the picture in my mind of a human system being developed in relation to a wetter environment and doing so in a way that seems to not have significantly diminished the water sources and possibly enhanced them.

I consider myself a hydrologist, but if I relied only on my university hydrology knowledge or books on hydrogeology in Hawai'i, I would not have an accurate idea as to what the mid- to higher-elevation water resources looked like in Kona before cattle, coffee, and the plantations. Modern sources generally note the lack of surface water in Kona, and focus on groundwater and the unique rainfall patterns. Only the larger, ephemeral/intermittent stream channels are noted on maps, the exception being in older hydrogeology journals that were based on actual fieldwork. These sources do note the presence of some marshy areas and springs, but are still limited in both spatial and temporal perspective. While my image of what the water landscape of my 'āina hānau (land of my birth) looked like remains incomplete, it would be grossly erroneous if I relied only on observations now and the literature readily available.

You do see important glimpses of the kahiko water sources and systems in some ethnographic studies, Hawaiian Kingdom documents such as land award testimonies, or, ironically, archeological or environmental studies required before development. For instance, an interview with now deceased kupuna Violet "Lei" Lincoln-Ka'elemakule Collins of Kona notes, "Tūtū Moku'ōhai had told us about punawai (springs) that were up in the forest about Ke'ei-uka. There were three pools up there, and they had something like 'auwai (irrigation channels) that were used to water agricultural plots below. It was blue, fresh water, and the larger pools didn't seem to have any bottom. We used to swim there all the time. . . . There were also 'o'opu[4] and 'ōpae[5] in the pools, and wild watercress grew along the pool banks."[6]

This brief interview extends the picture in my mind of North Kona into South Kona, and challenges our current, often generic, view of mauka Kona. The implication of 'o'opu and 'ōpae in those deep, mauka pools, far from the ocean with no perennial streams, expands and challenges our current knowledge set on those species and ecosystems.

This ʻike kūpuna (ancestral knowledge) is so critical, not widely recorded in existing literature, not normally taught in our education systems, and is quickly being lost. Just as it only took one generation for ʻōlelo Hawaiʻi to almost go extinct, it could similarly only take a generation for much ʻike ʻāina and ʻike wai to be lost if an intentional effort is not made to perpetuate it and add to it.

### The Rise of Central, Institutional Decision-Makers and Experts

I would argue that much of the water resource research and knowledge generated in Hawaiʻi over the last half a century or so has been largely driven by plantation agriculture, development, and conflict related to those. Most rainfall and stream data were collected as part of the plantation systems, and much of the hydrological and biological studies of streams were directly or indirectly driven in response to the degradation of streams and the conflict around plantation water systems, their diversions, and instream flow standards. Hawaiʻi also enjoyed a period of robust groundwater research as wells began to dominate the landscape, and today much of the new groundwater research is directly or indirectly being done in response to environmental degradation and conflict over groundwater pumping effects. That is not necessarily a bad thing, as much of that conflict needed to occur in order to address degradation of our natural resources and mismanagement of water.

However, having resource scarcity and conflict be a primary driver for much of the investment in water resource science will affect the nature and trajectory of a field. I don't think it's a coincidence that hydrologists and hydrogeologists are hard to find in this state, and that it is hard to keep them, especially in the state departments. It is often a thankless job that has the constant shadow of litigation and private and political agendas competing with the pursuit of new understandings of Hawaiʻi's water systems and associated biota. I often wish there was more "pure" study of ʻāina here in Hawaiʻi. What I mean is more research purely focused on the water and ʻāina rather than being laced with a bias toward one side or another of an issue or conflict. I am generalizing of course, and acknowledge all the seekers of knowledge—regardless if western or Indigenous systems of knowledge generation—who put the lands and waters of Hawaiʻi front and center in their work.

While our system of water resource management and laws allows for input and knowledge from all places, it has (in my opinion) moved in a direction such that a relatively small group of scientists, lawyers, and consultants dominate the water resource science and conflict resolution landscape. I am not making a value judgment or blaming anyone, just stating my opinion on the reality. After all, I would probably be lumped in with that group and have great respect for so many working hard in this field every day. Also, the application of science to solve problems is why my field of engineering exists, so it is perhaps natural that "issues" have driven research agendas.

The University of Hawai'i and the United States Geological Survey are the only sizable centers of water resource research in the state, and legal battles have greatly influenced our knowledge of water resources. That environment contributes to a privileging of certain kinds of data and knowledge sets, which by extension can lead to other kinds of data and knowledge sets being diminished and, in the worst case, phased out over time.

For instance, in most if not all of the important surface water cases today, the streamflow data (Q) is THE critical information for deliberation. Due to case precedents set over time and legal wranglings over "quality, objective data," the data must pretty much come from the USGS or a USGS-approved monitoring station or it will be contested. My point is not that USGS data collection is not top quality, but that over time we have created a worldview where the Q (flow) data set is the driver of decisions. These are numbers necessarily generated at a limited number of points on a waterbody and often over limited time periods. Though testimony is still a vital part of case hearings, at the end of the day I think those deliberating lean more on the hard data in their final verdicts. To be clear, I don't disagree that we must seriously consider the data resolving these cases, but merely pointing out that there has been a ripple effect on the types, sources, and culture of 'ike wai over time. In contrast, the knowledge and "traditional data set" of Kānaka 'Ōiwi was THE source of information in the Hawaiian Kingdom. It is what formed the basis for land and water deliberations in the nineteenth century.

This conversation is a complex one for me, as I personally value multiple forms of knowledge gathering and analysis. Modern data collection is invaluable as it is a practical way to get some level of spatial and temporal knowledge of a phenomenon when you don't have the benefit of a deep knowledge set based on prolonged observation over time. I do believe the rich, multilayered knowledge set that comes with deep observation over time can be superior for the multifaceted and nonlinear decisions we are often faced with.

However, because the number of people or communities with deep knowledge of our water or ʻāina resources is diminishing, our choices are limited. I myself have relied heavily on water resource data collection to inform research as well as decision-making during my tenure managing resources for Kamehameha Schools (KS) across the state. While I stand behind the need to continue to collect this data, my time in the field has made me a firm believer that just focusing on that data collection is limited in what it can tell us, and, more significantly, starts limiting our ability and capacity for greater observations, complex mental mapping of systems, intuition and ʻike pāpālua (double knowledge).

I remember being in the field on Kīlauea volcano in my twenties with a brilliant ʻōiwi scientist and mentor of mine, Dr. Jim Kauahikaua. We had all these fancy geophysical sensors, and as an eager scientist-in-training, I was hyperfocused on the readings and taking careful notes. As I was questioning some strange readings and what the sensor was saying, I remember him telling me, "Look up! Look at what is around you for the answer." His answer was multilayered, but the key point at that moment was to remind me that the "ground truthing" is important and right in front of you if you are open to it. Just like a child today learning about the ocean in Hawaiʻi from an iPhone but who may not bother to jump in the water with a diving mask. Auwe!

While I am not pretending to have come close to developing the ʻike wai as my ancestors did, I did develop enough wisdom to realize that I was failing in my job to understand and mālama (to care for) the water resources I was charged to steward if I relied only on the reams of reports in the KS files, contested case briefs, USGS studies, and what I could get from similar sources. These are all valuable and have their place, but there is so much I learned by simply "pani ka waha, hoolohe me ka pepeiao" while with wai and ʻāina itself as the kumu (teacher or source of knowledge). Not only did I come to realize the flaws and, in some cases downright errors, in many of these reports, but each time I feel I grew in my capacity to learn and appreciate the "anomalies" and uniqueness of each water in relation to the wahi (places) around it.

If I am supposed to be an "expert" on water in Hawai'i and only a fraction of my professional time working with water was spent immersed in deep learning at each site, imagine the 'ike wai of a Kanaka with a lifetime of knowing the entirety of an ecosystem and whose survival and identity were intertwined with it.

## The Characteristics of a "Healthy" Hawaiian Stream

When I moved back to Hawai'i after graduate school and started working in the field of water resource management and research, I came into it with an image of a healthy kahawai influenced by idealized ahupua'a posters and decades of stream research affected by legal conflict over streamflow. In a nutshell, in my mind, a Hawaiian stream without water year-round from top to bottom must be unhealthy, degraded by man, and someone must be held accountable for the diversion of water.

At a basic level, as a hydrologist, I knew all about the natural differences between perennial, intermittent, and ephemeral streams, but the buzz around stream diversions and conflict in Hawai'i had influenced my clarity and openness to the unique characteristics and circumstances for each stream. It was only after years of fieldwork that I started to shed my generalized notions and assumptions. If I didn't, it was clear that what I was seeing made no sense against the preconceived images and "stream stereotypes" that are prevalent.

For instance, there are natural and healthy intermittent or ephemeral Hawaiian streams that do not regularly connect to the ocean year-round. Such is the case now, and such was the case in ka wā kahiko (in times of old). On one hand, this would seem an obvious statement for anyone who studies hydrology, but in my experience working in Hawai'i, any stream that is not connecting to the ocean all the time becomes a source of potential conflict and mistrust. There are good, historical reasons for this reaction as there are far too many cases where it is or has been true that the stream has been greatly degraded by large-scale, man-made diversions and the like. These cases absolutely must be looked at and addressed properly.

My point in this section is to simply note the lessons learned, as they serve to inform honest and effective stewardship and decision-making. The complex factors of rainfall, geology and geomorphology, vegetation, diversion, nonnative species, and many more dimensions require a thought process that does not jump so quickly to conclusions on what the problem "must" be. For example, take a stream I worked with that had marginal presence of native species, was perennial, but was also significantly diverted—Oʻahu Stream.[7] After years of work to return a majority of diverted streamflow, there was and continues to be little positive effect on the population of native species. In Oʻahu Stream, it is my opinion that the population of alien species in the lower reaches of the stream and a choked muliwai are some of the key driving factors for the marginal presence of native stream life. One could argue that if the health of the stream ecosystem was the desired outcome, we should have diverted resources (time, people, money) to work in the lower reaches of the stream rather than the years of conflict resolution, design and engineering, and construction in the mauka reaches to increase streamflow. It could very well have a bigger impact on the native fish and shellfish of that place. Then again, I could be wrong again, and maybe that stream was never known for a strong diversity and population of native stream biota.

Take another stream, Hawaiʻi Stream, where an equally large majority of water was being diverted and it was clear that native stream animal populations were negatively impacted by the lower flow. When we returned a reasonably significant percentage of streamflow, we saw a strong, quick rebound in populations in the middle to upper reaches of the ahupuaʻa. However, it became quickly apparent that the stream naturally had heavily losing reaches makai (lower reaches of the stream). Even with a large percentage of water returned, this fairly large stream does not always reach the ocean. However, the biota population was some of the strongest I have seen in this district.

It is possible the intermittedness of Hawaiʻi Stream may actually be an advantage for this stream in terms of native versus nonnative fish. The top half of the ahupuaʻa always has water, and the adult ʻoʻopu populations there thrive. There are enough rain events (freshets) to connect to the ocean and allow spawning and recruitment. Flash floods seem to favor the native species while the nonnatives that are present in other streams along this coastline have not been able to establish themselves due to the periodic heavy flows, lack of constant flow in the lower reaches, and a rough offshore environment. The adapted native species may have a competitive advantage in this particular stream environment. There are no nonnative fish that we know of, and the presence of Tahitian prawn waxes and wanes. That being said, I am not arguing that more flow could not create more habitat for stream life in Hawai'i Stream. I am simply asking for us to recognize and discuss the unique characteristics and personalities of our kahawai.

To further highlight the tenacious and evolved survival traits of our 'ōiwi stream species, there is a stream, Maui Stream, I worked on for which I made initial judgments and was proven wrong. This was a heavily diverted stream when I started working on it, with little to no water in the middle to lower reaches except after large rain events. To exacerbate the situation, the makai reach was concreted for the last mile, with a very poor low-flow channel. When we returned water to this small stream, I thought it would be a stretch to expect the stream species to return. To my surprise, they reestablished themselves in about two years, after who knows how many decades of absence. Maui Stream seems to have more 'o'opu than O'ahu Stream despite having a fraction of overall flow and a mile of concreted channel. Clearly many other factors are at play that we need to learn from as we seek to restore our instream native species.

To summarize, we see robust stream populations in streams that do not reach the ocean all year. We see streams with strong, perennial flow with weak, almost nonexistent native populations. We also see streams with extremely challenging lower reach of concrete obstacles rebound quickly with but a fraction of the streamflow.

There are many more examples that continue to intrigue and puzzle me. Those are the streams that taught me the most and continue to teach me. However, to be able to accept the knowledge, we need to shed limiting generalizations and biases fueled by reports, charged court cases, and political agendas and instead focus on the complexity and subtleties of 'āina.

### Hahai no ka ua i ka ululā'au v. 2.0

The 'ōlelo no'eau "Hahai no ka ua i ka ululā'au—The rain follows the forest" is well-known in water resource and mālama 'āina circles. It describes the importance of forests and watersheds to water and hence life in Hawai'i. No argument there. What I want to do in this short section is to use the way this saying has been interpreted to dive deeper into the forest and ua/wai (rain/water) relationship, and challenge ourselves to take these 'ike kūpuna as starting points, not ending points, in our understanding of 'āina.

I have been in many interactions where this ʻōlelo noʻeau has been used to strictly declare that there is more rain where there is more native forest and, conversely, the decline of native forest is the reason for changes in rainfall. While I believe there are places where the forest or vegetation is directly linked to production of precipitation, in many cases I think it is more accurate to say that the forest is key to the effective and efficient capture, transport, storage, and delivery of rain spatially and temporally. That's a lot of words, but what does it mean?

Imagine two plots of land north of Hilo town, adjacent to each other. Both plots have the same elevation, steepness, and size. One is covered by more or less intact native forest for that place, and one is pastureland with sparse trees and vegetation. The amount of precipitation that falls on each plot is likely the same. Rainfall in a wet, windward area like Hilo district is driven primarily by trade winds, and it's unlikely that the vegetation on these two plots would have any effect. Same would go for two plots of ʻāina in Hāna, Kāneʻohe, or other wet, windward regions.

However, what happens to the rain that falls on those two plots is greatly affected by the presence or absence of native forest:

- capture of fog drip;
- slow dripping of rain off leaves and down tree trunks, and multiple levels of forest structure. This helps slow delivery of water to the ground and maximize capture in high-rainfall events. The opposite happens on the pasture where water quickly runs off;
- dissipation of raindrop energy by leaves and understory that minimizes erosion;
- absorption by moss, understory shrubs and grasses, and mulchy forest soils. This helps regulate the recharge of water while minimizing eroding runoff;
- strong reduction of evapotranspiration by, among other things, reduction in albedo/temperature and wind velocity transporting moisture from an area;
- creation of rich soil and complex geochemistry by roots and decomposing vegetation.

This is a very simplified view of complex biophysical dynamics. While the body of research is growing, we still do not have a solid corpus of work that fully articulates the critical nature of forests on the capture, delivery, storage, and timing of water across Hawai'i's watersheds. Plenty of opportunities for us to grow our 'ike!

If I can take us back to my opening mo'olelo about the punawai on my 'ohana property in mauka Kona—it is dry now. I believe the changes in the forest in that area are largely responsible. Such has been the case in many areas that were deforested.

Where forest cover and changes in vegetation may directly affect rainfall frequency and magnitude is where convective weather processes are the primary driver of precipitation. This happens primarily in the leeward districts, such as my home district of Kona, famous for the 'ōpua clouds that gather in the afternoons due to warm, moist air rising over the land surface at the mid-elevations. Both the moisture and flow dynamics of the air that becomes clouds and rainfall are affected by what is growing over the land surface. Another area would be the 'Ewa plain, where clouds that make it over the Ko'olau range dissipate on their way to the Wai'anae Coast. Change in land cover in the central plain could affect moisture dynamics in that region.

## The Energy-Water Challenge

There are interesting, often contrasting, conversations around when to use stream water and when to use groundwater in Hawai'i. On one hand, there is a strong desire to divert as little streamflow as possible, and in many communities an express desire that it be only for traditional Hawaiian crops. On the other hand, there are those who feel potable groundwater should not be used or "wasted" for agriculture or industrial uses. That creates a bit of a bind . . . especially when there is such a strong desire for Hawai'i to once again become food and fiber self-sufficient. I would put it out there that energy is one of the most important pieces to this conversation. In parts of Hawai'i, if energy was cheap or free, I would advocate for going 100 percent groundwater, even for agriculture.

Here's why. There are so many physical and social-political issues affecting our stream dynamics and I don't see it getting much better anytime soon. Diverting stream water takes more direct management, usually involves longer transmission systems (the stream can't be moved, after all), is generally more susceptible to the elements and natural disasters, much more variable in flow and quality, more directly impacts native flora and fauna (in my opinion), has to be treated for many uses due to stricter water quality regulations, and is a source of so much conflict. We use it because where we have large streams, we can access large amounts of water "for free" . . . in an energy-free way, to be specific.

Groundwater, on the other hand, is buffered from flora and fauna, is protected from natural disasters, is higher in quality, is more or less everywhere in some shape or form, and has a built-in storage mechanism in the form of the aquifer. It can also be scaled easily from a small, shallow well to a deep, large-capacity well. Wells can also be put where we need them, reducing transmission needs.

We don't use them as much because of the cost of energy and, secondarily in most locations, the cost to install the well. So, while it would not be the case for all locations and uses (lo'i, for instance), if energy were free, I think we would consider decommissioning some stream diversions and replacing them with wells close to the intended use area. The primary downside would be that if natural disasters cut off energy, then batteries and backup become important.

If we can somehow get to the point where renewables are actually reducing our cost of energy or providing excess for strategic water well pumping, this could help solve many water resource and stream ecology issues too. Ironically, at one point the sugar plantations were using the free bagasse energy from their factories to pump and push tremendous amounts of groundwater. Unfortunately, it was not to substitute for stream diversions but to take even more water for the very thirsty era of King Sugar.

### The Challenge of Submarine Groundwater Discharge

The issue of submarine groundwater discharge is coming to the forefront of water resource issues, and conflict. While the discharge of groundwater to the ocean has been recognized and studied for a very long time by hydrogeologists, the term "submarine groundwater discharge" (SGD) came onto the scene in the 1990s, primarily due to the work by Willard S. Moore, a geochemist at South Carolina University. Interestingly, I remember going to an awards ceremony at Woods Hole Oceanographic Institution while a graduate student in the late '90s, and one of the awards was to Dr. Moore for his "discovery" of the significant amount of groundwater inputs into the ocean.[8] Not to diminish his important work, but it was interesting to see the groundwater hydrologists in the room smirk at each other, as they had been aware of and studying discharge to the ocean for decades. It reminds us that it was not so long ago that most disciplines in science really did not communicate with each other, and the crossing of geology, chemistry, oceanography, biology, and hydrology has only really become normal in the last twenty years or so. It is good to see the increasing normalization of cross-disciplinary work, which is more relatable to holistic views of Indigenous cultures.

On a related note, while having a conversation with a Niʻihau person many years ago, I was telling her—with my elementary ʻōlelo Hawaiʻi skills—about my interest in water and specific research interests at that time. She went on to note how on Niʻihau, after a large rain event up mauka, they would expect a blooming of limu for harvest in a week or two. While the translation in my head may have been a bit inaccurate, I want to note how important and sophisticated her casual comment to me was. To recognize that rain falling in the upland is seeping into the ground, traveling underground with a specific travel time, picking up nutrients from the ground along the way, and fertilizing the nearshore environment to spur the growth of limu summarizes the work of multiple scientists and fields of study over the last decade or so. It influenced my own research while at UH in the early 2000s, and I am proud of the work of scientists like Dr. Dan Amato, who worked with me and then with Dr. Celia Smith to publish a paper on nutrients on limu growth so that we can begin to understand the geochemical and ecological implications of groundwater flow and discharge. Much good work continues on the topic at the University of Hawaiʻi and elsewhere.

In 2014, I was sitting in a conference room at the Sheraton Keauhou Hotel in Kona, listening in on a series of presentations related to the Keauhou groundwater aquifer controversy in West Hawaiʻi. At one point, a speaker put up graphs from the paper I had published with Dr. Amato and others, and began to explain how Duarte et al.'s work supported the assertion that groundwater pumping in Kona is negatively affecting the nearshore environment by affecting limu growth and, by extension, the food chain and traditional practices connected to that keystone species. I was surprised to hear this assertion, to say the least! Here are excerpts from the conclusion section of the paper that was being cited:

> In most cases the science is in its infancy for there is no definitive analysis available. This may be leading to, in our opinion, very dangerous legal and policy precedent. . . . It seems clear that for the

> very wet, windward sides of the islands, it would take a tremendous amount of pumping to significantly affect SGD fluxes and net impact on the nearshore water quality. Of course the locating of a well or well field can heighten or lessen impacts. For drier, leeward areas, there is a greater chance of nearshore effects. However, it is possible that the impact could be on the order of or less than that of naturally occurring weather patterns and seasonal fluctuations.[9]

While I had been thinking for a long time that SGD would become the next battleground for water conflict in Hawai'i, similar to instream flow standards for streams, it became very real for me that day. I think it is only the beginning and I believe it could be even more contentious, expensive, and divisive than the science and legal battles related to streams in Hawai'i.

Why? Because it is more complex and open to interpretation. Unlike with streamflow and quality effects on stream biota and cultural/ecological services:

- we cannot directly see or monitor the flow underground;
- we are limited in our understanding and mapping of the geologic structures that are paramount to understanding groundwater dynamics;
- the discharge to the ocean is happening very heterogeneously laterally across the shoreline boundary and at depth as one moves offshore;
- variability in temporal and spatial distribution of recharge often rivals or greatly masks influence of anthropogenic impacts;
- the science linking the nutrients in freshwater discharge across marine species and ecosystems is limited, but growing;
- we really don't have a baseline against which to judge impacts.

By contrast, for a stream system, we can readily see and study flow in the stream system, can assess the direct impacts on stream biota, have convenient recharge-streamflow relationships via hydrographs and the like, can measure water quality pretty easily, and in general we can study a given stream as a contained unit.

This is largely not the case for assessing the impact of, say, groundwater pumping on the nearshore environment. There is no doubt that large-scale groundwater pumping will have some kind of effect, but whether it is significant or not is very difficult to ascertain in most cases. From a flow quantity perspective, the variability in rainfall spatially and temporally/historically could be as, if not more, impactful on the nearshore environment.

From the perspective of nutrients' effects on biota, what is a "healthy baseline"? On one hand, we worry too much about nutrients in the water degrading habitat due to things like eutrophication, and on the other hand, we are waging battles that there are not enough nutrients going into the ocean now. To complicate things even further, in many cases it is not that the change in nutrients would destroy an ecosystem, but it would change it. One species may become more prevalent in one area than in another. This is, of course, already the case. It has changed over time, both naturally and because of man. Overfishing, selective fishing (shifting the population dynamics in favor of certain species), storm runoff, boats, nonnative species, sunscreen, and many other factors are already creating a complicated and challenging issue for us to deal with as island people. Honestly distinguishing the impact of SGD on the nearshore environment from everything else assailing Hawai'i's unique ecosystems is something that will challenge us for years.

### Choosing Truth over Agendas . . . Even Your Own

Few things have challenged me more in the water arena than maintaining integrity and honesty in speaking about and to water resources and issues. It has bothered me to see situations where people have decided what the "truth" is about a water resource or issue based on a political agenda or past history, with no genuine context or knowledge of the water resource or system. It is not that they are not entitled to their opinion or that the politics and historical context is not important, it is just that we often end up compromising something in the process—'ike wai.

I will use my home of Kona as an example. There is a part of me that feels a sense of loss due to the rapid changes and development in Kona. I personally support efforts to curb or limit development in places and to re-enfranchise Native Hawaiians in Kona. I will fight for that. However, I would not compromise the integrity of 'ike wai to wage that fight. If a water source was being threatened, then for sure! However, say there is enough water for a proposed development—I feel it would be wrong to manipulate data and science to fight the project on the basis of water impacts. I should still fight the project, but on grounds that don't compromise what I believe to be true.

Whether it is conventional agricultural versus traditional uses, or development versus impacts to the ecosystem, water is often used as an effective tool to do battle. It is so susceptible to divergent studies and interpretation of data, and so very easy to get off the path of objective search for knowledge. I don't want to leave behind knowledge that is so colored by politics and agendas that it is incomplete at best and, at worst, false. I feel we need to be better about creating safe spaces and mindsets to record and debate reality. So much in this world pushes us to "win the battle" and see who is "right." I guess there can be some value in that, but at this moment in time, I feel an approach that also values collective success and collaboration is needed. Even if battles need to happen sometimes, can we create a pu'uhonua (place of refuge) for 'ike where the weapons are left at the door?

## Ensuring Perpetuation of 'Ike Wai and Mālama Wai

Like with so much in life, pilina (relationships) are key to success and health. Relationships between people are key to effective, sustainable conflict resolution, decision-making, and stewardship. Pilina with our extended 'ohana in the kahawai, nahele, lo'i, loko i'a, mala, and kai. Pilina with 'āina and wai critical to understanding the nature and nuances of this place we live in.

My mind can't help but drift back to the ahupua'a and valley of Honokāne on Hawai'i Island. In reality, I have only spent a limited amount of time in that valley, but I was lucky in the lessons learned from that kahawai. For instance, the lessons on the dynamics of rain and stream response. One day we spent the day surveying the stream, which was at normal flow. In the early evening, it started to rain in the back of the valley and progressed makai. You could feel the changes in wind and moisture, and hear the changing streamflow as the night progressed. Lying in my cot that night, I found myself reflecting on the fact that our kupuna here, with years of pilina with this place, would have easily predicted the rain, the timing of the changes in river flow, as well as the time for the stream to recede (which we watched the next day). On another day, when the lower half of the stream was still dry, it again started to rain in the valley, and I was able to watch a dry stretch of stream slowly fill and creep makai. No amount of books and science can substitute for these lessons and productive questions generated by those times.

The concept of aloha ʻāina (love for the land)[10] takes many forms in the Hawaiʻi of today, for both Native Hawaiians and others. It seems to be in our iwi (bones), and infectious for anyone who is of this place or has been truly touched by ʻāina. However, I do think the development and perpetuation of ʻike wai takes intentional effort, and I do not think it can happen if we do not foster future generations that are pili with the waters of this place. In addition, I do not think wisdom to be effective stewards—to mālama wai—can be developed without working and engaging. Doing things. Making mistakes. Learning. Improving. Hana hou.

No amount of computer modeling, data collection, or fancy equipment can substitute for truly spending time in a place to listen, smell, see, taste, and feel with all senses. I truly believe it will be as sad as losing a language if future generations slip further from that relationship than my generation already has. Just like language, it can be recovered, and I think it will help heal not only our wai, but the health of lāhui (Hawaiian nation).

We live in a society now where a person can affect the outcome of an issue or conflict via vehicles like social media, regardless of how much they really know. How do we ensure some level of integrity of ʻike wai is not only preserved, but has a strong voice? How do we steer things in a direction such that it holds the level of influence it did in days past? How do we validate such ʻike wai and individuals? I don't know the full answer, but I know it partially lies in Ma ka hana ka ʻike wai. No hana, no ʻike. Thus, ensuring our children are pili (close) with ʻāina, kai, and wai is essential to producing robust knowledge, opportunities to mālama ʻāina,[11] and therefore producing ʻike wai that they will then pass on. And so the cycle continues, building on the past and innovating for times we do not yet know, but can be ready for.

## Notes

1. "In working one learns." Mary Kawena Pukui, *ʻŌlelo Noʻeau: Hawaiian Proverbs and Poetical Sayings* (Honolulu: Bishop Museum Press, 1983), #2088.
2. Carlos Andrade, *Hāʻena: Through the Eyes of the Ancestors* (Honolulu: University of Hawaiʻi Press, 2008): "the place that feeds your family, not only physically, but spiritually, mentally and emotionally."

3. ʻAukuʻu, black-crowned night heron, is an indigenous Hawaiian water bird that you usually only see near wetlands, streams, or coastal areas.
4. Mary Kawena Pukui and Samuel H. Elbert, *Hawaiian Dictionary* (Honolulu: University of Hawaiʻi Press, 1986), 290: "general name for fishes included in the families Elotridae, Gobiidae, and Blennidae."
5. Ibid., 291: "general name for Shrimp."
6. Violet Kuʻuleikeonaona Lincoln-Kaʻelemakule Collins, informal oral history interviews with Kepa Maly, March 5, 1996, at Huliheʻe Palace; April 9, 1996, at Hōlualoa; and May 23, 1996, at Kahaluʻu.
7. I will be using aliases for the stream examples given.
8. Willard S. Moore, "Large Groundwater Inputs to Coastal Waters Revealed by 226Ra Enrichments," *Nature* 380, no. 6575 (1996): 612–614.
9. Thomas Kaeo Duarte et al., "Optimal Management of a Hawaiian Coastal Aquifer with Nearshore Marine Ecological Interactions," *Water Resources Research* 46, no. 11 (2010).
10. Kamanamaikalani Beamer, "Tūtūʻs Aloha ʻĀina Grace," in *The Value of Hawaiʻi 2: Ancestral Roots, Oceanic Visions,* edited by Aiko Yamashiro and Noelani Goodyear-Kaʻōpua (Honolulu: University of Hawaiʻi Press, 2014), 11–17. One definition of "aloha ʻāina" could be a movement toward the union of culture and ecosystem. Noelani Goodyear-Kaʻōpua, Ikaika Hussey, and Erin Kahunawaikaʻala Wright, *Nation Rising* (Durham, NC: Duke University Press, 2014); see also Noenoe Silva: "where nationalism and patriotism tend to exalt the virtues of a people or a race, aloha ʻāina exalts the land" (Noenoe Silva, *Aloha Betrayed* [Durham, NC: Duke University Press, 2004]).
11. For more cultural and historical context, see Lilikalā Kameʻeleihiwa, *Native Land and Foreign Desires: Pehea Lā E Pono Ai?* (Honolulu: Bishop Museum Press, 1992).

## Bibliography

Andrade, Carlos. *Hāʻena: Through the Eyes of the Ancestors.* Honolulu: University of Hawaiʻi Press, 2008.

Beamer, Kamanamaikalani. "Tūtūʻs Aloha ʻĀina Grace." In *The Value of Hawaiʻi 2: Ancestral Roots, Oceanic Visions,* edited by Aiko Yamashiro and Noelani Goodyear-Kaʻōpua, 11–17. Honolulu: University of Hawaiʻi Press, 2014.

Duarte, Thomas Kaeo, Sittidaj Pongkijvorasin, James Roumasset, Daniel Amato, and Kimberly Burnett. "Optimal Management of a Hawaiian Coastal Aquifer with Nearshore Marine Ecological Interactions." *Water Resources Research* 46, no. 11 (2010).

Goodyear-Ka'ōpua, Noelani, Ikaika Hussey, and Erin Kahunawaika'ala Wright. *A Nation Rising: Hawaiian Movements for Life, Land, and Sovereignty.* Durham, NC: Duke University Press, 2014.

Kame'eleihiwa, Lilikalā. *Native Land and Foreign Desires: Pehea Lā E Pono Ai?* Honolulu: Bishop Museum Press, 1992.

Moore, Willard S. "Large Groundwater Inputs to Coastal Waters Revealed by 226Ra Enrichments." *Nature* 380, no. 6575 (1996): 612–614.

Pukui, Mary Kawena. *'Ōlelo No'eau: Hawaiian Proverbs and Poetical Sayings.* Honolulu: Bishop Museum Press, 1983.

Pukui, Mary Kawena, and Samuel H. Elbert. *Hawaiian Dictionary.* Honolulu: University of Hawai'i Press, 1986.

Silva, Noenoe. *Aloha Betrayed.* Durham, NC: Duke University Press, 2004.

CHAPTER 11

# Instream Flow Standards That Promote a Biocultural Landscape

*Ayron M. Strauch*

A biocultural landscape is a vision where the use of resources is integrated with the structure and function of ecosystems and promotes, rather than erodes, local cultural practices, ensuring that management actions consider the cumulative impact to the social-ecological system. In this context, "sustainable" is not just defined as the maximum rate of extraction for a given condition, such as the number of fish removed from a population per year; rather it is the totality of the repercussions of resource withdrawal, such as the biotic and abiotic interactions populations of aquatic species engage in over space and time and the cultural importance of those species. Humans have and will continue to modify their environment with continued dependence on ecosystem services. With a biocultural lens, we can more clearly articulate the consequences of natural resource use. By identifying these consequences, we can work to modify how management decisions play a role in shaping a more positive outcome. In the context of water in Hawai'i, management becomes more than an engineering issue, an ecological model, or the question of property rights. Rather, water is the central feature that connects the sky to the land, the land to the sea, the geology to the biology, gods to ancestors, and ancestors to the present day.

## Background

For thousands of years, humans have altered the natural course of rivers, springs, and streams to suit their needs.[1] From the modification of stream banks to support floodplain agriculture, the construction of aqueducts that gravity-fed water across barren deserts, and the water wheel that exploits the energy of flowing water, the human-water relationship has propelled civilizations forward.[2] The manipulation of water resources, important as it has been for improving health, hygiene, agricultural productivity, extraction of minerals, and energy independence over time, has also had profoundly negative consequences for aquatic environments. Both physical structures in streams and the removal of water from them can have consequences for the structure and function of aquatic ecosystems, the availability of water for domestic and municipal purposes, the interruption of cultural practices or recreational needs, and even transboundary political relationships.[3]

Historically, resource management science has focused on converting the Earth's life-support systems into commodities to be exploited in more and more efficient ways while minimizing environmental or social problems created by their depletion or mismanagement. While our understanding of natural systems has improved over time, the basic approaches to management have largely not, and a new science of resilience has emerged to fundamentally shift this behavior.[4] This shift requires first an appreciation of integrated systems and adaptive management with a focus on linkages across systems and feedback controls first promoted by Holling.[5] Second, there needs to be a shift to a sociocultural approach rather than a resource-commodity approach, building on the community-based resource management literature.[6] A biocultural landscape vision embraces how localized social systems developed management practices using local ecological knowledge to deal with the unpredictability of complex systems.[7]

## Flow Regimes

Describing what flowing water should look like is surprisingly difficult. Some streams have predictable seasonality in flow, in which snowmelt and winter rainfall drive peak flows in early summer, followed by predictably low late-summer flows. Alternatively, streamflow may be so dependent on rainfall-driven runoff that flow predictability beyond short time frames is poor. This is frequently the case for small watersheds with short, low-order streams and steep flow duration curves. In most cases, these two examples typify the distinction between streams in temperate latitudes and those in the humid tropics. While there are hundreds of statistical metrics used to describe streamflow,[8] no single metric is effective across large spatial or temporal scales. Hydrology in the tropics is directly related to patterns in climate (e.g., amount and distribution of rainfall) and geology (e.g., porosity, hydraulic conductivity) that drive vegetation growth, erosion, rate of stream channel formation, and downstream flow.[9] However, there are substantial gaps in our understanding of the hydrogeology of surface and groundwater resources that can result in a high degree of uncertainty in predicting how much water would be naturally flowing in any given location at a particular time.[10] Therefore, forecasting to what degree reductions in flow will affect the stream, much less the ecological patterns within the stream, is a major challenge.

## Hawaiian Science and Western Science

The notion that streams flow from mountain to sea is not new, just as the idea that streams at varying elevations and distances from the ocean support differing ecosystems. The Hawaiian concept of ma uka (toward the mountain) to ma kai (toward the ocean) flow, where wai (fresh water) transitions to kai (salt water), dates back centuries.[11] However, in western science, it wasn't until 1980 that Robin Vannote[12] codified the River Continuum Concept (RCC) as a model for water conservation. In short, the RCC is based on the theory that streamflow is in dynamic equilibrium among various physical and biological characteristics that influence the movement of water, such as stream width, water depth, velocity (which is dependent on gradient), sediment load, and riparian vegetation in a continuum from mountain to sea. The RCC provides a mechanism to understand and classify the waters of a stream by mapping the occurrence of indicator organisms. By providing a clear depiction of the natural patterns of aquatic communities, scientists and conservationists could demonstrate that a management action resulted in a change to a stream. This understanding provided a basis for developing instream flow standards.

Hawaiians understood that with wai came life, bounty, productivity, and reward. "Waiwai" (i.e., lots of water) is the Hawaiian word for wealth.[13] Flash floods signaled renewal with the flush in nutrients, a reorganization of the stream channel, the transport of overgrowth along stream banks, and the conveyance of aquatic species to the estuary and marine environment. The concepts of exploitation, conservation, release, and reorganization that make up the adaptive cycle in resilience theory codified these concepts for western science,[14] while they were understood in Hawaiian culture for generations. Winter et al.[15] coined the term "moku system" to describe the Hawaiian biocultural resource management system. The moku system integrates the complex cultural understanding of landscapes, the mosaic of social and political relationships, and mechanisms for managing resources for abundance across land, water, and marine ecosystems, by considering the dynamic integration of biotic and abiotic elements that build resilience.

Isolated tropical islands have relatively few endemic species. As a result, aquatic food webs are simpler, conspecific and interspecific interactions are more easily studied, and the effects of abiotic (e.g., geology, hydrology, temperature, dissolved oxygen, nutrients) gradients are more clearly articulated. Many tropical aquatic species are amphidromous,[16] and the ecology of Hawaiian species is summarized in greater detail elsewhere (see Kinzie and Ford[17] or Fitzsimons et al.[18]). Briefly, in Hawai'i, this includes five species of fish, two crustaceans, and two mollusks. They have adult forms that live and reproduce in fresh water, and larval forms that necessitate salt water. Like many R-selected species (i.e., high reproductive rate, low investment in rearing offspring), females produce millions of eggs that hatch and then need to get to the ocean quickly. Most amphidromous species in Hawai'i reproduce throughout the year, although some research suggests a peak with the onset of higher flows during the early wet season, possibly as a result of a greater proportion of females waiting to reproduce.[19] With the mountain shrimp 'ōpae kala'ole (*Atyoida bisulcata*), females protect their fertilized eggs until they can be successfully released under higher flow conditions. Higher flows are needed to ensure that the newly hatched larvae make it to the ocean within the one-to-three days before they are physiologically unable to survive in fresh water. Considering that each female may produce millions of eggs multiple times a year, the larvae form a critically important food resource for estuarine and marine food webs. In total, isolated tropical islands, because of their short watersheds and lack of species diversity, are ideal to study the RCC.

Scientists like to study things. They like to probe and prod, heat and cool, add and subtract, take samples and run tests, count and measure. But streams are long. Even in small tropical islands, a single stream may run ten miles or more from headwaters to ocean mouth across steep gradients. Walking up to a stream bank, one may be able to see a few hundred feet upstream and downstream. But from that singular place, very little of the stream can be understood. One cannot appreciate the spatial or temporal dynamics of energy (e.g., carbon or sunlight), nutrients, or habitat that affects the ever-changing composition of organisms. Nor how predator-prey relationships, competition for resources, or the temporary nature of substrates can influence the distribution of species across large scales. For example, gravel beds may be needed for successful breeding for a species but, due to the unique geology of a particular stream reach, they may be completely absent. To study ten miles of anything is challenging, but to study something that may, for the most part, be physically inaccessible, is mind-numbing. The advent of more sophisticated computing techniques and the widespread adoption of georeferencing the collection of both environmental and biological data have resulted in the more recent integration of ecological and geophysical sciences, producing the discipline of landscape ecology. By extension, riverscapes are the integration of landscape level features at the aquatic environment.[20] Recent research by Tingley et al.[21] has integrated local, upstream, and downstream characteristics to describe the composition of aquatic communities across Hawai'i.

We need to appreciate the complexities that make freshwater ecosystems in Pacific islands both unique and also great model systems for studying the consequences of altered flow regimes for biological communities as well as the cultural practices that are integrated within these social-ecological landscapes.

## Instream Flow Standards (Environmental Flows)

An instream flow standard is the amount or depth of water necessary to protect the social and ecological values of a stream. Two critical legal issues resolved by the Hawai'i Supreme Court led to the establishment of the State Water Code in Hawai'i: the McBryde decision (*McBryde v. Robinson*, 1973) and the Reppun decision (*Reppun v. Board of Water Supply*, 1982). Both decisions applied common property rights arguments to the use of water (whether riparian or instream). I will not get into the details of those cases but will simply state that they resulted in recognizing the appurtenant or prescriptive water rights of landowners or cultural practitioners riparian to the stream and that they were entitled to a reasonable amount of water to be used within the ahupua'a (Hawaiian land division). So technically, these two decisions resulted in the first instream flow standards.

In the 1980s, the U.S. Army Corps of Engineers and the Department of Energy pushed for the development of renewable hydropower to offset the high costs of importing fossil fuels to Hawai'i following the sixfold increase in crude oil from 1973 to 1980. A number of studies were commissioned to examine the feasibility of developing new hydropower on Kaua'i, Maui, and Hawai'i islands.[22] Like existing hydropower plants in Hawai'i, these would be run-of-the-river hydropower plants, where relatively small amounts of water were diverted into a penstock and dropped hundreds of feet in elevation to power a turbine. This type of hydropower is appealing because it requires less infrastructure and therefore lower capital costs and potential environmental impacts. These studies developed instream flow standards to quantify the lost habitat in the dewatered reaches before the tailrace returned water back to the stream. The use of the Physical Habitat and Biological Simulation Model (PHABSIM) was highly regarded for the development of environmental flows downstream of large-scale reservoir and hydropower projects on the mainland. This model relied on habitat suitability curves for specific species of interest, detailed channel surveys across varying flows to develop estimates of weighted usable area, and an understanding of the flow duration curve for the channel to simulate the degree to which altered hydrological regimes affected the available habitat for the species. While well intentioned, the model is not an ecological model and does not establish relationships between hydrology and biological goals such as abundance, reproductive output, or species diversity.

Further, many amphidromous species are relatively flexible in their habitat preferences in terms of depth, velocity, or substrate, and species exhibit a wide range of feeding, reproductive, or defensive behaviors that are interrelated to substrate, depth, or velocity.[23] While a species may prefer to feed on large bedrock surfaces with moderate velocities that support algal growth one–two feet underwater, that same species may prefer to reproduce in gravel beds under slower flow conditions in half a foot of water.[24] The PHABSIM model, and others like it, take a complicated multivariate issue and condense it down to a single flow value, which doesn't do the ecology justice. Use of such a model, while well-intentioned, isolates the management goal (weighted usable area) from a social-ecological perspective. Despite the great research conducted over the years, we have not conclusively demonstrated the linkage between hydrology and endemic amphidromous species ecology in Hawai'i.

In 1987, the Hawai'i state legislature established the Commission on Water Resource Management (Commission) and the legal framework (Hawai'i Revised Statues 174C, known as the Water Code) for determining instream flow standards for streams in Hawai'i.[25] The Commission was tasked with understanding the unique geohydrology of each stream and aquifer, and then examining everything and anything connected to the water. The Code defines instream uses as aquatic habitat for fish and other biota, ecosystem services such as the maintenance of riparian habitat, wetlands, estuaries, coastal and marine environments, implications for water quality, traditional and customary practices, aesthetic values, and recreational usage. Further, the Commission must also evaluate the reasonable and beneficial usage of water, whether it fits within other landscape-level protections including land ownership and management, state land use designations, current and future county development plans, and the alternatives available to its usage. As if that wasn't enough, much of the data needed to make a reasonable decision has either never existed (e.g., detailed understanding of reproductive potential across species and environments), outdated or nonexistent across large portions of Hawai'i (e.g., flow regimes), or limited in extent (e.g., consequences of riparian species invasion on aquatic ecosystems). Commission staff must also work with community groups, landowners, stakeholders, scientists, practitioners, and the public in general to develop these data sets. To recapture: the data necessary to make an informed decision needs an anthropologist, geologist, economist, hydrologist, biologist, conservationist, agronomist, engineer, statistician, and planner.

The Water Code states that the Commission must use the best available information to make its decision, but it also does not establish what constitutes that standard. Despite its passage in 1987, the Commission didn't even have a Stream Protection and Management (SPAM) branch organized until 2002; and in 2024, the SPAM branch was still not fully funded or staffed to meet the expectations of its mandate.

## Ma Uka to Ma Kai Perspectives

Water is critical for the practice of Hawaiian culture.[26] Briefly, the critical pieces that need to be understood are: (1) water is a central feature that connects creation to ancestry, ancestors to the present day, and today to the future; (2) water is central in the analogy between the island that gives life and the human body; (3) land and water are perpetuated through culture across space and time; and (4) there is sociocultural value in food resources from the mountain, estuary, and ocean, which were integrated together to produce some of the largest isolated communities in human history. The Hawaiian islands are all shield-cone volcanic islands with somewhat circular shapes, with the caldera in the middle as the source for progressive layers of lava, although some of the volcanoes merged to form singular islands. The northeast trade winds bring warm, moist air to the islands and get pushed upslope to the peak of the island. This rise in elevation coincides with a decrease in temperature, and once the dew point is reached, moisture precipitates from the sky as rainfall and fog drip. This is why the windward (literally facing the wind) watersheds are wet and the leeward watersheds are dry. The peak of the mountain becomes the source of water, the source of life. Water supports not just the flow of streams down the hillslope, but the transport of particulate and dissolved nutrients, larvae from aquatic species critical for food webs, and cold fresh water that stimulates estuarine and nearshore environments. Thus, the mountain peak can be envisioned as the navel, connecting the sky to the island as an umbilical cord connects the mother to the fetus. The streams carry the minerals and organic material that support productive ecosystems, just as veins carry the nutrients needed to support healthy bodies. It is the kuleana (responsibility) of people to mālama (care for) and aloha (respect) the ʻāina (land).

## Ma Uka

Ma uka (toward the upland or mountain) watershed habitat in Hawaiʻi is characterized by diverse, native-dominated cloud forest, with ʻōhiʻa and koa overstory and an understory of shrubs and herbs. The ʻōhiʻa tree plays a central role in Hawaiian culture, considered one of the kinolau or physical manifestations of Kū, one of the four principal Hawaiian akua (deities). The tree is the dominant native forest species across Hawaiʻi, supporting a diverse range of bird species that are also central to Hawaiian culture. One creation story is in the Kumulipo,[27] which explains that Hawaiians descend from akua and are physically related to all living things in Hawaiʻi, meaning that caring for the ʻōhiʻa forest is central to caring for ancestors. The ʻōhiʻa is the literal and figurative foundation of the watershed, supporting epiphytes and an understory that captures large quantities of moisture from the air, slowly recharging soil moisture and groundwater reserves. Streams at these elevations are generally small in size and in quantity of water, fed primarily from shallow springs and seepage that originate in these high-elevation forests. While not critical as habitat for amphidromous species, these streams support a variety of water-dependent damselfly and snail species, many of which are endemic and found nowhere else in the world.

When these small surface flows merge to form perennial streams, the increase in water accelerates the downward and outward erosion of the stream channel into the underlying basalt, forming the stream channel. Further downstream, the increase in runoff-driven peak flows results in an acceleration of energy available to transport sediment, big and small, downstream. This erosion results in higher gradient stream channels. Of course, rainfall, substrate age, and, by extension, island age all play a role in the degree to which a stream channel has eroded into the underlying parent material of the watershed. On younger substrates, the recent succession of lava flows can influence the movement of surface water to such a degree that water pours over, around, and under the basalt in no discernable stream channel. Often these streams end tens to hundreds of feet above sea level with a terminal waterfall, where no estuary environment can form. By contrast, some older stream channels on O'ahu and Kaua'i have meandering reaches, well-defined floodplains, and large estuaries. Moreover, the upslope movement of nickpoints, where low-elevation, low-gradient reaches transition sharply to high-elevation reaches, are easily discernible in many of these watersheds, a geologic phenomenon that is impossible to identify in younger watersheds. [28]

### The Stream in Between

The middle elevation stream reaches, where watersheds were sufficiently wide to support forest growth and human settlements, were the "Goldilocks" of stream channels: large enough in terms of watershed area and rainfall resulting in sufficient streamflow to support social and ecological systems, but small enough to not be a major flooding risk during extreme rainfall events. In these reaches, large numbers of freshwater fish (e.g., 'o'opu such as nākea) are available, po'owai can divert sufficient water to feed lo'i kalo, and the ocean is never beyond half a day's hike away. Restoring sufficient streamflow to promote the successful recruitment and survival of amphidromous species as well as the protection or restoration of cultural practices that depend on flowing water at this elevation is an important goal for the Commission.

### Ma Kai

The ma kai (toward the sea) reaches of a stream channel are particularly important from a biocultural landscape perspective. Estuaries support a diversity of species, ecosystem services, and cultural practices. Wide, low-gradient streams result in the blending of fresh water and salt water to produce a nutrient-rich, brackish environment with an abundant source of organic material as well as the nearly unlimited supply of larvae from upstream amphidromous species. Culturally important nearshore species such as āholehole (*Kuhlia sandvicensis*) and ʻamaʻama (*Mugil cephalus*) move in and out of this habitat. During low flows, this rich supply of nutrients and sediment settle out as the stream approaches the ocean, resulting in highly productive wetland habitat. Wetlands in Hawaiʻi support a wide range of endemic birds, invertebrates, crustaceans, mollusks, and fish while also providing critical breeding habitat for many marine species. The muliwai (estuary) environment sustains a diversity of limu (i.e., seaweed and algae) that is a culturally important food resource. Not only was the muliwai an important fishing area, but Hawaiians invented aquaculture that was integrated with the hydrological, ecological, and tidal seasons. The wide variety of loko iʻa (fishpond) management practices observed across the islands harnessed the varying surface and groundwater resources to reproduce the temperature, nutrients, and salinity conditions necessary to cultivate fish such as the ʻamaʻama. While urbanization, groundwater extraction, and neglect have negatively affected many loko iʻa, many community groups continue to push for their reconstruction, not just as a source of pride, but to return to the biocultural practices that sustained communities long before colonization. The Commission is beginning to incorporate the hydrological, ecological, and subsequently, cultural consequences of water resource management as they pertain to the estuary and nearshore environment into its decision-making.

### Ma Kai to Ma Uka

While water resource management typically looks at the gravity-driven movement of water from ma uka to ma kai, from an aquatic ecosystems perspective, habitat distribution starts in the nearshore and estuary environment and moves upstream, from ma kai to ma uka. The five endemic species of amphidromous fish in Hawai'i inhabit differing but frequently overlapping portions of the available stream reaches.[29] This differentiation is largely driven by each species' ability to navigate cascades and climb up rock faces at waterfalls. The estuary is an important habitat for the recruitment of postlarvae to stream ecosystems, as it provides ʻoʻopu (freshwater goby) the conditions necessary for the physical maturation that gives them the ability to climb streams. The least effective climber is the ʻoʻopu ʻakupa (*Eleotris sandwicensis*), which is only found in estuarine environments below the first cascade. The only slightly better climber is the ʻoʻopu naniha (*Stenogobius hawaiiensis*), which can navigate up to the first minor waterfall. ʻOʻopu nākea (*Awaous stamineus*) are good climbers, but their distribution is dependent on the distance inland from the coast to the first large (e.g., more than sixty feet) waterfall; as they grow, their larger size affects the physics of scaling waterfalls. By contrast ʻoʻopu nōpili (*Sicyopterus stimpsoni*) and ʻoʻopu alamoʻo (*Lentipes concolor*) remain relatively small and therefore remain excellent climbers, with alamoʻo, capable of navigating over of one thousand feet in vertical height. The nōpili and alamoʻo have uniquely fused pelvic fins and associated musculature that allows them to coordinate their body movements in such a way that they can scale up a wetted pathway along the outermost splash zone of a waterfall rock face. However, the similarities end there, as the alamoʻo uses a "powerburst" technique to propel itself upslope, while the nōpili uses an "inching" technique.[30] There may be overlap in species distribution at large scales among the three climbing species, but differences in substrate preferences is also known to affect their distribution: nākea generally prefer slower, deeper runs and pools with gravel substrates, while nōpili prefer faster, shallower riffles and runs with larger cobble and boulders. Nākea grow to be a larger, more robust adult, and at a particular size can no longer navigate taller waterfalls, while adult nōpili are smaller in stature and more effective climbers. However, even nōpili and alamoʻo face a declining window of opportunity to scale vertical rock faces along waterfalls as the physics behind propelling upward becomes an impossibility after a threshold of body weight is surpassed.[31] ʻŌpae kalaʻole (*Atyioda bisulcate*) are similarly adept at climbing waterfalls, although they can fully survive outside of the splash zone of the waterfall, which gives them a unique advantage for accessing high-elevation habitats.[32]

### Surface Water–Groundwater Interactions

Water cycles through the terrestrial, marine, and atmospheric environment, but the nuanced connectivity between surface water and groundwater resources and its temporal and spatial variability is not well understood.[33] While streams may start as rainfall in the highest reaches of the watershed, most perennial streamflow starts as spring flow, where groundwater that has infiltrated the soil at higher elevations runs into an aquitard, or a poorly permeable geologic formation that causes the water to flow laterally. This complexity is especially important to watersheds in Hawai'i, where rainfall is partitioned into groundwater recharge and runoff in relatively small, but geologically complex drainage areas.[34] The complexity is dependent on differences in permeability among geological formations, particularly as it relates to high-elevation vertical dike-impounded or horizontal perched water bodies. Hydrologic conductivity may be as high as hundreds to thousands of feet per day in permeable basalt, but water may move less than one foot per day through an aquitard. As a result, groundwater movement is heavily influenced by the presence of aquitards. Dike structures are a series of vertical aquitards sandwiching more permeable basalt formations. Water that recharges the more permeable basalt layers gets trapped, often at hundreds of feet in head.[35] This water then moves laterally toward areas of lower head. Horizontal aquitards, often formed by erosional deposition and ash layers between postshield lava flows, contribute to the perpendicular movement of water. When the laterally flowing water reaches a channel that has been incised into the landscape (such as in a gulch), the groundwater discharges as a spring into the surface or stream.

A watershed's underlying geology often varies across space and is heavily dependent on the underlying volcano's history. Some geologic formations are much more permeable than others. For example, in Haleakalā, the primary shield-building phase of the Honomanū volcanics is extremely permeable, and the basal groundwater body that is present in these volcanic layers is dependent on the Gyben-Herzberg principle that less dense fresh water floats on denser salt water. The later postshield Kula volcanic layers are more complex, with intermixed layers of permeable and impermeable basalt, soil, and ash. Groundwater can contribute to streamflow when channels incise the Kula volcanics, but that same stream may lose flow as it flows over the Honomanū volcanics. This lost surface water infiltrates into the groundwater body of the Honomanū volcanics, which may then support spring flow in lower elevation reaches and wetlands along the coast. Similarly, streams gain flow from springs emanating in dike zones at high elevations around the caldera or rift zones, but lose water (recharging the basal aquifer) as they flow over permeable volcanics at lower elevations. Much of the groundwater recharge that occurs from streamflow results in coastal spring flow and historically supported large amounts of wetland habitat. Unfortunately, this same area has seen the greatest amount of urbanization in the last hundred years, and these wetland habitats have almost entirely been lost. In areas where spring flow can support ecological or cultural practices, these benefits must be considered in the development of instream flow standards.

## Erosion of Socioecological Systems

The ahupua'a in Hawai'i represents a stage of land use and management where the island was divided up into slices, like a pie, with boundaries radiating from the piko, or the mountaintop, down to the coastline. The ahupua'a were at times a watershed, although watershed boundaries do not always match the ahupua'a boundaries. This level of land division was relatively simple to understand and demarcate, providing all that was necessary to make a community prosperous while avoiding conflicts over communal-type resources. The upland forests provided products with practical (e.g., wood and rocks for homes, tools, and canoes) and cultural (e.g., flowers, bird feathers, and plants for lei, capes, and crowns) value. Streams and springs supported lo'i kalo and loko i'a in the lower elevations for sustenance. And the nearshore waters provided fish, seafood, and limu. A ma uka to ma kai perspective is an important starting point for all present-day resource management issues, as it helps ground communities in their local environment and provides a sense of place. By understanding how the ahupua'a originally functioned (whether natural or not), resource managers can more clearly articulate the intended goals of resource management actions.

In many circumstances, restoring the ahupua'a to a previous state is not possible. Urbanization and land use changes have permanently altered the functioning of watersheds. This is clearly evident on O'ahu, where once healthy forests, lo'i kalo complexes, and coastal fishponds supported thousands of Hawaiians. However, urbanization has broken the natural water cycle, with roofs and concrete flushing rainfall into channelized streams out to the ocean as fast as possible, substantially reducing groundwater infiltration. It is not surprising that coastal and freshwater quality is poorest and stream habitat most limited in watersheds that are highly urbanized.[36] Every pulse of rainfall transports oils, pesticides, herbicides, fertilizers, cleaners, plastics, soils, and all manner of trash in our human-centric environment into streams and oceans. To make matters worse, Hawaiʻi has a large number of cesspools, releasing as much as fifty-five million gallons of sewage into groundwater systems each day.[37] These sources of bacterial and viral pollution lead to the high levels of water quality degradation.[38] The state is far behind in its goal to connect the hundreds of thousands of cesspools to appropriate sewage treatment facilities or upgrade them to more appropriate septic systems. In most major cities with functioning sewage systems, streams provide aesthetic and recreational value, boosting real estate prices and acting as focal points of interest. Not in urbanized Hawaiʻi, where a stream that runs through a neighborhood tends to be regarded as a drainage system and a nuisance rather than an ecosystem. Much of urban Honolulu has been built over streams and wetlands, with concrete walls around them, hiding them as hazardous blights to keep the community away.

The privatization of land ownership, specifically along the coast, has resulted in the destruction and abandonment of many loko iʻa throughout the islands. It wasn't until a comprehensive inventory and their designation as historic sites that the few remaining fishponds received protection.[39] Not only is the restoration and operation of fishponds physically demanding, but their function is threatened by shifts in nearshore freshwater inputs due to altered watershed hydrology, sea level rise affecting tidal patterns, and the invasion by nonnative fish, algal, and vegetation species.

However, if specific management goals are identified, steps can be taken to achieve intermediate aspects of ahupuaʻa restoration. For example, if the goal is to restore a functioning nearshore fishery or loko iʻa, then establishing clear goals related to the management of invasive estuarine vegetation (e.g., hau bush, mangrove), the control of land-based sources of sediment, or the careful reconstruction of fishpond walls are achievable.

### Limitations to Meeting Water Resource Management Goals

The Commission has made great strides in amending status-quo instream flow standards across the state. Either through formal petitions or staff-initiated amendments, there are over 85 instream flow standards for the 376 perennial streams and approximately 1,300 registered stream diversions. The majority of those amended instream flow standards were established by contested case proceedings (e.g., Waiāhole Petition, East Maui Petition, Nā Wai ʻEhā Petition, West Kauaʻi Petition). While contested cases provide a structured judicial forum for establishing relevant data and other information necessary for making management decisions, these proceedings can take decades to adjudicate. Further, once an instream flow standard has been established, implementation can be challenging, leaving the community with unanswered questions.

The Commission's small staff dedicated to the protection and management of streams (currently at seven filled positions for the state) limits its ability to gather and analyze relevant data, formulate recommendations, and implement decisions. The Commission has never been fully funded to fulfill the mandate of the Water Code. The list of responsibilities for the instream use section is daunting, with a mandate that is statewide in duties and obligations. Staff make flow measurements, maintain stream-monitoring equipment, conduct ecological surveys, respond to and investigate complaints, hold meetings, follow up with Commission orders, manage water use reporting, coordinate large-scale datasets, provide expertise to other state agencies, and develop data sets necessary to make management decisions. Further, the high degree of specialization and breadth of knowledge needed for staff means there is a tremendous loss of institutional knowledge when one staff member leaves.

### Positive Outlook

The state has made incremental strides toward implementing a complete water resource monitoring program. The recent publication of the Statewide Monitoring Needs Assessment[40] articulates what Commission staff have been parlaying to the state legislature, to community members, to stakeholders, and to the counties for years: that informed management that is adaptive to future conditions requires appropriate monitoring. We cannot know what we do not measure. With an increase in general funds, the Commission has now added or reestablished ten additional USGS stream monitoring stations in the last seven years, with the focus on the poorly monitored regions of the state. Since 2002, the Commission has funded four regional low-flow studies to better quantify the availability of surface water during dry periods and to improve our understanding of surface water and groundwater interactions: East Maui, Nā Wai ʻEhā, West Maui, and Southeast Kauaʻi. Moreover, a statewide program to develop low-flow statistics at ungagged locations is ongoing. However, the complete implementation of the statewide monitoring program to understand current and future trends in water availability is limited by state funding constraints, and the Commission continues to grapple with limited staff and budget to implement the Water Code.

Using a biocultural lens has not replaced the need to make quantitative evaluations of hydrological processes to support the development of instream flow standards. Rather, it supports the assessment of instream values and the integration of cultural and ecological processes. Since 2017, Commission staff have been engaged with communities and agencies statewide to develop instream flow standards without the need for lengthy contested case proceedings. In total, nine streams or tributaries in West Maui, fourteen streams in East Hawaiʻi, five streams on Molokaʻi, two streams on Kauaʻi, and one stream on Oʻahu have amended instream flow standards in this manner. While each amended instream flow standard represents years of work, community support has enhanced the process and aided data collection. It is this coproduction of data, with a focus on the biology, ecology, and cultural values, from ma uka to ma kai, that has boosted the Commission's effectiveness at establishing reasonable, meaningful instream flow standards.

## Notes

1. I will hereafter refer to all flowing water as streams, which will include rivers, creeks, brooks, and all other nomenclature.
2. David Deming, "The Aqueducts and Water Supply of Ancient Rome," *Groundwater* 58, no. 1 (2020): 152–161.
3. L. De Stefano et al., "Assessment of Transboundary River Basins for Potential Hydro-Political Tensions," *Global Environmental Change* 45 (2017): 35–46.
4. F. Berkes and C. Folke, "Linking Social and Ecological Systems for Resilience and Sustainability," in *Linking Social and Ecological Systems: Management Practices and Social Mechanisms for Building Resilience,* edited by F. Berkes and C. Folke (New York: Cambridge University Press, 1998), 1–26.
5. C. S. Holling, *Adaptive Environmental Assessment and Management* (London: Wiley Press, 1978).

6. L. Gunderson, C. S. Holling, and S. Light, eds., *Barriers and Bridges to the Renewal of Ecosystems and Institutions* (New York: Columbia University Press, 1995).
7. A. M. Strauch and A. M. Alemdom, "Traditional Water Resource Management and Water Quality in Rural Tanzania," *Human Ecology* 39 (2011): 93–106.
8. N. LeRoy Poff et al., "The Natural Flow Regime," *BioScience* 47, no. 11 (1997): 769–784.
9. Ellen Wohl et al., "The Hydrology of the Humid Tropics," *Nature Climate Change* 2 (2012): 655–662.
10. Ibid.
11. D. Kapua'ala Sproat, "Water," in *The Value of Hawaii: Knowing the Past, Shaping the Future,* edited by C. Howes and J. K. Kamakawiwo'ole Osorio (Honolulu: University of Hawai'i Press, 2010), 187–194.
12. R. L. Vannote et al., "The River Continuum Concept," *Canadian Journal of Fisheries and Aquatic Sciences* 37 (1980): 130–137.
13. D. K. Sproat, "Wai through Kānāwai: Water for Hawai'i's Streams and Justice for Hawaiian Communities," *Marquette Law Review* 95 (2011): 127–211.
14. C. S. Holling, L. Gunderson, and D. Ludwig, "In Quest of a Theory of Adaptive Change," in *Panarchy: Understanding Transformations in Human and Natural Systems,* edited by L. H. Gunderson and C. S. Holling (Washington, DC: Island Press, 2002), 3–24.
15. Kawika Winter et al., "The *Moku* System: Managing Biocultural Resources for Abundance within Social-Ecological Regions in Hawai'i," *Sustainability* 10, no. 3 (2018): 554.
16. R. M. McDowall, "On Amphidromy, a Distinct Form of Diadromy in Aquatic Organisms," *Fish and Fisheries* 8, no. 1 (2007): 1–13.
17. R. A. Kinzie III and J. I. Ford, "Population Biology in Small Hawaiian Streams," Technical Report no. 147 (Honolulu: Water Resources Research Center, University of Hawai'i at Mānoa, 1982).
18. J. Michael Fitzsimons, M. G. McRae, and R. T. Nishimoto, "Behavioral Ecology of Indigenous Stream Fishes in Hawai'i," in *Biology of Hawaiian Streams and Estuaries: Proceedings of the Symposium on the Biology of Hawaiian Streams and Estuaries, Hilo, Hawai'i, 26–27 April 2005,* edited by N. L. Evenjhuis and J. M. Fitzsimons, Bishop Museum Bulletin in Cultural and Environmental Studies 3 (Honolulu: Bishop Museum Press, 2007), 11–21.
19. R. A. Kinzie III, "Evolution and Life History Patterns in Freshwater Gobies," *Micronesica* 30 no. 1 (1997): 27–40.

20. Kurt D. Fausch et al., "Landscapes to Riverscapes: Bridging the Gap between Research and Conservation of Stream Fishes," *Bioscience* 52, no. 6 (2002): 483–498.
21. R. W. Tingley III et al., "Identifying Natural Catchment Landscape Influences on Tropical Stream Organisms: Classifying Stream Reaches of the Hawaiian Islands," *Hydrobiologia* 826 (2019): 67–83.
22. U.S. Army Corps of Engineers, "National Hydroelectric Power Resources Study: Regional Assessment: Alaska and Hawaii," vol. 23, September 1981.
23. Fitzsimons, McRae, and Nishimoto, "Behavioral Ecology."
24. Kinzie, "Evolution and Life History Patterns."
25. I will avoid the distinction between interim instream flow standards and instream flow standards for now.
26. D. K. Sproat, "Wai Through Kānāwai: Water for Hawai'i's Streams and Justice for Hawaiian Communities," *Marquette Law Review* 95 (2011): 127–211.
27. See, e.g., Martha Warren Beckwith, *The Kumulipo: A Hawaiian Creation Chant* (Honolulu: University of Hawai'i Press, 1972).
28. L. W. Raming, K. X. Whipple, and A. M. Strauch, "Limits to Knickzone Retreat and Bedrock River Incision on the Hawaiian Islands," *Earth Surface Processes and Landforms* 49, no. 6 (2024): 1914–1931.
29. Kinzie, "Evolution and Life History Patterns."
30. H. L. Schoenfuss and R. W. Blob, "Kinematics of Waterfall Climbing in Hawaiian Freshwater Fishes (Gobiidae): Vertical Propulsion at the Aquatic-Terrestrial Interface," *Journal of Zoology* 261, no. 2 (2003): 191–205.
31. T. Maie, H. L. Schoenfuss, and R. W. Blob, "Ontogenic Scaling of Body Proportions in Waterfall-Climbing Gobiid Fishes from Hawaii and Dominica: Implications for Locomotor Function," *Copeia* 3 (2007): 755–764.
32. Strauch et al., "Population Response to Connectivity Restoration of High Elevation Tropical Stream Reaches in Hawai'i," *Conservation Science and Practice* 4 (2022): e12836.
33. Izuka et al., "Volcanic Aquifers of Hawai'i—Hydrogeology, Water Budgets, and Conceptual Models," U.S. Geological Studies Scientific Investigations Report 2015–5164, 2020.
34. I am ignoring canopy and soil evaporation and plant transpiration for this discussion.
35. A hydrologic term referencing the elevation of water above a reference point.

36. A. M. D. Brasher, "Impacts of Human Disturbances on Biotic Communities in Hawaiian Streams," *Bioscience* 53, no. 11 (2003): 1052–1060.
37. M. Mezzacapo et al., "Hawai'i's Cesspool Problem: Review and Recommendations for Water Resources and Human Health," *Journal of Contemporary Water Research and Education* 170 (2020): 35–75.
38. L. M. Economy et al., "Rainfall and Streamflow Effects on Estuarine *Staphylococcus aureus* and Fecal Indicator Bacteria Concentrations," *Journal of Environmental Quality* 48, no. 6 (2019): 1711–1721.
39. *Hawaiian Fishpond Study: Islands of Hawai'i, Maui, Lāna'i, and Kaua'i,* prepared for the Hawai'i Coastal Zone Management Program (Honolulu: DHM Inc., 1990).
40. C. L. Cheng et al., "Water-Resource Management Monitoring Needs," State of Hawai'i: U.S. Geological Survey Scientific Investigations Report 2020–5115, 2021.

## Bibliography

Anthony, S. S., C. D. Hunt Jr., A. M. D. Brasher, L. D. Miller, and M. S. Tomlinson. "Water Quality on the Island of Oahu, Hawaii, 1999–2001." *U.S. Geological Survey Circular* 1239 (2004).

Beck, C. A. "Hydroelectric Power in Hawaii: A Report on the Statewide Survey of Potential Hydroelectric Sites." Report for the Department of Planning and Economic Development, State of Hawai'i, 1981.

Beckwith, Martha Warren. *The Kumulipo: A Hawaiian Creation Chant.* Honolulu: University of Hawai'i Press, 1972.

Benstead, J. P., J. G. March, C. M. Pringle, and F. N. Scatena. "Effects of Low-Head Dam and Water Abstraction on Migratory Tropical Stream Biota." *Ecological Applications* 9, no. 2 (1999): 656–668.

Berkes, F., and C. Folke. 1998. "Linking Social and Ecological Systems for Resilience and Sustainability." In *Linking Social and Ecological Systems: Management Practices and Social Mechanisms for Building Resilience,* edited by F. Berkes and C. Folke, 1–26. New York: Cambridge University Press, 1998.

Brasher, A. M. D. "Impacts of Human Disturbances on Biotic Communities in Hawaiian Streams." *Bioscience* 53, no. 11 (2003): 1052–1060.

Brasher, A. M. D. "Monitoring the Distribution and Abundance of Native Gobies ('o'opu) in Waikolu and Pelekunu Streams on the Island of Moloka'i." PCSU Technical Report 113. Honolulu: Cooperative National Park Resources Studies Unit, University of Hawai'i at Mānoa, 1996.

Cebrián-Piqueras, M. A., A. Filyushkina, D. N. Johnson, V. B. Lo, M. D. López-Rodríquez, H. March, E. Oteros-Rozas, C. Peppler-Lisbach, C. Quintas-Soriano, C. M. Raymond, I. Ruiz-Mallén, C. J. van Riper, Y. Zinngrebe, and T. Plieninger. "Scientific and Local Ecological Knowledge, Shaping Perceptions towards Protected Areas and Related Ecosystem Services." *Landscape Ecology* 35 (2020): 2549–2567.

Cheng, C. L., S. K. Izuka, J. J. Kennedy, A. G. Frazier, and T. W. Giambelluca. "Water-Resource Management Monitoring Needs." State of Hawai'i: U.S. Geological Survey Scientific Investigations Report 2020–5115, 2021.

Couret, C. L., Jr. "The Biology and Taxonomy of a Freshwater Shrimp, *Atyid bisulcata* Randall, Endemic to the Hawaiian Islands." Master's thesis, University of Hawai'i at Mānoa, 1976.

Deming, David. "The Aqueducts and Water Supply of Ancient Rome." *Groundwater* 58, no. 1 (2020): 152–161.

De Stefano, L., Jacob D. Petersen-Perlman, Eric A. Sproles, Jim Eynard, and Aaron T. Wolf. "Assessment of Transboundary River Basins for Potential Hydro-Political Tensions." *Global Environmental Change* 45 (2017): 35–46.

Economy, L. M., T. N. Wiegner, A. M. Strauch, J. D. Awaya, and T. Gerken. "Rainfall and Streamflow Effects on Estuarine *Staphylococcus aureus* and Fecal Indicator Bacteria Concentrations." *Journal of Environmental Quality* 48, no. 6 (2019): 1711–1721.

Erős, T., and W. H. Lowe. "The Landscape Ecology of Rivers: From Patch-Based to Spatial Network Analyses." *Current Landscape Ecology Reports* 4 (2019): 103–112.

Fausch, Kurt D., Christian E. Torgersen, Colden V. Baxter, and Hiram W. Li. "Landscapes to Riverscapes: Bridging the Gap between Research and Conservation of Stream Fishes." *Bioscience* 52, no. 6 (2022): 483–498.

Fitzsimons, J. M., M. G. McRae, and R. T. Nishimoto. "Behavioral Ecology of Indigenous Stream Fishes in Hawai'i." In *Biology of Hawaiian Streams and Estuaries: Proceedings of the Symposium on the Biology of Hawaiian Streams and Estuaries, Hilo, Hawai'i, 26–27 April 2005,* edited by N. L. Evenhuis and J. M. Fitzsimons, 11–21. Bishop Museum Bulletin in Cultural and Environmental Studies 3. Honolulu: Bishop Museum Press, 2007.

Gingerich, S. B., and R. H. Wolff. "Effects of Surface-Water Diversions on Habitat Availability for Native Stream Macrofauna, Northeast Maui, Hawaii." U.S. Geological Survey Scientific Investigations Report 2005–5213, 2005.

Gunderson, L., C. S. Holling, and S. Light, eds. *Barriers and Bridges to the Renewal of Ecosystems and Institutions*. New York: Columbia University Press, 1995.

*Hawaiian Fishpond Study: Islands of Hawai'i, Maui, Lāna'i, and Kaua'i*. Prepared for the Hawai'i Coastal Zone Management Program. Honolulu: DHM Inc., 1990.

Holling, C. S. *Adaptive Environmental Assessment and Management*. London: Wiley Press, 1978.

Holling, C. S., L. Gunderson, and D. Ludwig. "In Quest of a Theory of Adaptive Change." In *Panarchy: Understanding Transformations in Human and Natural Systems*, edited by L. H. Gunderson and C. S. Holling, 3–24. Washington, DC: Island Press, 2002.

Izuka, S. K., J. A. Engott, K. Rotzoll, M. Bassiouni, A. G. Johnson, L. D. Miller, and A. Mair. "Volcanic Aquifers of Hawai'i—Hydrogeology, Water Budgets, and Conceptual Models." U.S. Geological Studies Scientific Investigations Report 2015–5164, 2020.

Kinzie, R. A., III. "Evolution and Life History Patterns in Freshwater Gobies." *Micronesica* 30, no. 1 (1997): 27–40.

Kinzie, R. A., III. "Reproductive Biology of an Endemic, Amphidromous Goby, *Lentipes concolor*, in Hawaiian Streams." *Environmental Biology of Fishes* 37 (1993): 257–268.

Kinzie, R. A., III, and J. I. Ford. "Population Biology in Small Hawaiian Streams." Technical Report no. 147. Honolulu: Water Resources Research Center, University of Hawai'i at Mānoa, 1982.

Lau, L. S., and J. F. Mink. *Hydrology of the Hawaiian Islands*. Honolulu: University of Hawai'i Press, 2006.

Maie, T., H. L. Schoenfuss, and R. W. Blob. "Ontogenic Scaling of Body Proportions in Waterfall-Climbing Gobiid Fishes from Hawaii and Dominica: Implications for Locomotor Function." *Copeia* 3 (2007): 755–764.

McDowall, R. M. "On Amphidromy, a Distinct Form of Diadromy in Aquatic Organisms." *Fish and Fisheries* 8, no. 1 (2007): 1–13.

Mezzacapo, M., M. J. Donohue, C. Smith, A. El-Kadi, K. Falinski, and D. T. Lerner. "Hawai'i's Cesspool Problem: Review and Recommendations for Water Resources and Human Health." *Journal of Contemporary Water Research and Education* 170 (2020): 35–75.

National Hydropower Study. U.S. Army Corps of Engineers. Final Draft Report, January 1981.

Poff, N. LeRoy, J. David Allan, Mark B. Bain, James R. Karr, Richard E. Sparks, and Julie C. Stromberg. "The Natural Flow Regime." *BioScience* 47, no. 11 (1997): 769–784.

Radtke, R. L., R. A. Kinzie III, and D. J. Shafer. "Temporal and Spatial Variation in Length of Larval Life and Size at Settlement of the Hawaiian Amphidromous Goby *Lentipes concolor.*" *Journal of Fish Biology* 59, no. 4 (2001): 928–938.

Raming, L. W., K. X. Whipple, and A. M. Strauch. "Limits to Knickzone Retreat and Bedrock River Incision on the Hawaiian Islands." *Earth Surface Processes and Landforms* 49, no. 6 (2024): 1914–1931.

Schoenfuss, H. L., T. A. Blanchard, and D. G. K. Kuamo'o. "Metamorphosis in the Cranium of Postlarval *Sicyopterus stimpsoni,* an Endemic Hawaiian Stream Goby." *Micronesica* 30 (1997): 93–104.

Schoenfuss, H. L., and R. W. Blob. "The Importance of Functional Morphology for Fishery Conservation and Management: Applications to Hawaiian Amphidromous Fishes." In *Biology of Hawaiian Streams and Estuaries: Proceedings of the Symposium on the Biology of Hawaiian Streams and Estuaries, Hilo, Hawai'i, 26–27 April 2005,* edited by N. L. Evenhuis and J. M Fitzsimons, 125–141. Bishop Museum Bulletin in Cultural and Environmental Studies 3. Honolulu: Bishop Museum Press, 2007.

Schoenfuss, H. L., and R. W. Blob. "Kinematics of Waterfall Climbing in Hawaiian Freshwater Fishes (Gobiidae): Vertical Propulsion at the Aquatic-Terrestrial Interface." *Journal of Zoology* 261, no. 2 (2003): 191–205.

Sproat, D. K. "Wai through Kānāwai: Water for Hawai'i's Streams and Justice for Hawaiian Communities." *Marquette Law Review* 95 (2011): 127–211.

Sproat, D. K. "Water." In *The Value of Hawai'i: Knowing the Past, Shaping the Future,* edited by C. Howes and J. K. Kamakawiwo'ole Osorio, 187–194. Biography Monographs. Honolulu: University of Hawai'i Press, 2010.

Strauch, A. M., and A. M. Alemdom. "Traditional Water Resource Management and Water Quality in Rural Tanzania." *Human Ecology* 39 (2011): 93–106.

Strauch, A. M., R. A. MacKenzie, C. P. Giardina, and G. L. Bruland. "Climate Driven Changes to Rainfall and Streamflow Patterns in a Model Tropical Island Hydrological System." *Journal of Hydrology* 523 (2015): 160–169.

Strauch, A. M., R. W. Tingley III, J. Hsiao, P. B. Foulk, T. C. Frauendorf, R. A. MacKenzie, and D. M. Infante. "Population Response to Connectivity Restoration of High Elevation Tropical Stream Reaches in Hawai'i." *Conservation Science and Practice* 4 (2022): e12836.

Tingley, R. W., III, D. M. Infante, R. A. MacKenzie, A. R. Cooper, and Y.-P. Tsang. "Identifying Natural Catchment Landscape Influences on

Tropical Stream Organisms: Classifying Stream Reaches of the Hawaiian Islands." *Hydrobiologia* 826 (2019): 67–83.

U.S. Army Corps of Engineers. "National Hydroelectric Power Resources Study: Regional Assessment: Alaska and Hawaii." Vol. 23. September 1981.

Vannote, R. L., G. W. Minshall, K. W. Cummins, J. R. Sedell, and C. E. Cushing. "The River Continuum Concept." *Canadian Journal of Fisheries and Aquatic Sciences* 37 (1980): 130–137.

Winter, K. B., K. Beamer, M. B. Vaughan, A. M. Friedlander, M. H. Kido, A. N. Whitehead, M. K. H. Akutagawa, N. Kurashima, M. P. Lucas, and B. Nyberg. "The *Moku* System: Managing Biocultural Resources for Abundance within Social-Ecological Regions in Hawaiʻi." *Sustainability* 10 (2018): 1–19.

Wohl, Ellen, Ana Barros, Nathaniel Brunsell, Nick A. Chappell, Michael Coe, Thomas Giambelluca, Steven Goldsmith, Russell Harmon, Jan M. H. Hendrickx, James Juvik, Jeffrey McDonnell, and Fred Ogden. "The Hydrology of the Humid Tropics." *Nature Climate Change* 2 (2012): 655–662.

CHAPTER 12

# Hierarchy of Water Management in Hawai'i

*Kamanamaikalani Beamer, Kawena Elkington, and Pua Souza*

This section provides a brief summary of the hierarchy of water management in Hawai'i as well as several illustrations to assist the reader in understanding Hawai'i's water management structure (see figure 12.1 for a basic hierarchy). For a more thorough and detailed analysis, please see chapters 3 and 4.

Authority for water management in Hawai'i begins with constitutional provisions that dictate Hawai'i's trust relationship, management, and conservation of water resources. The primacy of the Public Trust Doctrine in Hawai'i is traced back to the cultural and historical significance of water in Hawaiian society and the codifications of these principles in the Hawaiian Kingdom. These frameworks provide the context for adaptive water management today. The 1978 constitutional convention enshrined the Public Trust Doctrine in Hawai'i's Constitution. Article XI, §1 of Hawai'i's Constitution states that "all public natural resources are held in trust by the State for the benefit of the people." Article XI, §7 references water specifically, including the duty to "protect, control, and regulate the use of Hawai'i's water resources for the benefit of its people." Together, they define the Public Trust Doctrine as "a fundamental principle of constitutional law in Hawai'i."[1] Today, water management in Hawai'i is complex and largely influenced by unique geographic and climatic conditions, as well as Hawai'i's history and culture.

Article XI, §7 of the Constitution also facilitates the establishment of a water resources agency to administer water regulation: "The legislature shall provide for a water resources agency which, as provided

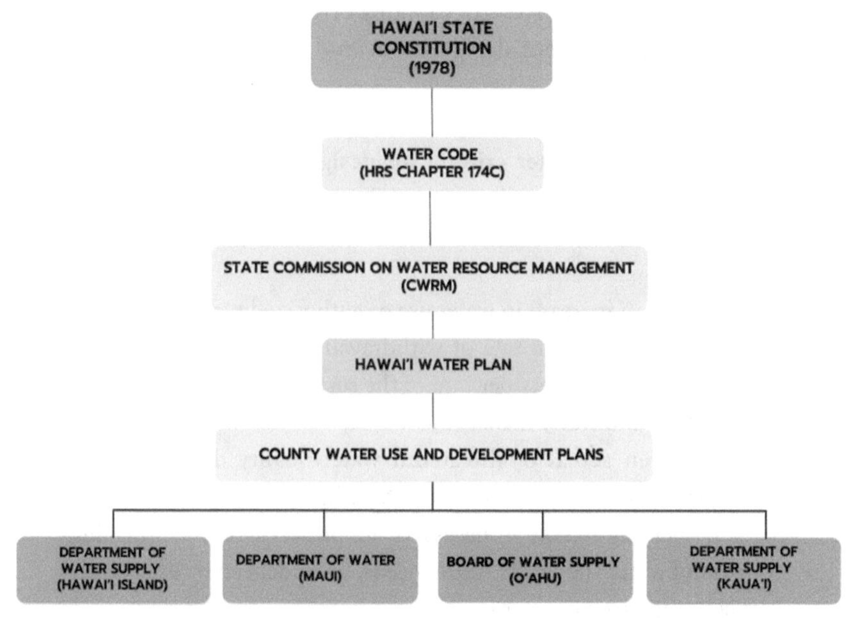


**Figure 12.1** Institutions involved in Hawai'i water management.

by law, shall set overall water conservation, quality and use policies; define beneficial and reasonable uses; protect ground and surface water resources, watersheds and natural stream environments; establish criteria for water use priorities while assuring appurtenant rights and existing correlative and riparian uses and establish procedures for regulating all uses of Hawaii's water resources."[2] This set the beginning stages for the State Commission on Water Resource Management (CWRM, or "Water Commission"), which was created by the 1987 Hawai'i state legislature when it enacted the State Water Code. The Commission's mission is "to protect and manage the waters of the State of Hawai'i for present and future generations."[3] The commission is responsible for regulating water use, promoting conservation, setting instream flow standards, sustainable yields for groundwater aquifers, and protecting Hawai'i's water resources. The Water Commission also has a heightened level of administrative duty and authority when it designates surface and groundwater management areas (WMA). Areas are designated because of a determined resource need for more active management by the Commission. Non-water management areas are *not* subject to the same level of regulation as WMAs. The State Water Code dictates criteria for groundwater and surface water management designation, and these criteria are quoted verbatim below.

> **§174C-44 Groundwater criteria for designation.** In designating an area for water use regulation, the commission shall consider the following:
>
> 1. Whether an increase in water use or authorized planned use may cause the maximum rate of withdrawal from the ground water source to reach ninety per cent of the sustainable yield of the proposed ground water management area;
> 2. There is an actual or threatened water quality degradation as determined by the department of health;
> 3. Whether regulation is necessary to preserve the diminishing ground water supply for future needs, as evidenced by excessively declining ground water levels;
> 4. Whether the rates, times, spatial patterns, or depths of existing withdrawals of ground water are endangering the stability or optimum development of the ground water body due to upconing or encroachment of salt water;
> 5. Whether the chloride contents of existing wells are increasing to levels which materially reduce the value of their existing uses;
> 6. Whether excessive preventable waste of ground water is occurring;
> 7. Serious disputes respecting the use of ground water resources are occurring; or
> 8. Whether water development projects that have received any federal, state, or county approval may result, in the opinion of the commission, in one of the above conditions.
>
> Notwithstanding an imminent designation of a ground water management area conditioned on a rise in the rate of ground water withdrawal to a level of ninety per cent of the area's sustainable yield, the commission, when such level reaches the eighty per cent level of the sustainable yield, may invite the participation of water users in the affected area to an informational hearing for the purposes of assessing the ground water situation and devising.[4]

**[§174C-45] Surface water criteria for designation.** In designating an area for water use regulation, the commission shall consider the following:

1. Whether regulation is necessary to preserve the diminishing surface water supply for future needs, as evidenced by excessively declining surface water levels, not related to rainfall variations, or increasing or proposed diversions of surface waters to levels which may detrimentally affect existing instream uses or prior existing off stream uses;
2. Whether the diversions of stream waters are reducing the capacity of the stream to assimilate pollutants to an extent which adversely affects public health or existing instream uses; or
3. Serious disputes respecting the use of surface water resources are occurring.[5]

Not all of Hawai'i is under designation. Figures 12.2–12.4 are maps of designated water management areas that show the areas that are designated today.

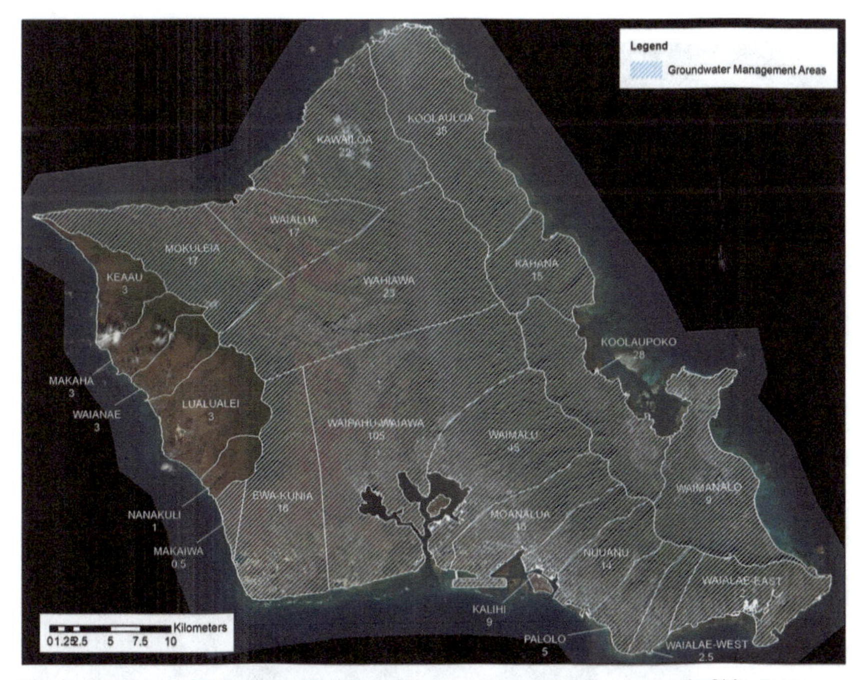


**Figure 12.2** Hawai'i State Water Commission's designated water management areas for O'ahu. Note that only a select few areas across Hawai'i are designated and more actively managed. Map created by Hawai'i Commission on Water Resource Management hydrologist Ayron M. Strauch, PhD.

**Figure 12.3** Hawai'i State Water Commission's designated water management areas for Moloka'i. Note that only a select few areas across Hawai'i are designated and more actively managed. Map created by Hawai'i Commission on Water Resource Management hydrologist Ayron M. Strauch, PhD.

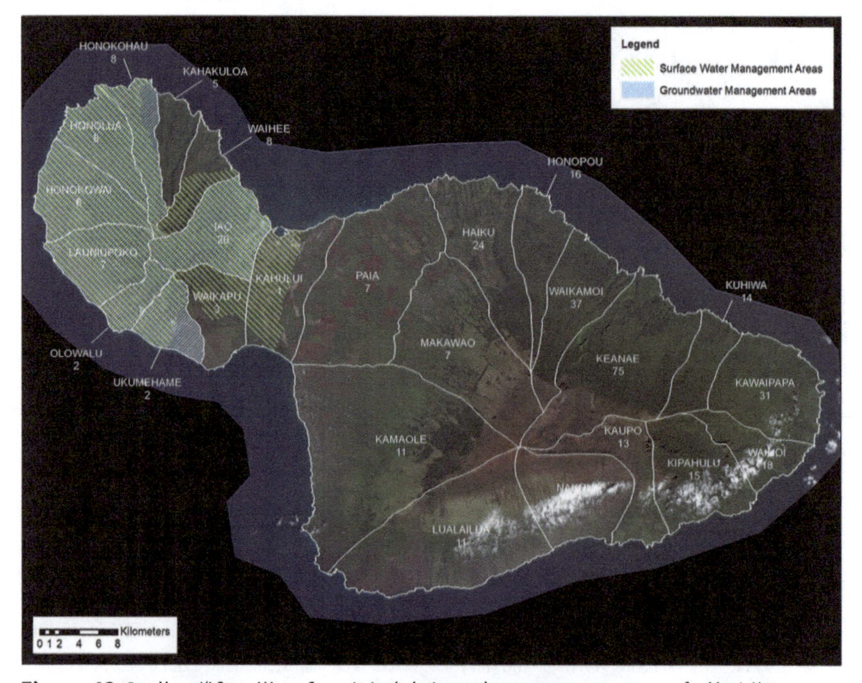


**Figure 12.4** Hawai'i State Water Commission's designated water management areas for Maui. Note that only a select few areas across Hawai'i are designated and more actively managed. Map created by Hawai'i Commission on Water Resource Management hydrologist Ayron M. Strauch, PhD.

"Sustainable yield" refers to the maximum amount of water that can be withdrawn from a particular water source while still maintaining its long-term availability and the ecological health of the surrounding environment. This is one primary tool used by the CWRM to ensure groundwater is sustained for future generations, although it has been contested in some areas across Hawai'i, and there are reasons for the Commission to update their sustainable yield calculations.[6] To determine the sustainable yield, the CWRM considers various factors, including the natural recharge rates of water sources, the existing and projected water demands, and the environmental needs of streams, rivers, and groundwater-dependent ecosystems. They also take into account the hydrological characteristics of each region, including rainfall patterns, recharge estimates, geology, and other relevant factors. Future estimates need to take into account the requirements for traditional and customary rights in calculating sustainable yields. The goal of managing groundwater resources based on sustainable yield is to ensure a balance between water supply and demand, protect the long-term viability of water sources, and prevent overextraction or depletion of the aquifer. When done properly, this approach should promote the sustainable use of water while preserving the ecological integrity and the ability of ecosystems to function properly. By setting sustainable yields, the CWRM aims to make informed decisions about water allocation, permitting, and management practices, ensuring the long-term availability and sustainability of water resources for Hawai'i's future. Figures 12.5–12.10 are maps of Hawai'i's aquifer systems and present sustainable yields.

The Hawai'i State Water Code outlined in the Hawai'i Revised Statutes, chapter 174C, is administered by the Hawai'i Department of Land and Natural Resources, which is responsible for enforcing its provisions and ensuring that water resources in Hawai'i are managed sustainably. The code covers a wide range of topics related to water management, including the allocation of water rights, and the regulation of water use. It also acknowledges the importance of a comprehensive water resources planning program to address water supply and conservation needs. The resulting Hawai'i Water Plan, which was initially adopted in 1990, serves as a guiding document for implementing this policy. The plan consists of five main parts:

1. A Water Resource Protection Plan prepared by the Commission on Water Resource Management;
2. A Water Quality Plan prepared by the Department of Health;

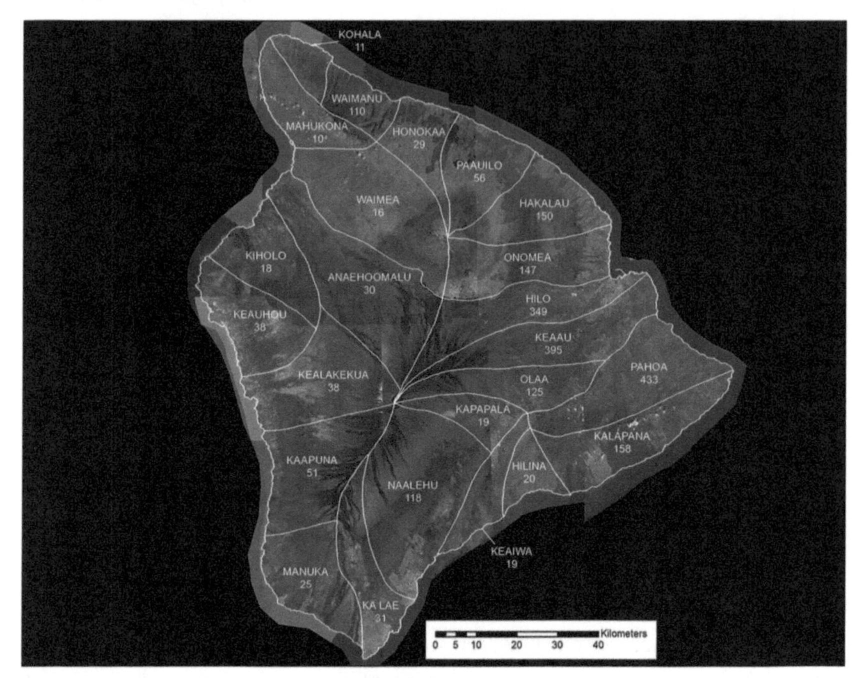


**Figure 12.5** Aquifer boundaries and sustainable yields of Hawai'i Island. Map created by Hawai'i Commission on Water Resource Management hydrologist Ayron M. Strauch, PhD.

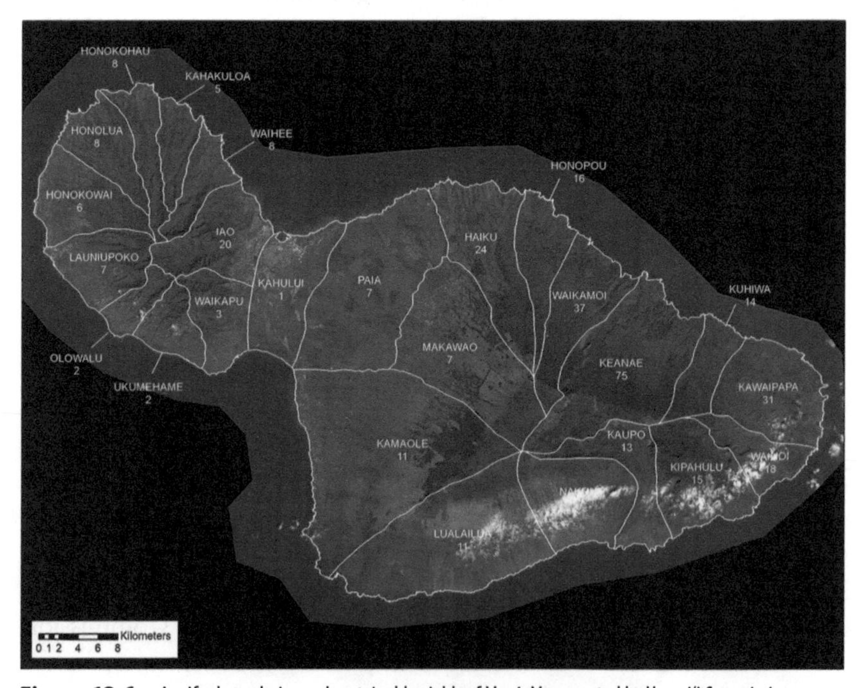


**Figure 12.6** Aquifer boundaries and sustainable yields of Maui. Map created by Hawai'i Commission on Water Resource Management hydrologist Ayron M. Strauch, PhD.

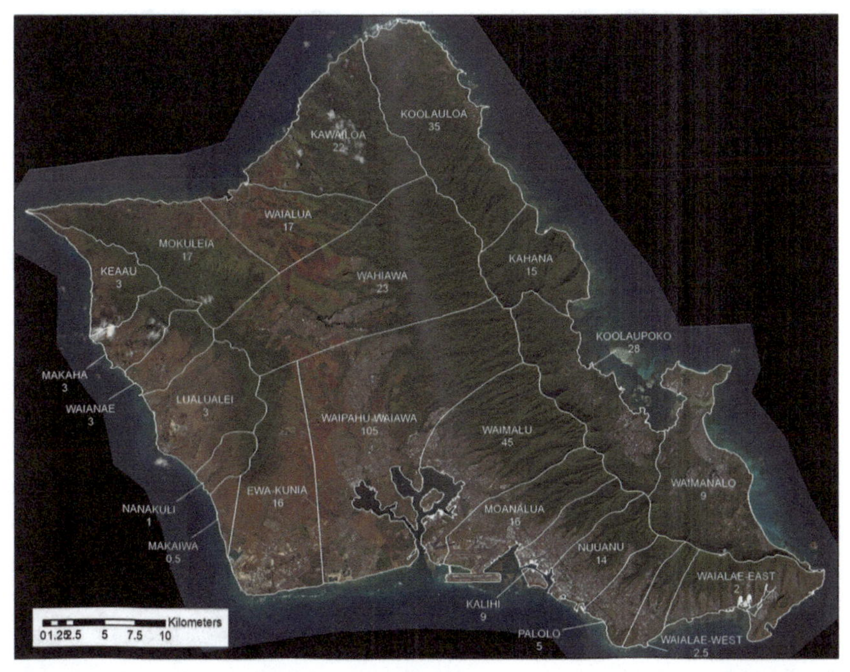


**Figure 12.7** Aquifer boundaries and sustainable yields of O'ahu. Map created by Hawai'i Commission on Water Resource Management hydrologist Ayron M. Strauch, PhD.

**Figure 12.8** Aquifer boundaries and sustainable yields of Kaua'i. Map created by Hawai'i Commission on Water Resource Management hydrologist Ayron M. Strauch, PhD.

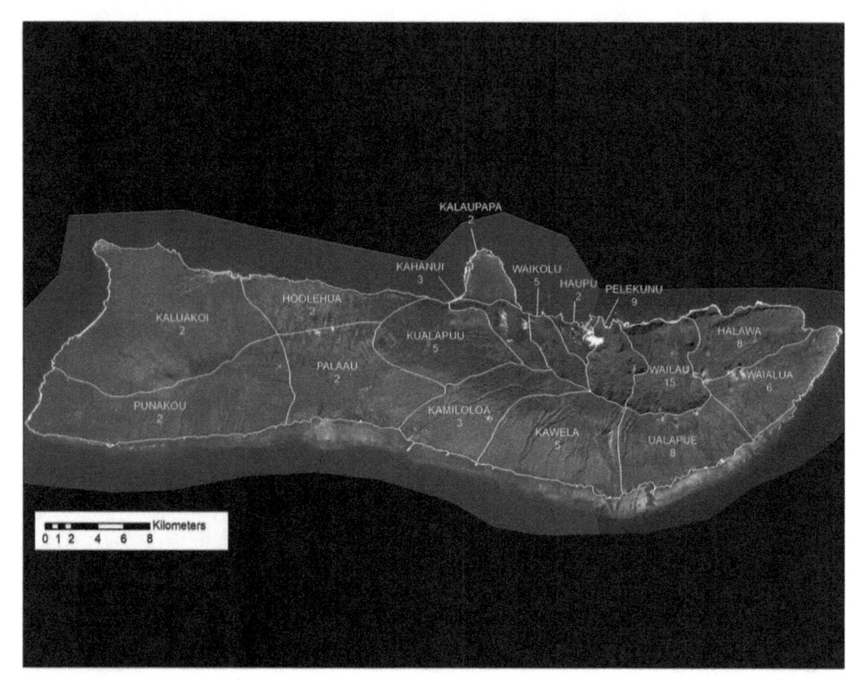


**Figure 12.9** Aquifer boundaries and sustainable yields of Moloka'i. Map created by Hawai'i Commission on Water Resource Management hydrologist Ayron M. Strauch, PhD.

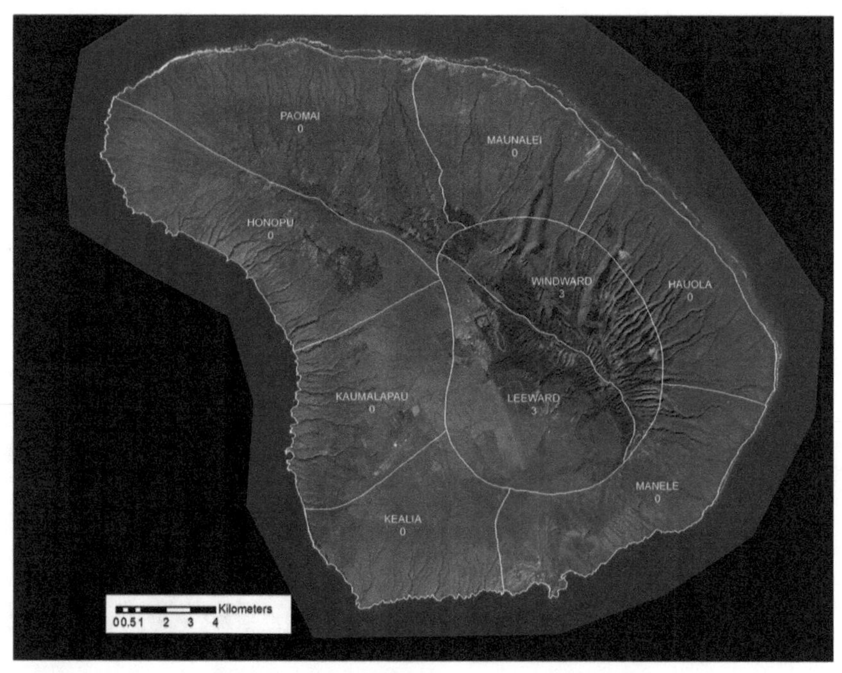


**Figure 12.10** Aquifer boundaries and sustainable yields of Lana'i. Map created by Hawai'i Commission on Water Resource Management hydrologist Ayron M. Strauch, PhD.

3. A State Water Projects Plan prepared by the Engineering Division of the Department of Land and Natural Resources;
4. An Agricultural Water Use and Development Plan prepared by the Department of Agriculture; and
5. A Water Use and Development Plans prepared by each county.

The Water Resource Protection Plan and Water Quality Plan serve as the legal and policy framework for the development, conservation, and use of water resources. The State Water Projects Plan and Agricultural Water Use and Development Plan provide information on statewide water needs and development plans. These components are integrated into the County Water Use and Development Plans, which allocate water use and development for each county. While the CWRM is the trustee over all of Hawai'i's water resources, there are several county-level agencies with responsibility for water management, including the Department of Water Supply on the island of Hawai'i, the Department of Water on Maui, the Board of Water Supply on O'ahu, and the Department of Water Supply on Kaua'i (see figure 12.1).

## Notes

1. *Waiāhole Combined Contested Case Hearing* ("*Waiāhole I*"), 94 Hawai'i 97, 132, 9P.3d 409, 444 (2000); D. Kapua'ala Sproat, *Ola I Ka Wai: A Legal Primer for Water Use and Management in Hawai'i* (Honolulu: Ka Huli Ao Center for Excellence in Native Hawaiian Law, 2009).
2. Hawai'i Constitution, art. XI, sec. 7.
3. State of Hawai'i, Department of Land and Natural Resources: Commission on Water Resource Management. https://dlnr.hawaii.gov/cwrm/aboutus/. Accessed June 23, 2023.
4. Hawai'i Revised Statutes, sec. 174C-44.
5. Hawai'i Revised Statutes, sec. 174C-45.
6. For more information, see Jonathan L. Scheuer and Bianca K. Isaki, "Protecting Water While It Is below Ground: The Unsustainability of 'Sustainable Yield,'" in Scheuer and Isaki, *Water and Power in West Maui* (Lahaina: North Beach-West Maui Benefit Fund, 2021), 131–149.

## Bibliography

Scheuer, Jonathan L., and Bianca K. Isaki. "Protecting Water While It Is below Ground: The Unsustainability of 'Sustainable Yield.'" In Scheuer and Isaki, *Water and Power in West Maui,* 131–149. Lahaina: North Beach-West Maui Benefit Fund, 2021.

Sproat, D. Kapua'ala. *Ola I Ka Wai: A Legal Primer for Water Use and Management in Hawai'i.* Honolulu: Ka Huli Ao Center for Excellence in Native Hawaiian Law, 2009.
*Waiāhole Combined Contested Case Hearing* ("*Waiāhole I*"), 94 Hawai'i 97, 132, 9P.3d 409, 444 (2000).

PART IV

# LET WAI FLOW

Part 4, "Let Wai Flow," is deeply inspired by the eight years I served as a commissioner on the Hawai'i State Commission on Water Resource Management (CWRM). In truth, the impetus for this book came while I was in the first year of my second term as a commissioner. Administering and upholding Hawai'i's Water Plan and the State Water Code, the CWRM is one of the most important entities responsible for securing Hawai'i's water future. However, it is interesting to note here that there is often little visibility given to the work of this indispensable group. This last part of *Waiwai: Water and the Future of Hawai'i* opens with my final remarks, given at the end of my last meeting as a commissioner, after having served for eight years on the CWRM. These eight years held many moments of both personal and professional growth, expanding my pilina to wai to places I had never imagined. I continue to believe in the work of this commission and most importantly want to emphasize the power of community organizing as essential for Hawai'i's water future.

CHAPTER 13

# Final Remarks as a Commissioner, June 21, 2021

*Kamanamaikalani Beamer*

Since the establishment of the Commission, in roughly thirty years of work, from the 1980s until 2012, the Commission set updated instream flow standards for sixteen times in a total of ten streams.

Between the years of 2013 [to] our present meeting in June 2021, our commission has set updated instream flow standards forty-eight times for a total of forty-six streams all across our islands.

These streams are: Wailuku, Waikapū, Waiakoali, Kawaikōī, Kauakinana, Kōke'e, Waiahulu, Koai'e, Waimea, 'Āinakō, Ukumehame, Olowalu, Launuipoko, Kaua'ula, Honopo, Hanehoi, Waiakamoi, Wahinepe'e, Puohokamoa, Haipua'ena, Punalau, Honomanu, Nua'ailua, Pi'inau, Palahulu, Waiokamilo, Wailuanui, East Wailua'iki, West Wailua'iki, Kōpili'ula, Puaka'a, Waiohu'e, Wai'a'aka, Pa'akea, Kapa'ula, Hanawī, Makapipi, Kahoma, Kanahā, Waiaama, Lāwa'i, Honokōhau, Kaluanui, Honolua, Waioli, and Ka'ie'ie.

This is an effort to celebrate, and I want to thank the communities that petitioned our commission, our staff, and especially Dr. Ayron Strauch, our deputies, Bill Tam, Jeff Pearson, Kaleo Manuel, Chair Suzanne Case and William Ailā, and lastly, our commission members who began to establish policy for our commission to set Interim Instream Flow Standards prior to communities petitioning us to do so. I want to acknowledge our present commission, Mike, Neil, Paul, Wayne—it has been a pleasure to serve with you, and I have learned so much from each of you. Your voluntary commitment is something I will always remember. Additionally, I would like to mahalo commissioners Jonathan Starr, whom I had the pleasure of working with and learning alongside, as well as the commissioners who have now passed, Milton Pavao and Bill Balfour. It is a significant achievement to have in eight years nearly quadrupled the output of all previous commissions over the last nearly thirty years in the setting of IIFS. And I do believe we have restored these streams while also balancing reasonable and beneficial use for agriculture and economics. However, there are also significant matters that require our urgent attention in this area; we must continue to rebalance wai across our islands, enforce IIFS we have implemented, and thoughtfully listen to the voices and concerns of the community.

Our fiduciary duty as Kahuwai is to implement the public trust to protect and manage the waters of Hawai'i for present and future generations. In implementing and upholding the Public Trust, we must also continue to earn the trust of our communities. I know this work is hard and adverse. There have been times when some of you have seen us get challenged and held responsible for the actions or inactions of previous commissions, or the deeds of plantation companies whose infrastructure has emptied community streams for generations. We even face the paradoxical situation of the illegal overthrow of the Hawaiian Kingdom, the occupation, and lingering issues of justice and sovereignty over resources and water. This work comes with kuleana, and we must carry these burdens. And while they are significant, they pale in comparison to the pain, the intergenerational trauma, carried by many communities who have witnessed their traditional and customary practices diminished because the lifeline of their community, the stream, has been emptied for generations. And these communities responded by carrying that trauma and facing the injustice head-on in the courts and through the processes mandated by this commission, sometimes fighting for over twenty years.

To me, that is tremendous kuleana and difficult and adverse work to continue. So, when we meet a community in an open hearing, we can keep in mind that we are perhaps one of many faces they have encountered on their path to return water and life to their community. Whether we like it or not, we carry the burden of balancing reasonable and beneficial usages of water and the legacy of historical injustice. And I think one of the most moving and inspiring developments I've witnessed has been a resurgence of these communities of aloha 'āina all across our islands as we have seen the next generation of leadership emerge all across the Pae ʻĀina to better mālama our wai, in Waiʻoli, in West Maui, East Maui, Na Wai ʻEhā, West Kauaʻi, Wailua, Waimea, Keauhou, and Heʻeia.

Ma ka lohe ʻia ka leo makuahine mai nā pua o Hawaiʻi, ma kēia ahaʻōlelo, ʻo ka haʻaheo wale kēia, A ʻo ka maʻa mau keia no nā hanauna hou, Puka ka ʻŌlelo!

But one thing I have learned in these eight years is we have the chance to play a role in restorative justice. And I believe there are few greater callings than having the chance to restore life to a community; the people, the plants, the ʻoʻopu, the hīhīwai, the ʻōpae, the aquifer itself are all resurrected when we restore a stream. So we have done forty-eight times in the last eight years. I hope this commission can complete all the roughly four hundred streams over the next decade. We now have the processes and experience in place to restore every stream in our islands within the next fifteen years.

We also have done significant work in the areas of groundwater and moved the Commission toward a more integrated one-water approach over the last eight years. I hope the Commission continues on this path, so that our approaches toward integrated management can be more actualized across our islands. I believe this to be of critical importance for mitigating the climate crisis and careful management of our aquifers and streams. In reality, like the Mele of Kane tells us, the engineered silos of groundwater, surface water, and stormwater are really only one water. And we actually don't manage the resource as much as we manage human use and consumption, and respect for the resource.

I hope the Commission will continue to work toward the understanding and protection of groundwater-dependent ecosystems as these areas are of critical importance for mitigating climate change, and toward the protection of traditional and customary practices, as well as overall nearshore marine ecosystems health.

Lastly, the most difficult issues for me to leave unresolved are the issues in Red Hill and the leaking underground fuel storage tanks. This has been the most serious threat to our drinking water in one of the most important aquifers on Oʻahu. We have had hearings and requested updates, but I continue to be deeply concerned about the existential threat the inaction of the Navy possesses to the drinking water of Oʻahu and the public trust resources of this commission. The fact that there was another recent leak in the past months after all of the proposed and supposed modifications to the fuel tanks calls for urgent action to be taken by this commission. If there were any other user of water who has already damaged and continues to pose this kind of threat to our drinking water, this would have been urgently resolved. I implore this commission to be bold to resolve this issue for the future of Oʻahu's groundwater.

In closing, I would like to end with a story. One that to me shows the potential of this commission, which is composed of stellar individuals from diverse backgrounds all of whom possess some significant experience in water resource management, to come to common ground and find integrated and reasonable and beneficial solutions to the challenges we face. As we all know, water is the basis of life; it is the common resource that will always need to exist.

Upon entering this room for my first commission meeting, under a completely different composition of this commission, I was greeted by commissioner Balfour. I knew he had served several years on the Commission and had spent his life as an employee for the plantations. Bill was a strong, burly guy. I walk into this room and Bill says, "Hey, son, how old are you?" I answer, "Thirty-five." He says to me, "Oh, I am old enough to be your father, in fact I am old enough to be your grandfather, maybe your great-grandfather." To which I was dumbfounded. Bill initiated this conversation for three meetings in a row, until finally I knew I had to respond. So finally, by the third meeting, I figured it out. I just recited a bit of my genealogy to him and assured him he wasn't any of these people but, "Thank you, Bill." And after that, he no longer indulged me in this line of questioning.

And so for the first few years, he and I, I think, disagreed on a number of issues, but I also tried to maintain aloha for Bill, and as time passed, we developed some respect for each other's positions, and I learned more about his knowledge of place and resource from being on the ground managing the ditches. I also taught Bill some about the Mahele, the public trust and ancestral Hawaiian water resource management, as well as lingering historical justice issues. Near the end of Bill's service on the Commission, many took note of his open support of public trust responsibilities as well as the serious issues of waste allegations that we had been working through at the time. In many ways Bill's metamorphosis was instrumental in some of our decision-making at the time. And I share this story with everyone today because I believe that we all have that chance to work toward restorative justice while balancing reasonable and beneficial usages; no matter our previous professional backgrounds, or political affiliations, we can learn new ways and learn from each other if we keep our aloha and put our fiduciary duty to wai first.

E like me ka olelo o ko kakou wahi kupuna, he wai e ola, he wai e mana e ola no ea, a hiki i ke aloha aina hope loa.

# Epilogue

## What Will We Do? Seven Actions for the Future of Water in Hawai'i

*Kamanamaikalani Beamer*

Wherever you are right now, please stand up. Find the closest location for you to experience wai. If you are at your desk, take a walk to the nearest water fountain. If you are at home, place this book down on the closest table or shelf and get into the shower, jump into the ocean, or, if you are lucky enough to live by a stream, go for it. But while you are immersing yourself in this experience, do nothing else but feel and taste the water. I am serious. This exercise is essential. I will be here when you get back.

Now that you have had a moment with your wai and you felt its precious caress, I hope you might imagine the twenty-five years it likely took for those raindrops to percolate through our ʻāina and into the aquifer before they were pumped from the ground and into your pipes. Or perhaps, if you found the time to immerse your body into the ocean, did you look around and ponder where the nearest muliwai or spring source might be? Did you think about the hundreds of plantation water diversions that continue to dry streams? Whatever the case may be, the authors of this book ask you to make a weekly or even daily practice of reconnecting with wai. If you already do it, maikaʻi, fantastic! If not, consider this request a gift for your benefit and ours—a practice to help restore a community of people who value and revere our precious wai, with a charge to do better for its and all of our future.

And the future can seem overwhelming and impossible at times. As Hawaiʻi experiences the convergence of multiple crises, the future often looks bleak. In many ways, securing Hawaiʻi's water future in the midst of the multiple crises discussed in this book is one of the most critical interventions we can make for the betterment of the generations to come. This book has been an attempt to intervene in the future course of water management for Hawai'i. We offered a collection of thoughtful essays written by a handful of Hawai'i experts and community organizers who have advocated for a better water future for Hawai'i. It is one critical resource toward understanding the multifaceted management of Hawai'i's water resources. We identified key challenges facing our islands—issues like sea level rise, changing precipitation and weather patterns, the linear economic systems of waste, siloed water management regimes, historical injustice, the role of power and politics in the management of water, and the existential threats to aquifer systems in places like Red Hill caused by the U.S. military.

We also identified regions of hope for our future. The critical importance and regenerative potential of the public trust doctrine to guide decision-making. The role that ancestral and local knowledge can play in restoring our relationships to wai—in redesigning future management regimes and the crucial reforms needed for the future of our islands. We can see incredible progress through emerging data supporting the return of ecosystems' health when we restore Hawai'i's streams, demonstrating that we must educate and collectively steward our water resources.

The question of the future of water in Hawai'i is essentially a question about the future of life in our islands. It is a question that forces us to ask not only if we will continue to exist on our islands but also the ways in which we as a society *want* and *should* live in one of the most precious places on our planet. No human is responsible for the creation of water. Holding a patent or bottling and selling a product does not equate to creating water. Our ancestors understood, and many continue to know, that wai, as a manifestation of Kāne, is a sacred gift for the life and future of our world. In earlier times, our ancestors developed governance structures, laws, and customs to mimic the water cycle to care for this precious resource and to benefit human and natural systems. These are not fanciful statements or idealistic accounts of an Indigenous society. They are models for how future socioeconomic systems must change to confront the climate crisis. If we could restore, resurrect, and implement even a portion of this knowledge system, we could exponentially improve Hawai'i's water future. If more could have the courage to get educated and speak our truths to power, we might be able to enact policies that mitigate catastrophic climate impacts and forge a more equitable and just society. If we made the time and opened our minds, we could meet with and learn from highly skilled 'ōiwi community leaders who continue to fight for water justice in our islands. If we elected courageous leaders who were willing to risk potential short-term political challenges for the longer-term greater good of society, we could do more than stop the U.S. Navy from leaking jet fuel into O'ahu's aquifer. If we commit, as a society, to develop an economic system and policies to make the wealth of our society more like the water cycle, circularly redistributing resources and value in regularly established intervals, we could ensure a more just and equitable society. We have the knowledge base, collective desire, and ancestral models to achieve these goals while we organize and build capacity. None of these things are out of reach or impossible. In fact, these innovations are only made more necessary because of the multiple crises of our times. From this vantage point, the only thing that is radical or extreme is *not* to adapt, to simply continue on the present course toward destruction.

A famous Hawaiian song begins with the line "Where I live there are rainbows." We understand that these rainbows are only possible because of the glistening of light that shines through the storm. Rainbows make the dull and gray, the uncertainty felt in the rumble of the thunder, the adverse and seemingly endless torrents metamorphize into robust radiant colors, signaling an end to the storm—gifting hope for a future state that lies beyond. The famed chant "He Mele no Kāne" that I discussed and translated in the introduction of this book mentions multiple forms of rainbows and their role in being an ouli, an omen in the water cycle.

Rainbows reveal themselves in the shape of a pi'o, an arch. The pi'o of the rainbows, the bowing, reaching, and inclusion—the bending and refracting of light through water to illuminate the truth, to make visible what is often unseen. The revelation for our eyes to witness color and hope where there once was the gloom of the storm is an important metaphor for these times. We need to see the truth for the future of water in Hawai'i. This book is one resource for gathering truth. But most importantly, What can we see and do together? How can we effectively utilize the resources and people's power to support real-world change for our islands? To mālama our place and people? To create systems that bend and bow to share abundance and to be like light shining through water to illuminate truth? The future of water in Hawai'i needs us to organize; *what, then, will we do?*

## Six Actions for the Future of Water in Hawai'i

1. ***Ola I Ka Wai:*** First and foremost cherish and defend the wai that surrounds you. Mahalo and embrace the concept of Kāneikawaiola, "the life-giving waters of Kāne." In your daily life, find gratitude that the generations before you cared for our resources enough to make it possible for you to drink water today. Remember that none of us create water, but we have a duty to protect it for future generations. Speak of and advocate for the Public Trust Doctrine in Hawai'i. Ensure that legislators, decision-makers, and appointed officials understand and exercise their sacred duty to uphold the public trust. Realize that your duty to wai is not just for human consumption but for all life on our planet.
2. ***Let Wai Flow:*** Do all you can to restore streams and groundwater-dependent ecosystems to enhance the lives of our communities and to protect biocultural resources. There are nearly four hundred streams across Hawai'i. Over ninety streams have been addressed through the work of the Hawai'i State Commission on Water Resource Management, but we have many more to go! You can support community groups across our islands like Hui o Nā Wai Eha, Moloka'i Nō Ka Heke, Hui Kaloko-Honokohau, and Pō'ai Wai Ola, to name a few who have been advocating for these efforts in their communities. Let's ensure an IIFS is set for every stream in Hawai'i and flow restored to all streams in the next five years.
3. ***Reduce and Reuse:*** Support sustainable agricultural use of water. Push for graywater and reused water to be used for lawns and in toilets, and where drinking water is being wasted. Can you believe we are flushing away our most precious resource?
4. ***Restore Waiwai:*** Support efforts to reestablish a regenerative economy in Hawai'i based on the Ancestral Circular Economy discussed in chapter 1. Check out 'Āina Aloha Economic

Futures (www.ainaalohafutures.com) and Doughnut Economics (www.doughnuteconomics.org) for examples and inspiration. Recognize that the linear extractive economies that have been in place in the islands—the plantation economy, militarism, unfettered tourism, and overdevelopment have all adversely impacted our wai. We can do better! Help us make the political and social changes necessary to achieve a better economy for our people and place.

5. ***Kiaʻi Wai:*** Speak and act for future generations' rights to experience life, water, and Kāneikawaiola in our islands. Confront and challenge business-as-usual policies and structures that harm our wai. This includes shutting down Red Hill, and questioning the structures that waste and damage water for future generations. Consider supporting advocacy efforts of hui like Oʻahu Water Protectors and Kaʻohewai.
6. ***Kahu Wai:*** Support Landback and the de-occupation of Hawaiʻi and the lands of indigenous peoples around the world. At present, it is estimated that "indigenous people make up 5 percent of the world's population but protect 80% of global biodiversity."[1] Natural systems will benefit from increased indigenous governance over the world's natural resources. In Hawaiʻi, confronting the legacy of the illegal overthrow, the occupation of Hawaiian lands, and the U.S. military's negligence over our resources is necessary for a more just and regenerative society.
7. ***Be like Wai:*** Recognize deeper insights into wai and ourselves. I am sharing some insights on issues I have come to know. Losing a parent is hard. For everyone who has lost a mother or father, I want to send you my aloha and wish I could give you a giant hug. Years can pass, and the hurt can surprisingly hit as strongly as the first waves of grief. It could be a song, a smell, a dream, or something you see that recalls a moment shared with your loved one. It can put you in a spin sometimes. I think it's all natural, and perhaps it is how it should be. To love is to be open and vulnerable. Perhaps the relationship and love shared between a parent and keiki is the most vulnerable and foundational relationship we can share. If you have a friend who has lost a parent, they will likely come out of that loss forever changed. I still feel like I am coming to know the person I am becoming, having

> lost my birth mom. I am so grateful for all of the time we had, and even more so for being a father and trying to parent in the ways I find to be the most healing, loving, but also firm and predictable. Even at forty-five years old, I am reborn in so many ways. Our life experiences and the filter through which we see our lives are the nourishment that determines who we become. These are like rainbows. Can we recognize the light refracting through water in our own lives in the shape of the piʻo? We are like the wai that is constantly flowing to reach the sea, only to change form and return as rain. Every raindrop is this truth. We are wai. Be like wai and nourish those around you. Detoxify and purify yourself every time you change form. Leave behind the toxins and ills and know you are reborn like every drop of wai.

*Waiwai: Water and the Future of Hawaiʻi* is merely the introduction; what we do from today forward will be the story we write through our collective action. OLA I KA WAI!

## Notes

1. Gleb Raygorodetsky, "Indigenous Peoples Defend Earth's Biodoversity—But They're in Danger," *National Geographic*, November 16, 2018, https://www.nationalgeographic.com/environment/article/can-indigenous-land-stewardship-protect-biodiversity-.

## Contributors

**Kamanamaikalani Beamer, PhD,** is a full professor and the inaugural Dana Naone Hall Endowed Chair in Hawaiian Studies, Literature, and the Environment at Hawaiʻinuiākea School of Hawaiian Knowledge at the University of Hawaiʻi, Mānoa. He serves a dual appointment in the Hawaiʻinuiākea School of Hawaiian Knowledge and in the William S. Richardson School of Law as part of the Ka Huli Ao Center for Excellence in Native Hawaiian Law. Dr. Beamer has multiple publications in the study of aloha ʻāina and the circular economy. In 2021, he concluded two consecutive terms as a commissioner on the Hawaiʻi State Water Resource Management Commission. After eight years of service, his accomplishments include restoring water to forty-eight streams across Hawaiʻi and challenging the U.S. Navy on issues at Red Hill. In addition to his academic and government service, Dr. Beamer is a father, organizer, and activist in movements to advance aloha ʻāina and water justice.

**T. Kāʻeo Duarte, PhD,** is the vice president of community and ʻāina resiliency at Kamehameha Schools (KS), where he leads a team with kuleana for 360,000 acres of ʻĀina Pauahi (KS land holdings) on five islands. Prior to joining KS, he was a faculty member at the University of Hawaiʻi at Mānoa, where he specialized in Indigenous resource management and hydrological sciences. He holds a bachelor of science degree in civil engineering from Princeton University, and a doctorate in environmental engineering from the Massachusetts Institute of Technology.

Born and raised in Koʻolaupoko, Oʻahu, **Kawena Elkington** is pursuing a PhD in the department of Geography and Environment. In her studies, Kawena focuses on the intersection of Indigenous knowledge and economic development trends in Hawaiʻi. This interest stems from an overall goal of engaging with economic activity from a Kanaka ʻŌiwi perspective to build capacity and install generational sustainability.

She received a BA in Indigenous resource management and an MA in Hawaiian studies from the University of Hawai'i at Mānoa.

**Charles H. Fletcher III, PhD,** is a geologist serving as interim dean of the School of Ocean and Earth Science and Technology, University of Hawai'i at Mānoa. He is professor of earth sciences and chairperson of the Honolulu Climate Change Commission. His research focuses on observing and modeling coastal system evolution under past, present, and future sea level change. He teaches earth science, climate change, and coastal geology. Chip is a prominent public speaker and contributor to local and national media. He has been a principal advisor in funding and awarding over thirty graduate research degrees in earth and planetary sciences, and has received a number of teaching, research, and community service awards.

**Sharde Mersberg Freitas** is a Native Hawaiian mother, daughter, 'āina justice warrior, hula practitioner, and community organizer, and currently resides on Hawai'i Island. Together with her husband, they raise their five children ma ka 'ōlelo Hawai'i. Sharde was born and raised on the west side of O'ahu, and spent a lot of time during her formative years with her Tūtū and Papa. Sharde has published articles in public health and Native Hawaiian well-being, the intersection of 'āina and health, and occasionally posts on her blog.

**Thomas W. Giambelluca, PhD,** is the director of Water Resources Research Center and professor of geography and environment at the University of Hawai'i at Mānoa. Dr. Giambelluca received BS and MA degrees in geography from the University of Miami, and a PhD in geography from the University of Hawai'i at Mānoa. His research is focused on ecohydrology and land-atmosphere interaction, including the effects of biological invasions on water, soils, carbon exchange, and the hydrology of montane cloud forests under changing global climate, with a focus on tropical environments. One aspect of his work aims to improve understanding of Hawai'i's climate: how it has changed over the past century, how it is likely to change in the future, and how the changes have affected and will affect hydrological processes and terrestrial ecosystems. He is currently leading efforts to build the Hawai'i Mesonet, a statewide network of telemetered weather stations to support research, resource management, and extreme event warning capabilities.

**Ikaika Lowe** is from Lā'iewai, Ko'olauloa, O'ahu, and is pursuing an MA in Hawaiian studies at the University of Hawai'i at Mānoa, where he also received his BA degree in Hawaiian studies. Ikaika's research interests focus on water rights and management in the Hawaiian Kingdom period and Māhele land research.

The board of directors of the **O'ahu Board of Water Supply (BWS)** appointed Ernest Y. W. Lau as the tenth manager and chief engineer of the BWS. As manager, he is responsible for the overall strategic direction and management of the BWS, with a focus on furthering the department's mission to provide a safe and dependable water supply, now and into the future. He assumed this position on February 1, 2012. Ernest previously served as the administrator of the Public Works Division under the State Department of Accounting and General Services, where he oversaw the planning, coordinating, directing, and controlling of a statewide program of engineering, architectural, and construction services. He previously worked as deputy director of the State Commission on Water Resource Management, Department of Land and Natural Resources, where he worked collaboratively with the commission members to set policies and make decisions in accordance with the State Water Code. He also served as the manager and chief engineer of the Kauai Department of Water from 1996 to 2003, and as deputy manager from 1995 to 1996. Prior to that position, he worked for the Board of Water Supply, City and County of Honolulu, for more than fourteen years as an engineer in long-range planning and water systems planning.

**Pua Souza** is from Honomaka'u, Kohala, and is currently pursuing a PhD in curriculum and instruction at the University of Hawai'i at Mānoa. She is a graduate of Hawai'inuiākea School of Hawaiian Knowledge, where she received her bachelor's degree in Hawaiian studies. Pua also holds a master's degree in social work from Myron B. Thompson School of Social Work. Her research focuses on centering conceptualizations of 'Ōiwi relationality and positionality within the context of 'Ōiwi agency as a means to assert instructional philosophies that promote practices of aloha ʻāina and community care. She currently works as a graduate research assistant in the Pōʻai Ke Aloha ʻĀina Research Lab at Kamakakūokalani Center for Hawaiian Studies.

Born and raised on the North Shore of Kauaʻi, **D. Kapuaʻala Sproat, JD,** is the director of the Ka Huli Ao Center for Excellence in Native Hawaiian Law, and professor of Native Hawaiian and environmental law at the William S. Richardson School of Law, University of Hawaiʻi at Mānoa. Before the University of Hawaiʻi, Sproat spent nine years as an attorney in the mid-Pacific office of Earthjustice. She earned her BA from Mills College in political, legal, and economic analysis, and her JD from the University of Hawaiʻi's William S. Richardson School of Law in 1998, with an environmental law certificate. In 2015, Sproat was appointed to the Board of Water Supply's board of directors.

**Ayron M. Strauch, PhD,** is a hydrologist with the State of Hawaiʻi Commission on Water Resource Management and affiliated graduate faculty at the University of Hawaiʻi at Mānoa Department of Natural Resources and Environmental Management. Dr. Strauch received BA degrees in biology and anthropology from Washington University in St. Louis, and interdisciplinary MS and PhD degrees in biology with a certificate in water systems, science, and society from Tufts University in Boston. His research focused on ecohydrology and human-water interactions across socioecological systems, including the interactions among geological, ecological, and social systems on water resources. He works closely with university faculty and students across the state to improve our understanding of Hawaiʻi's freshwater resources, freshwater biota, and management of water as a public trust resource. He is currently leading efforts to establish instream flow standards statewide that protect traditional and customary practices, aquatic ecosystems, and balance instream and non-instream uses. Dr. Strauch has led efforts to expand the monitoring of freshwater resources to support water resource management, research that improves watershed modeling, and our understanding of extreme hydrological conditions such as flooding and drought.

**William Tam, Esq.,** is a water and natural resources lawyer who has practiced law in Hawaii since 1976. After working as a Legal Aid lawyer in Lihue, Waianae, and Honolulu, he served seventeen years as a deputy attorney general for the Hawaii Board of Land and Natural Resources and the Hawaii Commission on Water Resources Management. He represented the state in major natural resources (ocean, energy, and water) litigation, including the state's successful defense in the U.S. Supreme Court in *Robinson v. Ariyoshi*. He was the principal co-author of the 1987 Hawaii Water Code and counsel in the Waiahole contested case hearings. Later, he successfully litigated Clean Water Act cases including the $5 Billion Federal Consent Decree requiring Honolulu to upgrade its municipal wastewater treatment plants and sewer system. He was co-counsel for the Hawaiian Homes Commission in the U.S. Supreme Court in *Rice v. Cayetano* (2000). From 2010 to 2015, he was deputy director of the Hawaii Commission on Water Resources Management and the Department of Land and Natural Resources (DLNR), and administrator of DLNR's Division of Aquatic Resources. He taught Hawaiian land law and administrative law as an adjunct professor at the William S. Richardson School of Law, University of Hawai'i at Mānoa. He worked extensively on the application of the public trust doctrine in Hawai'i. He graduated from Wesleyan University (1970) and Boston University School of Law (1976).

**Wayne Tanaka** is the current executive director for the Sierra Club of Hawai'i, one of the longest-standing institutional advocates for the closure of the U.S. Navy's Red Hill Bulk Fuel Storage Facility. Born and raised on the island of O'ahu, Wayne grew up exploring the tide pools of Wāwāmalu, the streams of Kāne'ohe and He'eia, and the ridgelines of the Ko'olau mountain range. An attorney and engineer by training, Wayne served for nearly a decade in the public policy program at the Office of Hawaiian Affairs prior to his position at the Sierra Club. His passion for native ecosystems and community-based resource management has also led him to serve on the board of directors of the Conservation Council for Hawai'i, and on the founding board of the then–newly minted Kua'āina Ulu 'Auamo. He has authored and co-authored book chapters and essays on konohiki fishing rights, Papahānaumokuākea, nearshore fisheries management, Indigenous food sovereignty, and the intersection of race and politics in Hawai'i, among other topics.

**Mahina Tuteur** is from Ko'olaupoko, O'ahu, and is currently pursuing a PhD in Indigenous politics at University of Hawai'i at Mānoa. She is a graduate of the William S. Richardson School of Law, with certificates in environmental and Native Hawaiian law. She has worked at several nonprofit, state, and federal agencies and currently serves on the State Environmental Advisory Council. In addition to her work in the Pōʻai ke Aloha ʻĀina research lab, Mahina is a post-juris doctor research and teaching fellow at Ka Huli Ao Center for Excellence in Native Hawaiian Law, where she assists with various scholarship projects aimed at evolving the law and advancing justice for Kānaka Maoli and other Indigenous peoples.

# Index

Note: Page numbers in *italics* refer to illustrative matter.